Advances in Intelligent Systems and Computing

Volume 735

Series editor

Janusz Kacprzyk, Polish Academy of Sciences, Warsaw, Poland
e-mail: kacprzyk@ibspan.waw.pl

The series "Advances in Intelligent Systems and Computing" contains publications on theory, applications, and design methods of Intelligent Systems and Intelligent Computing. Virtually all disciplines such as engineering, natural sciences, computer and information science, ICT, economics, business, e-commerce, environment, healthcare, life science are covered. The list of topics spans all the areas of modern intelligent systems and computing.

The publications within "Advances in Intelligent Systems and Computing" are primarily textbooks and proceedings of important conferences, symposia and congresses. They cover significant recent developments in the field, both of a foundational and applicable character. An important characteristic feature of the series is the short publication time and world-wide distribution. This permits a rapid and broad dissemination of research results.

More information about this series at http://www.springer.com/series/11156

Ajith Abraham · Abdelkrim Haqiq
Azah Kamilah Muda · Niketa Gandhi
Editors

Innovations in Bio-Inspired Computing and Applications

Proceedings of the 8th International Conference on Innovations in Bio-Inspired Computing and Applications (IBICA 2017) Held in Marrakech, Morocco, December 11–13, 2017

 Springer

Editors
Ajith Abraham
Machine Intelligence Research Labs
Auburn, WA
USA

Abdelkrim Haqiq
Faculty of Sciences and Techniques
Hassan 1st University
Settat
Morocco

Azah Kamilah Muda
Faculty of Information and Communication
   Technology
Universiti Teknikal Malaysia Melaka
Durian Tunggal, Melaka
Malaysia

Niketa Gandhi
Machine Intelligence Research Labs
Auburn, WA
USA

ISSN 2194-5357   ISSN 2194-5365 (electronic)
Advances in Intelligent Systems and Computing
ISBN 978-3-319-76353-8   ISBN 978-3-319-76354-5   (eBook)
https://doi.org/10.1007/978-3-319-76354-5

Library of Congress Control Number: 2018935891

Printed on acid-free paper

This Springer imprint is published by the registered company Springer International Publishing AG part of Springer Nature
The registered company address is: Gewerbestrasse 11, 6330 Cham, Switzerland

Preface

Welcome to the Proceedings of the 8th International Conference on Innovations in Bio-Inspired Computing and Applications (IBICA 2017). IBICA 2017 was organized in conjunction with 13th International Conference on Information Assurance and Security (IAS 2017). Conferences were held at Mogador Hotels & Resorts, Marrakech, Morocco, during December 11–13, 2017. IBICA–IAS 2017 conferences are jointly organized by the Machine Intelligence Research Labs (MIR Labs), USA, and Faculty of Sciences and Techniques, Hassan 1st University, Settat, Morocco.

The aim of IBICA is to provide a platform for world research leaders and practitioners, to discuss the "full spectrum" of current theoretical developments, emerging technologies, and innovative applications of bio-inspired computing. Bio-inspired computing is currently one of the most exciting research areas, and it is continuously demonstrating exceptional strength in solving complex real-life problems.

The themes of the contributions and scientific sessions range from theories to applications, reflecting a wide spectrum of the coverage of bio-inspired computing, intelligent systems, and its applications. IBICA 2017 received submissions from over 18 countries, and each paper was reviewed by at least 5 reviewers in a standard peer-review process. Based on the recommendation by 5 independent referees, finally 33 papers were accepted for publication in the proceedings published by Springer-Verlag.

Many people have collaborated and worked hard to produce a successful IBICA 2017 conference. First and foremost, we would like to thank all the authors for submitting their papers to the conference, for their presentations and discussions during the conference. Our thanks go to Program Committee members and reviewers, who carried out the most difficult work by carefully evaluating the submitted papers. Our special thanks to Oscar Castillo (Tijuana Institute of Technology, Tijuana, Mexico) and Alexander Gelbukh (Instituto Politécnico Nacional, Mexico City, Mexico) for the exciting plenary talks.

We express our sincere thanks to special session chairs and organizing committee chairs for helping us to formulate a rich technical program.

Abdelkrim Haqiq
Ajith Abraham
IBICA 2017 - General Chairs

IBICA 2017 Organization

Honorary Chairs

Ahmed Nejmeddine   President of Hassan 1st University, Settat, Morocco

Houssine Bouayad   Acting Dean of FST, Hassan 1st University, Settat, Morocco

General Chairs

Abdelkrim Haqiq   GREENTIC, FST, Hassan 1st University, Settat, Morocco

Ajith Abraham   MIR Labs, USA

General Co-chairs

Layth Sliman   EFREI, Paris, France

Adel M. Alimi   University of Sfax, Tunisia

PC Co-chairs

Antonio J. Tallón-Ballesteros   University of Seville, Spain

Nelishia Pillay   University of Pretoria, South Africa

Millie Pant   Indian Institute of Technology Roorkee, India

Huang Dijiang   Arizona State University, USA

Nizar Rokbani   University of Sousse, Tunisia

Ghita Mezour   International University of Rabat, Morocco

Advisory Board

Albert Zomaya	The University of Sydney, Australia
Bruno Apolloni	University of Milano, Italy
Hideyuki Takagi	Kyushu University, Japan
Imre J. Rudas	Óbuda University, Hungary
Janusz Kacprzyk	Polish Academy of Sciences, Poland
Javier Montero	Complutense University of Madrid, Spain
Krzysztof Cios	Virginia Commonwealth University, USA
Mario Koeppen	Kyushu Institute of Technology, Japan
Patrick Siarry	Université Paris-Est Créteil, France
Salah Al-Sharhan	Gulf University of Science and Technology, Kuwait
Sebastian Ventura	University of Cordoba, Spain
Vincenzo Piuri	Università degli Studi di Milano, Italy

Publication Chairs

Azah Kamilah Muda	UTeM, Malaysia
Niketa Gandhi	Machine Intelligence Research Labs, USA

Web Service

Kun Ma	University of Jinan, China

Publicity Chair

Brahim Ouhbi	ENSAM, Moulay Ismail University, Meknès, Morocco

Organizing Chairs

Jaouad Dabounou	FST, Hassan 1st University, Settat, Morocco
Mohamed Hanini	FST, Hassan 1st University, Settat, Morocco
Mohamed Chakraoui	Multidisciplinary Faculty of Khouribga, Morocco

Organizing Committee

Youmna El Hiss	FST, Hassan 1st University, Settat, Morocco
Ayman Hadri	FST, Hassan 1st University, Settat, Morocco
Amine Maarouf	xHub, Technopark, Casablanca, Morocco
Ahmed Boujnoui	FST, Hassan 1st University, Settat, Morocco
Hamid Taramit	FST, Hassan 1st University, Settat, Morocco
Adnane El Hanjri	FST, Hassan 1st University, Settat, Morocco
El Mehdi Kandoussi	FST, Hassan 1st University, Settat, Morocco

International Program Committee

H. El Bakkali	University Mohammed V Rabat ENSIAS, Rabat, Morocco
Abdelkrim Haqiq	Hassan 1st University, Morocco
Abderrahim Beni Hssane	Chouaib Doukkali University, El Jadida, Morocco
Ajith Abraham	Machine Intelligence Research Labs, USA
Alan Barton	Carleton University, Canada
Alberto Cano	University of Córdoba, Spain
Antonio J. Tallón Ballesteros	Universidad de Sevilla, Spain
Aswani Cherukuri	VIT University, India
Azah Kamilah Muda	Universiti Teknikal Malaysia Melaka, Malaysia
Brahim Ouhbi	Université Moulay Ismail, Morocco
Enrique Dominguez	Universidad de Málaga, Spain
Francisco Martine	National Institute of Astrophysics, Optics and Electronics, France
Ghizlane Orhanou	Faculté des Sciences, Université Mohammed V-Agdal, Rabat, Morocco
Haresh Suthar	Panipat Institute of Engineering & Technology, India
Josu Ceberio	University of the Basque Country, Spain
Julio Cesar Nievola	Pontifícia Universidade Católica do Paraná, Brazil
Julio Ponce	Universidad Autónoma de Aguascalientes, Mexico
Kang Tai	Nanyang Technological University, Singapore
Katsuhiro Honda	Osaka Prefecture University, Japan
Kelemen Arpad	University of Maryland, USA
Leticia Hernando	The University of the Basque Country, Spain
Lin Wang	Jinan University, China
Lubna Gabralla	Sudan University of Science and Technology, Sudan

Additional Reviewers

Abdelali El Bouchti	Bournemouth University, UK
Arun Kumar Sangaiah	VIT University, Tamil Nadu, India
Brahim Ouhbi	Moulay Ismaïl University, Meknès, Morocco
El Moukhtar Zemmouri	Moulay Ismaïl University, Meknès, Morocco
Ghizlane Orhanou	Mohammed V University of Rabat, Morocco
Hanini Mohamed	Hassan 1st University, Morocco
Kamal Oudidi	National School of Computer and Systems Analysis, Rabat, Morocco
Kusum Deep	Indian Institute of Technology Roorkee, India
Mohamed Hanini	Hassan 1st University, Morocco
Mohamed Moughit	Hassan 1st University, Morocco
Mohammed Ridouani	GREENTIC/EST, UH2C, Casablanca, Morocco
Oussama Mjihil	Hassan 1st University, Morocco
Said El Kafhali	Hassan 1st University, Morocco
Sambit Bakshi	National Institute of Technology Rourkela, India
Trivedi Kishor	Duke University, NC, USA
Yassine Maleh	Hassan 1st University, Morocco
Yassine Sadqi	University Sultan Moulay Slimane, Beni Mellal, Morocco
Omar Iraqui	University of Milan, Milano, Italy
Jesus Benito-Picazo	University of Malaga, Málaga, Spain
Prashant K. Gupta	South Asian University, India

Contents

Contents

Dynamic Parameter Adaptation Using Interval Type-2 Fuzzy Logic in Bio-Inspired Optimization Methods

Oscar Castillo[✉], Frumen Olivas, and Fevrier Valdez

Tijuana Institute of Technology, Calzada Tecnologico s/n, Tomas Aquino, 22379 Tijuana, Mexico
{ocastillo,fevrier}@tectijuana.mx, frumen@msn.com

Abstract. In this paper we perform a comparison of the use of type-2 fuzzy logic in two bio-inspired methods: Ant Colony Optimization (ACO) and Gravitational Search Algorithm (GSA). Each of these methods is enhanced with a methodology for parameter adaptation using interval type-2 fuzzy logic, where based on some metrics about the algorithm, like the percentage of iterations elapsed or the diversity of the population, we aim at controlling their behavior and therefore control their abilities to perform a global or a local search. To test these methods two benchmark control problems were used in which a fuzzy controller is optimized to minimize the error in the simulation with nonlinear complex plants.

Keywords: Interval type-2 fuzzy logic · Ant Colony Optimization
Gravitational Search Algorithm · Dynamic parameter adaptation

1 Introduction

Bio-inspired optimization algorithms can be applied to most combinatorial and continuous optimization problems, but for different problems need different parameter values, in order to obtain better results. There are in the literature, several methods aim at modeling better the behavior of these algorithms by adapting some of their parameters [18, 19], introducing different parameters in the equations of the algorithms [4], performing a hybridization with other algorithm [17], and using fuzzy logic [5–9, 14, 16].

In this paper a methodology for parameter adaptation using an interval type-2 fuzzy system is presented, where on each method a better model of the behavior is used in order to obtain better quality results.

The proposed methodology has been previously successfully applied to different bio-inspired optimization methods like BCO (Bee Colony Optimization) in [1], CSA (Cuckoo Search Algorithm) in [3], PSO (Particle Swarm optimization) in [5, 7], ACO (Ant Colony Optimization) in [6, 8], GSA (Gravitational Search Algorithm) in [9, 16], DE (Differential Evolution) in [10], HSA (Harmony Search Algorithm) in [11], BA (bat Algorithm) in [12] and in FA (Firefly Algorithm) in [15].

The algorithms used in this research are ACO (Ant Colony Optimization) from [8] and GSA (Gravitational Search Algorithm) from [9], each one with dynamic parameter adaptation using an interval type-2 fuzzy system. Fuzzy logic proposed by Zadeh in [20–22] help us to model a complex problem, with the use of membership functions and fuzzy rules, with the knowledge of a problem from an expert, fuzzy logic can bring tools to create a model and attack a complex problem.

The contribution of this paper is the comparison between the bio-inspired methods which use an interval type-2 fuzzy system for dynamic parameter adaptation, in the optimization of fuzzy controllers for nonlinear complex plants. The adaptation of parameters with fuzzy logic helps to perform a better design of the fuzzy controllers, based on the results which are better than the original algorithms.

2   Bio-Inspired Optimization Methods

ACO is a bio-inspired algorithm based on swarm intelligence of the ants, proposed by Dorigo in [2], where each individual helps each other to find the best route from their nest to a food source. Artificial ants represent the solutions to a particular problem, where each ant is a tour and each node is a dimension or a component of the problem. Biological ants use pheromone trails to communicate to other ants which path is the best and the artificial ant tries to mimic that behavior in the algorithm.

Artificial ants use probability to select the next node using Eq. 1, where with this equation calculate the probability of an ant k to select the node j from node i.

$$P_{ij}^k = \frac{[\tau_{ij}]^\alpha [\eta_{ij}]^\beta}{\sum_{l} \in N_i^k [\tau_{il}]^\alpha [\eta_{il}]^\beta}, \quad if\ j \in N_i^k \tag{1}$$

The components of Eq. 1 are: P^k is the probability of an ant k to select the node j from node i, τ_{ij} represents the pheromone in the arc that joins the nodes i and j and η_{ij} represents the visibility from node i to node j, with the condition that node j must be in the neighborhood of node i. Also like in nature the pheromone trail evaporates over time, and the ACO algorithm uses Eq. 2 to simulate the evaporation of pheromone in the trails.

$$\tau_{ij} \leftarrow (1 - \rho)\tau_{ij}, \quad \forall (i,j) \in L \tag{2}$$

The components of Eq. 2 are: τ_{ij} representing the pheromone trail in the arc that joins the nodes i and j, ρ represents the percentage of evaporation of pheromone, and this equation is applied to all arcs in the graph L.

There are more equations for ACO, but these two equations are the most important in the dynamics of the algorithm, also these equations contain the parameters used to model a better behavior of the algorithm using an interval type-2 fuzzy system.

GSA proposed by Rashedi in [13], is a population based algorithm that uses laws of physics to update its individuals, more particularly uses the Newtonian law of gravity and the second motion law. In this algorithm each individual is considered as an agent, where each one represent a solution to a problem and each agent has its own mass and can move to another agent. The mass of an agent is given by the fitness function, agents with bigger mass are better. Each agent applies some gravitational force to all other agents, and is calculated using Eq. 3.

$$F_{ij}^d(t) = G(t) \frac{M_{pi}(t) \times M_{aj}(t)}{R_{ij}(t) + \varepsilon} (x_j^d(t) - x_i^d(t)) \tag{3}$$

The components of Eq. 3 are: F_{ij}^d is the gravity force between agents i and j, G is the gravitational constant, M_{pi} is the mass of agent i or passive mass, and M_{aj} is the mass of agent j or active mass, R_{ij} is the distance between agents i and j, ε is an small number used to avoid division by zero, x_j^d is the position of agent j and x_i^d is the position of agent j.

The gravitational force is used to calculate the acceleration of the agent using Eq. 4.

$$a_i^d(t) = \frac{F_i^d(t)}{M_{ii}(t)} \tag{4}$$

The components of Eq. 4 are: a_i^d is the acceleration force of agent i, F_i^d is the gravitational force of agent i, and M_{ii} is the inertial mass of agent i.

In GSA the gravitational constant G from Eq. 3, unlike in real life here it can be variable and is given by Eq. 5.

$$G(t) = G_0^{-\alpha t/T} \tag{5}$$

The components of Eq. 5 are: G is the gravitational constant, G_0 is the initial gravitational constant, α is a parameter defined by the user of GSA and is used to control the change in the gravitational constant, t is the actual iteration and T is the total number of iterations. To control the elitism GSA uses Eq. 6 to allow only the best agents to apply their force to other agents, and in initial iterations all the agents apply their force but Kbest will decrease over time until only a few agents are allowed to apply their force.

$$F_i^d(t) = \sum_{j \in Kbest, j \neq 1} rand_i F_{ij}^d(t) \tag{6}$$

The components of Eq. 6 are: F_i^d is the new gravity force of agent i, *Kbest* is the number of agents allowed to apply their force, sorted by their fitness the best *Kbest* agent can apply their force to all other agents, in this equation j is the number of dimension of agent i.

3   Methodology for Parameter Adaptation

The optimization methods involved in this comparison have dynamic parameter adaptation using interval type-2 fuzzy systems, and each of these adaptations are described in details for ACO in [8] and for GSA in [9]. The way in which this adaptation of parameters was performed is as follows: first a metric about the performance of the algorithms needs to be created, in this case the metrics are a percentage of iteration elapsed described by Eq. 7 and the diversity of individuals described by Eq. 8, then after the metrics are defined we need to select the best parameters to be dynamically adjusted, and this was done based on experimentation with different levels of all the parameters of each optimization method.

$$Iteration = \frac{Current\ Iteration}{Maximum\ of\ Iterations} \tag{7}$$

The components of Eq. 7 are: *Iteration* is a percentage of the elapsed iterations, *current iteration* is the number of elapsed iterations, and *maximum of iterations* is the total number iterations set for the optimization algorithm to find the best possible solution.

$$Diversity(S(t)) = \frac{1}{n_s} \sum_{i=1}^{n_s} \sqrt{\sum_{j=1}^{n_x} \left(x_{ij}(t) - \bar{x}_j(t)\right)^2} \tag{8}$$

The components of Eq. 8 are: *Diversity(S)* is a degree of dispersion of the population S, n_s is the number of individuals in the population S, n_x is the number of dimensions in each individual from the population, x_{ij} is the j dimension of the individual i, *tested x_j* is the j dimension of the best individual in the population. After the metrics are defined and the parameters selected, a fuzzy system is created to adjust just one parameter, and with this obtain a fuzzy rule set to control this parameter, and for all the parameters we need to do the same, and at the end only one fuzzy system will be created to control all the parameters at the same time combining all the created fuzzy systems. The proposed methodology for parameter adaptation is illustrated in Fig. 1, where it has the optimization method, which has an interval type-2 fuzzy system for parameter adaptation.

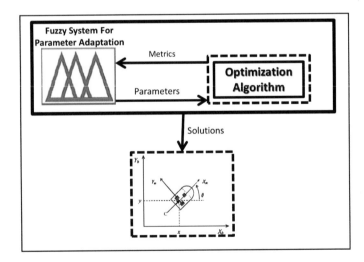

Fig. 1. General scheme of the proposal for parameter adaptation

Figure 1 illustrates the general scheme for parameter adaptation, in which the bio-inspired optimization algorithm is evaluated by the metrics and these are used as inputs for the interval type-2 fuzzy system, which will adapt some parameters of the optimization algorithm based on the metrics and the fuzzy rules. Then this method with parameter adaptation will provide the parameters or solutions for a problem, in this case the parameters for the fuzzy system used for control. The final interval type-2 fuzzy systems for each optimization method are illustrated in Figs. 2 and 3 respectively, for ACO and GSA correspondingly. Each of these fuzzy systems has iteration and diversity as inputs, with a range from 0 to 1 using the Eqs. 7 and 8 correspondingly to each input, and two

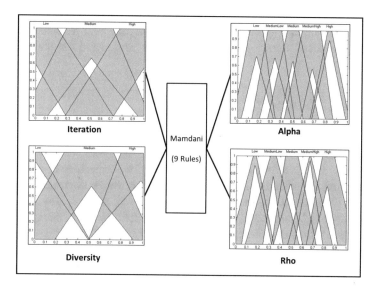

Fig. 2. Interval type-2 fuzzy system for parameter adaptation in ACO

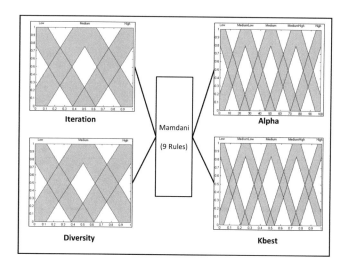

Fig. 3. Interval type-2 fuzzy system for parameter adaptation in GSA

outputs but these differs from each optimization method because each one has its own parameters to be dynamically adjusted.

The interval type-2 fuzzy system from Fig. 2 has two inputs and two outputs, the inputs are granulated into three type-2 triangular membership functions and the outputs into five type-2 triangular membership functions, and nine rules, in this case the parameters to be dynamically adjusted over the iterations are α *(alpha)* and ρ *(rho)* from Eqs. 1 and 2 respectively, both with a range from 0 to 1.

The interval type-2 fuzzy system from Fig. 3 has *iteration* and *diversity* as inputs with three type-2 triangular membership functions and two outputs, which are the parameters to be adjusted in this case, α *(alpha)* with a range from 0 to 100 and *Kbest* from 0 to 1, each output is granulated into five type-2 triangular membership functions with a fuzzy rule set of nine rules. The parameters α *(alpha)* and *Kbest* are from Eqs. 5 and 6 respectively.

4   Problems Statement

The comparison of ACO and GSA is through the optimization of a fuzzy controller from two different non-linear complex plants, where these two problems use a fuzzy system for control. The first problem is the optimization of the trajectory of an autonomous mobile robot and the objective is to minimize the error in the trajectory, the robot has two wheeled motors and one stabilization wheel, it can move in any direction. The desired trajectory is illustrated in Fig. 4, where first the robot must start from point (0, 0) and it needs to follow the reference using the fuzzy system from Fig. 5 as a controller. The reference illustrated in Fig. 4 helps in the design of a good controller because it uses only nonlinear trajectories, to assure that the robot can follow any trajectory.

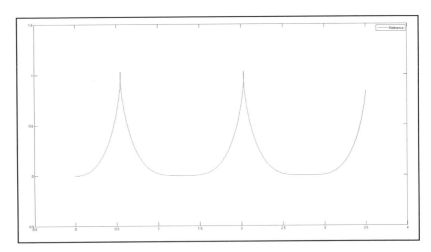

Fig. 4. Trajectory for the autonomous mobile robot

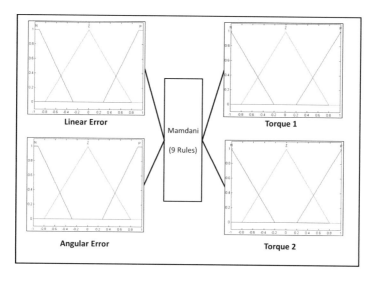

Fig. 5. Fuzzy controller for the autonomous mobile robot

The fuzzy system used for control illustrated in Fig. 5, and uses the linear and angular errors to control the motorized wheels of the robot. In this problem the optimization methods will aim at finding better parameters for the membership functions, using the same fuzzy rule set. The second problem is the automatic temperature control in a shower, and the optimization method will optimize the fuzzy controller illustrated in Fig. 6, which will try to follow the flow and temperature references. The fuzzy system used as control is illustrated in Fig. 6 and has two input variables, temperature and flow; a fuzzy rule set of nine rules and two outputs cold and hot is presented. The fuzzy system

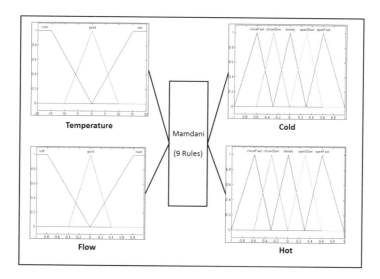

Fig. 6. Fuzzy controller for the automatic temperature control in a shower

uses the inputs and with the fuzzy rules to control the open-close mechanism of the cold and hot water.

5   Simulations, Experiments and Results

The optimization methods were applied to the optimization of the membership functions of the fuzzy system used as controllers for the two problems described in Sect. 4. Using the parameters from Table 1, each method was applied to both problems. In the case of the problem of the trajectory of an autonomous mobile robot there are 40 points to be search for all the membership functions, and in the problem of the automatic temperature control in a shower there are 52 points. The methods to be compared are: the original ACO method, ACO with parameter adaptation, original GSA method and GSA with parameter adaptation.

Table 1. Parameters for each optimization method

Parameter	Original ACO	ACO with parameter adaptation	Original GSA	GSA with parameter adaptation
Population	30	30	30	30
Iterations	50	50	50	50
α (Alpha)	1	Dynamic	40	Dynamic
β (Beta)	2	2		
ρ (Rho)	0.1	Dynamic		
Kbest			Linear decreasing from 100% to 2%	Dynamic
G_0			100	100

The parameters from Table 1 are a challenge for the optimization methods, because there are only 50 iterations to found the best possible fuzzy controller for each problem. This is a good manner to show the advantages of the proposed methodology for parameter adaptation using an interval type-2 fuzzy system. Table 2 contains the results of applying all the optimization methods to the optimization of the fuzzy controller for an autonomous mobile robot, the average is from 30 experiments (with Mean Square Error (MSE)) and results in bold are best, the 30 experiments means that each method was

Table 2. Results of the simulations with the robot problem

MSE	Original ACO	ACO with parameter adaptation	Original GSA	GSA with parameter adaptation
Average	0.4641	**0.0418**	36.4831	15.4646
Best	0.1285	**0.0048**	10.4751	3.2375
Worst	0.9128	**0.1276**	76.0243	30.8511
Standard deviation	0.2110	**0.0314**	15.8073	8.6371

applied to the fuzzy controller optimization for 30 times resulting in 30 different fuzzy controllers for each method.

From Table 2 the optimization method that obtains better results is ACO with parameter adaptation using the proposed methodology with an interval type-2 fuzzy system, also it can be seen that the results of GSA with parameter adaptation are better that the original GSA, but ACO is better. The results in Table 3 are from applying all the methods to optimize the fuzzy controller for the automatic temperature control in a shower, the average is from 30 experiments (with the Mean Square Error (MSE)) and also the results in bold are best, same as the first problem the 30 experiments means that each method was applied to the optimization of the fuzzy controller and obtaining 30 different fuzzy controller for each method.

Table 3. Results of the simulations with the shower problem

MSE	Original ACO	ACO with parameter adaptation	Original GSA	GSA with parameter adaptation
Average	0.6005	0.4894	3.8611	**0.1151**
Best	0.5407	0.3980	1.9227	**0.0106**
Worst	0.9036	0.5437	6.5659	**0.3960**
Standard deviation	0.0696	0.0378	1.0860	**0.0913**

From the results in Table 3 in this case GSA with parameter adaptation using the proposed methodology using an interval type-2 fuzzy system can obtains better results than the other methods. Also it can be seen that ACO with parameter adaptation can obtain better results than the original ACO method and the original GSA method.

6   Statistical Comparison

The Z-test is a tool to prove that the methods with parameter adaptation can obtain on average better results than its counterparts the original methods, also to know what method is better on certain problem by comparing its results with all of the other methods. The comparison between the methods is using the statistical test Z-test, using the parameters from Table 4 and the results of the comparisons are in Tables 5 and 6 for the robot and shower problems, respectively.

Table 4. Parameters for the statistical Z-test

Parameter	Value
Level of significance	95%
Alpha (α)	5%
Alternative hypothesis (H_a)	$\mu_1 < \mu_2$ (claim)
Null hypothesis (H_0)	$\mu_1 \geq \mu_2$
Critical value	-1.645

Table 5. Results of the Z-test for comparison in the robot problem

μ_2 μ_1	Original ACO	ACO with parameter adaptation	Original GSA	GSA with parameter adaptation
Original ACO		10.8415	**−12.4795**	**−9.5098**
ACO with parameter adaptation	**−10.8415**		**−12.6269**	**−9.7803**
Original GSA	12.4795	12.6269		6.3911
GSA with parameter adaptation	9.5098	9.7803	**−6.3911**	

Table 6. Results of the Z-test for comparison in the shower problem

μ_2 μ_1	Original ACO	ACO with parameter adaptation	Original GSA	GSA with parameter adaptation
Original ACO		7.6813	**−16.4115**	23.1516
ACO with parameter adaptation	**−7.6813**		**−16.9950**	20.7332
Original GSA	16.4115	16.9950		18.8264
GSA with parameter adaptation	**−23.1516**	**−20.7332**	**−18.8264**	

The results in Table 5 are using the parameters in Table 4 for the Z-test, where it claims that a method (μ_1) has on average better results (we are comparing errors, so minimum is better) than the other method (μ_2), in Tables 5 and 6 the first column correspond to the methods as μ_1 and the first row correspond to the methods as μ_2, also we are not comparing the same method with itself, results in bold means that there are enough evidence to reject the null hypothesis.

From the results in Table 5, which correspond to the optimization of a fuzzy controller for the trajectory of an autonomous mobile robot, there is enough evidence that ACO method with parameter adaptation can obtain on average better results than all of the other methods. There is enough evidence that the original ACO method can obtain on average better results than the original GSA and GSA with parameter adaptation. There is also enough evidence that GSA with parameter adaptation can obtain on average better results than the original GSA method.

From the results in Table 6, which correspond to the optimization of a fuzzy controller for the automatic temperature control in a shower, there is enough evidence that GSA with parameter adaptation can obtain on average better results than all of the other methods. There is enough evidence that ACO with parameter adaptation can obtain on average better results than the original ACO method and the original GSA method. There is also enough evidence that the original ACO method can obtain on average better results than the original GSA method.

7 Conclusions

The optimization of a fuzzy controller is a complex task, because require the search of several parameters in infinite possibilities in the range of each input or output variables. The bio-inspired optimization methods help in the search because is guided by some kind of intelligence, from swarm intelligence or from laws of physics and can make a better search of parameters. With the inclusion of a fuzzy system in this case an interval type-2, the bio-inspired methods can search even in a better way, because is guided by the knowledge of an expert system that model a proper behavior in determined states of the search, in the beginning improves the global search or exploration of the search space and in final improves the local search or the exploitation of the best area found so far of the entire search space. From the results with the MSE there is clearly that ACO with parameter adaptation has the best results in the robot problem, and GSA with parameter adaptation has the best results in the shower problem, but with the statistical test it confirm these affirmations. The statistical comparison shows that the methods with parameter adaptation are better than their counterparts the original methods. Also ACO is a better method with the robot problem, but GSA is better in the shower problem.

References

1. Amador-Angulo, L., Castillo, O.: Statistical analysis of type-1 and interval type-2 fuzzy logic in dynamic parameter adaptation of the BCO. In: 2015 Conference of the International Fuzzy Systems Association and the European Society for Fuzzy Logic and Technology (IFSA-EUSFLAT 2015). Atlantis Press, June 2015
2. Dorigo, M.: Optimization, learning and natural algorithms. Ph.D. thesis, Dipartimento di Elettronica, Politechico di Milano, Italy (1992)
3. Guerrero, M., Castillo, O., Garcia, M.: Fuzzy dynamic parameters adaptation in the Cuckoo Search Algorithm using fuzzy logic. In: 2015 IEEE Congress on Evolutionary Computation (CEC), pp. 441–448. IEEE, May 2015
4. Hongbo, L., Ajith, A.: A fuzzy adaptive turbulent particle swarm optimization. Int. J. Innov. Comput. Appl. $1(1)$, 39–47 (2007)
5. Melin, P., Olivas, F., Castillo, O., Valdez, F., Soria, J., Garcia, J.: Optimal design of fuzzy classification systems using PSO with dynamic parameter adaptation through fuzzy logic. Exp. Syst. Appl. $40(8)$, 3196–3206 (2013)
6. Neyoy, H., Castillo, O., Soria, J.: Dynamic fuzzy logic parameter tuning for ACO and its application in TSP problems. In: Castillo, O., Melin, P., Kacprzyk, J. (eds.) Recent Advances on Hybrid Intelligent Systems. Studies in Computational Intelligence, vol. 451, pp. 259–271. Springer, Heidelberg (2012)
7. Olivas, F., Valdez, F., Castillo, O., Melin, P.: Dynamic parameter adaptation in particle swarm optimization using interval type-2 fuzzy logic. Soft. Comput. $20(3)$, 1057–1070 (2016)
8. Olivas, F., Valdez, F., Castillo, O., Gonzalez, C., Martinez, G., Melin, P.: Ant colony optimization with dynamic parameter adaptation based on interval type-2 fuzzy logic systems. Appl. Soft Comput. 53, 74–87 (2016)
9. Olivas, F., Valdez, F., Castillo, O., Melin, P.: Interval type-2 fuzzy logic for dynamic parameter adaptation in a modified gravitational search algorithm. Eng. Appl. Artif. Intell. (2017, under review)

10. Ochoa, P., Castillo, O., Soria, J.: Differential evolution with dynamic adaptation of parameters for the optimization of fuzzy controllers. In: Castillo, O., Melin, P., Pedrycz, W., Kacprzyk, J. (eds.) Recent Advances on Hybrid Approaches for Designing Intelligent Systems. Studies in Computational Intelligence, vol. 547, pp. 275–288. Springer, Cham (2014)

11. Peraza, C., Valdez, F., Castillo, O.: An improved harmony search algorithm using fuzzy logic for the optimization of mathematical functions. In: Melin, P., Castillo, O., Kacprzyk, J. (eds.) Design of Intelligent Systems Based on Fuzzy Logic, Neural Networks and Nature-Inspired Optimization. Studies in Computational Intelligence, vol. 601, pp. 605–615. Springer, Cham (2015)

12. Perez, J., Valdez, F., Castillo, O., Melin, P., Gonzalez, C., Martinez, G.: Interval type-2 fuzzy logic for dynamic parameter adaptation in the bat algorithm. Soft Comput. **21**(3), 667–685 (2016)

13. Rashedi, E., Nezamabadi-Pour, H., Saryazdi, S.: GSA: a gravitational search algorithm. Inf. Sci. **179**(13), 2232–2248 (2009)

14. Shi, Y., Eberhart, R.: Fuzzy adaptive particle swarm optimization. In: Proceeding of IEEE International Conference on Evolutionary Computation, Seoul, Korea, pp. 101–106. IEEE Service Center, Piscataway (2001)

15. Solano-Aragón, C., Castillo, O.: Optimization of benchmark mathematical functions using the firefly algorithm with dynamic parameters. In: Castillo, O., Melin, P. (eds.) Fuzzy Logic Augmentation of Nature-Inspired Optimization Metaheuristics. Studies in Computational Intelligence, vol. 574, pp. 81–89. Springer, Cham (2015)

16. Sombra, A., Valdez, F., Melin, P., Castillo, O.: A new gravitational search algorithm using fuzzy logic to parameter adaptation. In: 2013 IEEE Congress on Evolutionary Computation (CEC), pp. 1068–1074. IEEE, June 2013

17. Taher, N., Ehsan, A., Masoud, J.: A new hybrid evolutionary algorithm based on new fuzzy adaptive PSO and NM algorithms for distribution feeder reconfiguration. Energy Convers. Manag. **54**, 7–16 (2012)

18. Valdez, F., Melin, P., Castillo, O.: Evolutionary method combining particle swarm optimization and genetic algorithms using fuzzy logic for decision making. In: IEEE International Conference on Fuzzy Systems, pp. 2114–2119 (2009)

19. Wang, B., Liang, G., Chan Lin, W., Yunlong, D.: A new kind of fuzzy particle swarm optimization fuzzy_PSO algorithm. In: 1st International Symposium on Systems and Control in Aerospace and Astronautics, ISSCAA 2006, pp. 309–311 (2006)

20. Zadeh, L.: Fuzzy sets. Inf. Control **8**, 338–358 (1965)

21. Zadeh, L.: Fuzzy logic. IEEE Comput. **8**, 83–92 (1965)

22. Zadeh, L.: The concept of a linguistic variable and its application to approximate reasoning—I. Inform. Sci. **8**, 199–249 (1975)

Reducing Blackhole Effect in WSN

Sana Akourmis[1(✉)], Youssef Fakhri[1,2], and Moulay Driss Rahmani[1]

[1] LRIT, Research Unit Associated with the CNRST (URAC29),
Faculty of Sciences, University Mohammed V-Agdal, Rabat, Morocco
sakourmis@gmail.com, mrahmani@fsr.ac.ma
[2] LaRIT LAboratory, Faculty of Sciences, University Ibn Tofail,
Kenitra, Morocco
fakhri@uit.ac.ma

Abstract. The open nature of wireless medium, low processing power, wireless connectivity, changing topology, limited resources and hostile deployment of nodes makes WSN easy for outsider attackers to interrupt the legitimate traffic. This leads to various types of security threats among them active black hole attack in which all received data packets are dropped immediately after giving false routing information in order to attract the traffic towards itself. It is a denial of service attacks effective on the network layer. In this paper Black hole attack has being simulated on AODV routing protocol and a method called idsAODV has being tested by controlling in the "recvReply" function of AODV routing protocol if the RREP (Route Reply) message is arrived for itself. If it did not, the node forwards the message to its neighbor nodes, and if it did the RREP function is changed by RREP caching mechanism to identify the faked RREP message coming from malicious node. The simulation result shows that by using this method in WSN environment the packet loss was reduced under promiscuous mode.

Keywords: WSNs · AODV · IDS (intrusion detection system)
RREQ (route request) · RREP (route reply) · Black hole attack
Security attack · NS2.35

1 Introduction

Wireless sensor network is a typical network which has positive and negative points like other technologies. It is composed by nodes which are characterized by low energy constrained, low cost and low power. Multi-hop routes are needed to relay data from the monitored region to one or more gateway nodes to collect and transmit environmental data in an autonomous manner and in large scale to build a global view of this monitored region [1]. These relations between nodes are restricted to their communication range which allows them to connect each other to the destination node and subsequently their links to other nodes can be modified because of mobility that often changes the network topology, but due to these inherent characteristics like insecure communication, common transmission medium, broadcasting mechanism and simplicity of routing protocol helps the adversary to interrupt the network by simply being within radio range which can help this adversary node to intercept the transmitted data and afterwards, it can

© Springer International Publishing AG, part of Springer Nature 2018
A. Abraham et al. (Eds.): IBICA 2017, AISC 735, pp. 13–24, 2018.
https://doi.org/10.1007/978-3-319-76354-5_2

access to the network and transform the routing protocol by making the network prone to many types of attacks that targets different layers in OSI model [16] and interrupt the network operations through mechanisms such as data fabrication, selective forwarding, and packet drops which is the case of our black hole attack. Meanwhile, the physical layer, Mac layer and network layer plays a major role in routing mechanism of ad hoc networks [2, 3]. Therefore, for sensor node application, transport layer protocols can provide session control and reliability. This last is especially needed when the system plans to be accessed through internet. Moreover, in the network layer the variety of attacks differs either by adding or by modifying some parameters of routing message; such as hop count or sequence number or by not forwarding the packets. Thus, in the case of black hole attack, the adversary node attract the neighbor node by using greater sequence number, less hop and false route reply and never broadcasts the received RREQ as it is required by route discovery process in the AODV routing protocol for example. To do this, the black hole node announces itself that it has the most recent route to the destination. So Source node sends packets passing by this node and that node immediately drops them which fool the source node. This results in reduction of packet delivery ratio; because the intruder has to disable the send which is mentioned before by not broadcasting the RREQs that receives from the intermediate nodes. Thus, in this type of network each node has to listen the packets transmitted by its own neighbor nodes and black hole attack snoops on its neighbor to find which node is preparing to send this RREQ, in this way the black hole node propagates a RREP for any received RREQ pretending that it has a direct path to the destination [6, 7]. Many wireless routing protocols such as DSDV, DSR, HWMP, and AODV are vulnerable to this type of attacks. To summarize other existing attacks which target different layer of nodes in WSN, Table 1 shows these attacks.

Table 1. Different attacks existing in WSN [2–4]

Attacks	Corresponding layer
Denial of service, tampering, radio interference, physical capture	Physical layer
Unfair attacks, energy depletion, jamming, collision, traffic manipulation, exhaustion, unfairness	Data link layer
Message altering or false message, modification and replication on routing information, information disclosure, sinkhole attack, gray hole attack, selective forwarding attack, sybil attack, wormhole, hello flood attack, sending data to node out transmission range, neglect and greed, homing, misdirection	Network layer
Running out of memory, not synchronized attack, session hijacking, packet injection attack, jellyfish attack	Transport layer attack
Data gathering, task distribution, target tracking, repudiation, attacks on reliability, aggregation based attacks	Application layer attacks

In WSN environment especially in hostile deployment where human control is not always present, this type of network can be easily compromised for example by capturing a signal at any time, or by compromising a node in the network that remains the most damaging attack in WSN. In this way, attacks in WSN are classified as active, passive external and internal. In the active attack, the attacker exploits the wicked environment link in the security protocol to project attacks like replaying attack or packet modification, snooping, eaves dropping, traffic analysis, monitoring, etc. In passive attack, the access to information is obtained by the attacker without being detected such as wormhole, black hole, gray hole, information disclosure, resource consumption, routing attacks.

This last, is among the most difficult attacks to detect. In the other hand the external attacker is represented by external attacker node which has no rights to access the network and in the case of internal attack, the attacker deploys malicious node to compromise the sensor nodes and gets control of the network in order to takes authorization to access the network [3]. Consequently, the security goals in WSN can also be classified into four categories, namely Confidentiality, integrity, authentication and availability (CIAA). Hence, to transmit data securely between sensor nodes, secure communication is mandatory for this type of network. This work focus especially on the black hole attack launched by adversaries on AODV routing protocol which occurs in the network layer on wireless sensor networks and also an existing method which are called IDS has been tested in this work to mitigate the effect of this threat. We have taken three scenarios in our experiment which is created by 25 nodes. Firstly, we simulated sensor network without black hole attack and in second scenario, we launched black hole attack with different number of malicious node i.e. from 1 to 5 in order to observe the impact of malicious nodes in the network's performance especially on packet loss. Finally, we tested the proposed technique to evaluate the network's performance. The simulation results show that the proposed technique is able to reduce the impact of black hole attack in sensor network. This last will be against parameters such as packet loss, average throughput, average energy consumption, end-to-end delay under various scenarios based on NS2 simulator.

2   Related Work

The most important IDS systems can be classified into two categories: network-based detection (NIDS) and host-based intrusion detection.

In the former, malicious actions and attacks are done with the help of neighboring nodes by their cooperation between each other. In the latter, data is collected through the log's files of the Hope Rating System that runs on the node.

Many researchers have suggested various protocols with higher safety to defend WSN against security attacks. However, each attack has specific defense objects, and is unable to defend against particular attacks. Table 2 shows different proposed preventions against Black hole attack proposed by researchers.

Table 2. Literature summary table

Title	Objective	Methods	Description
Sheela et al. [7]	Detecting black hole attacks in wireless sensor networks using mobile agent	The primary goal of agent is to detect the black hole nodes by giving information of one node to its neighboring nodes in the network	This method requires transmission of redundant copies of a packet from each source node (SN)
Karakehayov [8]	Using REWARD to detect team black-hole attacks in wireless sensor networks	REWARD to (Receive, Watch, Redirect). This approach with the help of two broadcast messages; MISS and SAMBA to identify Black Hole nodes	This technique is very expensive – for a network with n black hole nodes
Lou et al. [9]	H-SPREAD: a hybrid multipath scheme for secure and reliable data collection in wireless sensor networks	(H-SPREAD) to improve both security and reliability. The new scheme is based on a distributed N-to-1 multipath discovery protocol which is able to find multiple node-disjoint paths from every sensor node to the base station simultaneously in one route discovery process	The proposed multipath discovery protocol is very efficient, with less than one message per path found
Weerasinghe et al. [10]	Preventing cooperative black hole attacks in mobile ad hoc networks	DRI (Data Routing Information) is used to keep track of past routing experience among mobile nodes in the network and crosschecking of RREP messages from intermediate nodes by source nodes	The main drawback of this technique is that mobile nodes have to maintain an extra database
Banerjee et al. [11]	Detection/removal of cooperative black and gray hole attack in mobile ad-hoc networks	This mechanism is capable of detecting and removing the malicious nodes launching these two types of attacks	False positive may occur in this mechanism and the algorithm may report that a node is misbehaving, when in fact it is not

(*continued*)

Table 2. (*continued*)

Title	Objective	Methods	Description
Marti et al. [12]	Mitigating routing misbehavior in mobile ad hoc networks	Watchdog and Pathrater use observation-based techniques to detect misbehaving nodes, and report observed misbehavior back to the source of the traffic	This technique is imperfect due to collision in routes, limited transmission power and partial dropping
Sasikala and Vallinayagam [14]	Secured intrusion detection system in mobile ad hoc network using RAODV	In this proposed Intrusion Detection algorithm used the protocols RAODV and AODV for measuring the efficiencies of the network security. A trust relationship is established based on a dynamic evaluation of the sender's "secure IP". RAODV gives the alarm to the neighboring nodes and also the performance time is increased when compared with AODV protocol. In RAODV protocol malicious nodes are detected	The proposed RAODV provides better security to data packets for sparse and significant security for denser medium. It provides better security compared to other protocols like AODV
Nasser and Chen [15]	"Enhanced intrusion detection system for discovering malicious nodes in mobile ad hoc networks"	Introduction of intrusion detection system called ExWatchdog. This system is an extension to the Watchdog by its ability to discover malicious nodes which can partition the network by falsely reporting other nodes as misbehaving	Compared to Watchdog, ExWatchdog increases the throughput by up to 11%. ExWatchdog solves a fatal problem of Watchdog

(*continued*)

Table 2. (*continued*)

Title	Objective	Methods	Description
Gurung and Saluja [16]	Mitigating Impact of Blackhole Attack in MANET	ANB-AODV (Anti Near Blackhole-AODV) is created to mitigate black hole attack in manet. In this approach, when sender broadcast the RREQ packet, it will wait for reply. The source node will get first reply from malicious node provided the malicious node is near to source node and acquire the data packet and it will not forward the packet to the destination	The proposed approach is effective in improving the performance of the network. When ANB-AODV and AFB-AODV protocol is used then there is decrease in the packet loss thus improving the performance of network
Khamayseh et al. [17]	A new protocol for detecting black hole nodes in ad hoc networks	The proposed protocol modifies the behavior of the original AODV by including this technique: Every node is provided with a data structure referred as trust table. This table is responsible for holding the addresses of the reliable nodes and the RREP is extended with an extra field called trust field. The source node sends its data only if the RREP is propagated by a reliable node. Otherwise it waits for further RREP	The main priority of the protocol is to send the data through reliable route. The protocol need to be supported by a technique to eliminate the black hole node from the network
Buchegger and Boudec [13]	performance analysis of the CONFIDANT protocol: cooperation of nodes-fairness in dynamic ad hoc networks	CONFIDANT detects misbehaving nodes by means of observation and more aggressively informs other nodes of this misbehavior through reports sent around the network	This scheme can be beneficial for fast misbehavior detection

3 Technique of AODV-IDS

The flowchart shows that the algorithms for IDS-AODV will discard the first RREP packet from malicious node which are fault node and choose second coming RREP packet from destination, it also find another path to destination. Because, generally the first route reply will be from the adversary which is black hole node with high destination sequence number (DSN) and is stored as the first entry in the RR-Table.

Then it compares the first destination sequence number with the source node sequence number, and the node is malicious node, if there exist more difference between them. So remove that entry from the RR-Table (Fig. 1).

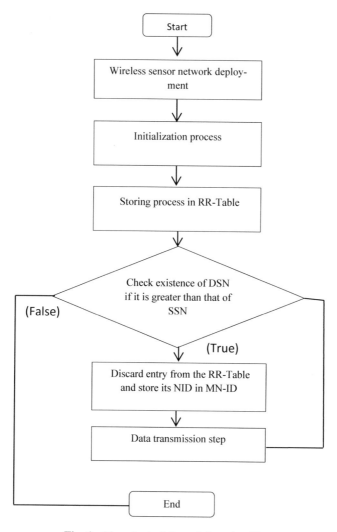

Fig. 1. Flowchart of the existing algorithm

4   Simulation Environment

The performance of AODV routing protocol in WSN is analyzed against parameters such as packet loss, average throughput in kbps, energy in joule, end-to-end delay under various scenarios using NS-2(v-2.35) simulator. Simulations are run on 10 seeds and average of the obtained parameter values are used for final analysis and comparison. An improved version of random waypoint model is used as the model of node mobility. The chosen parameters for simulation are presented in Table 3.

Table 3. Experimental setup

Simulation time	200.0 s
Topology	Mobile
Node placement	Random
Terrain dimension	800 × 550
Antenna model	OmniAntenna
Number of nodes	5, 10, 15, 20, 25
MAC layer	802.11
Routing protocols	AODV, BLACKHOLEAODV, IDSAODV
Radio propagation model	TwoRayGround
Traffic model	Constant bit rate
Packet size	256
Traffic rate	0.1 mbps
Number of malicious nodes	0 to 10
Transmission range	250 m
Observation parameters	PDF, end-to-end delay, throughput, energy

5   Results and Discussions

A simulation study was performed to evaluate the performance of WSN in presence of attacks using metrics such as packet loss, throughput, end to end delay and average energy. Results in Figs. 2 and 3 have been obtained in existence of 5 attackers and IDS

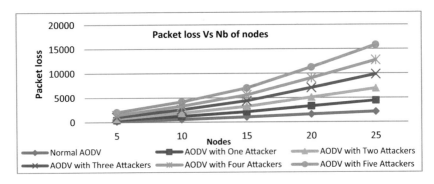

Fig. 2. Impact of black hole attack on packet loss

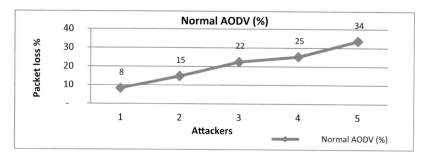

Fig. 3. Percentage of packet loss of AODV under different number of attackers

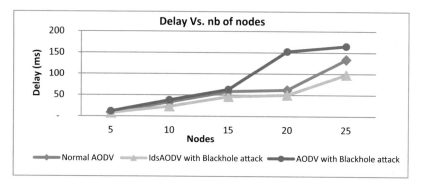

Fig. 4. Impact of blackhole attack on the average E-E delay.

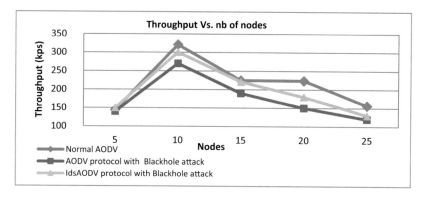

Fig. 5. Impact of blackhole attack on the network throughput.

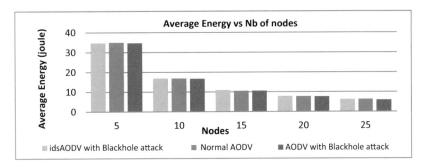

Fig. 6. Impact of blackhole attack on the average energy.

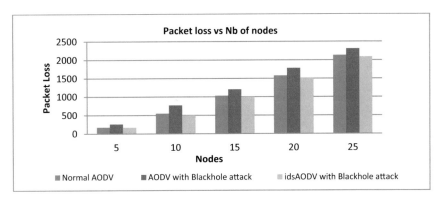

Fig. 7. Impact of blackhole attack on packet loss.

technique was also used in presence of black hole attacker node to check the network's performance.

Figure 7 shows how the modified protocol IDSAODV decreases the number of dropped packets compared to the original protocol AODV for a network attacked by one black hole. As shown by Figs. 2 and 3 the number of dropped packets increases as the number of malicious nodes increases from one to five, the black hole hole has the chance to provide in more connections and to drop more packets, for this reason the number of dropped packets keeps increasing.

The results show that throughput increases in the presence of Ids compared to original protocol and it is clear from the graph in Figs. 4 and 5 that the performance of the original protocol decreases dramatically when the network is attacked by a black hole node.

Figure 6 shows the energy consumption which is proportional to the number of nodes. Therefore, when the number of nodes increases, energy consumption decreases automatically.

6 Conclusion and Future Work

In this study an attempt has been made to find impact of malicious node in AODV routing protocol in WSN environment under different density of node with number of malicious attack. Result shows that packet loss and delay of normal AODV is much better than AODV with malicious attack. Under malicious attack AODV drops more packets with increase of number of attacks. It is concluded that performance of routing protocol (AODV) degrades by introducing malicious nodes is increased under attack and the throughput increases. To reduce the impact of this attack we have introduced IDS solution which is a slightly modified version of AODV protocol. Based on the results obtained, this solution can reduce the packet dropped in the network. The advantage of using this approach is that IDSAODV has minimum modification in AODV protocol and it does not need any additional overhead, and also the implementation does not make any modification in packet format. In future work, we will investigate other types of attacks on network layer such as wormhole attack on AODV protocol and we will try to find how we can reduce the effect of this attack on the network, either by testing the same technique or by introducing another response system and improve the performance of the network.

References

1. Yick, J., Mukherjee, B., Ghosal, D.: Wireless sensor network survey. Comput. Netw. **52**(12), 2292–2330 (2008)
2. Kaplantzis, S., Shilton, A., Mani, N., Sekercioglu, Y.A.: Detecting selective forwarding attacks in wireless sensor networks using support vector machines. In: 3rd International Conference on Intelligent Sensors, Sensor Networks and Information, ISSNIP 2007, pp. 335–340 (2007)
3. Mamatha, G.S., Sharma, D.S.: Network layer attacks and defense mechanisms in MANETS-a survey. Int. J. Comput. Appl. **9**(9), 12–17 (2010)
4. Sen, J.: Routing security issues in wireless sensor networks: attacks and defenses. In: Tan, Y. K. (ed.) Sustainable Wireless Sensor Networks. InTech, Rijeka (2010)
5. Sheela, D., Srividhya, V.R., Asma, B.A., Chidanand, G.M.: Detecting black hole attacks in wireless sensor networks using mobile agent. In: International Conference on Artificial Intelligence and Embedded Systems (ICAIES), pp. 15–16 (2012)
6. Karakehayov, Z.: Using REWARD to detect team black-hole attacks in wireless sensor networks. In: Workshop on Real-World Wireless Sensor Networks (REALWSN 2005), pp. 20–21, June 2005
7. Lou, W., Kwon, Y.: H-SPREAD: a hybrid multipath scheme for secure and reliable data collection in wireless sensor networks. IEEE Trans. Veh. Technol. **55**(4), 1320–1330 (2006)
8. Weerasinghe, K.G.H.D., Fu, H.: Preventing cooperative black hole attacks in mobile ad hoc networks: simulation implementation and evaluation (2008)
9. Banerjee, S.: Detection/removal of cooperative black and gray hole attack in mobile ad-hoc networks. In: Proceedings of the World Congress on Engineering and Computer Science, pp. 22–24, October 2008
10. Marti, S., Giuli, T.J., Lai, K., Baker, M.: Mitigating routing misbehavior in mobile ad hoc networks. In: Proceedings of the 6th Annual International Conference on Mobile Computing and Networking, pp. 255–265. ACM, August 2000

11. Buchegger, S., Le Boudec, J.Y.: Performance analysis of the CONFIDANT protocol. In: Proceedings of the 3rd ACM International Symposium on Mobile Ad Hoc Networking and Computing, pp. 226–236. ACM, June 2002

12. De Mazieux, A.D., Gauthier, V., Marot, M., et al.: Etat de l'art sur les réseaux de capteurs. Rapport de recherche INT N05001RST GET/INT UMR, vol. 5157 (2005)

13. Kumar, K.V., Somasundaram, K.: Detection of black hole attacks in manets by using proximity set method. Int. J. Comput. Sci. Inf. Secur. **14**(3), 136 (2016)

14. Sasikala, S., Vallinayagam, M.: Secured intrusion detection system in mobile ad hoc network using RAODV. In: Proceedings Published in International Journal of Computer Applications (IJCA). ISSN 975: 8887

15. Nasser, N., Chen, Y.: Enhanced intrusion detection system for discovering malicious nodes in mobile ad hoc networks. In: IEEE International Conference on Communications, ICC 2007. IEEE (2007)

16. Gurung, S., Saluja, K.K.: Mitigating impact of blackhole attack in MANET. In: International Conference on Recent Trends in Information, Telecommunication and Computing, ITC (2014)

17. Khamayseh, Y., et al.: A new protocol for detecting black hole nodes in ad hoc networks. Int. J. Commun. Netw. Inf. Secur. **3**(1), 36 (2011)

Minimum Spanning Tree in Trapezoidal Fuzzy Neutrosophic Environment

Said Broumi[1(✉)], Assia Bakali[2], Mohamed Talea[1], Florentin Smarandache[3], and Vakkas Uluçay[4]

[1] Laboratory of Information Processing, Faculty of Science Ben M'Sik, University Hassan II, Sidi Othman, B.P 7955, Casablanca, Morocco
broumisaid78@gmail.com, taleamohamed@yahoo.fr
[2] Ecole Royale Navale, Boulevard Sour Jdid, B.P 16303, Casablanca, Morocco
assiabakali@yahoo.fr
[3] Department of Mathematics, University of New Mexico, 705 Gurley Avenue, Gallup, NM 87301, USA
fsmarandache@gmail.com, smarand@unm.edu
[4] Department of Mathematics, Gaziantep University, 27310 Gaziantep, Turkey
vulucay27@gmail.com

Abstract. In this paper, an algorithm for searching the minimum spanning tree (MST) in a network having trapezoidal fuzzy neutrosophic edge weight is presented. The network is an undirected neutrosophic weighted connected graph (UNWCG). The proposed algorithm is based on matrix approach to design the MST of UNWCG. A numerical example is provided to check the validity of the proposed algorithm. Next, a comparison example is made with Mullai's algorithm in neutrosophic graphs.

Keywords: Neutrosophic sets · Trapezoidal fuzzy neutrosophic sets
Score function · Neutrosophic graph · Minimum spanning tree

1 Introduction

In 1998, Smarandache [1] proposed the concept of neutrosophic set (NS) from the philosophical point of view, to represent uncertain, imprecise, incomplete, inconsistent, and indeterminate information that are exist in the real world. The concept of neutrosophic set generalizes the concept of the classic set, fuzzy set, and intuitionistic fuzzy set (IFS). The major differences between the IFS and neutrosophic set (NS) are the structure of the membership functions, the dependence of the membership functions, and the constraints in the values of the membership functions. A NS has a triple-membership structure which consists of three components, namely the truth, falsity and indeterminacy membership functions, as opposed to the IFS in which information is described by a membership and non-membership function only. Another major difference is the constraint between these membership functions. In a NS, the three membership functions are independent of one another and the only constraint is that the sum of these membership functions must not exceed three. This is different from the IFS where the

© Springer International Publishing AG, part of Springer Nature 2018
A. Abraham et al. (Eds.): IBICA 2017, AISC 735, pp. 25–35, 2018.
https://doi.org/10.1007/978-3-319-76354-5_3

values of the membership and non-membership functions are dependent on one another, and the sum of these must not exceed one. To apply the concept of neutrosophic sets (NS) in science and engineering applications, Smarandache [1] initiated the concept of single-valued neutrosophic set (SVNS). In a subsequent paper, Wang et al. [2], studied some properties related to SVNSs. We refer the readers to [3, 11, 13–15] for more information related to the extensions of NSs and the advances that have been made in the application of NSs and its extensions in various fields. The minimum spanning tree problem is one of well–known problems in combinatorial optimization. When the edge weights assigned to a graph are crisp numbers, the minimum spanning tree problem can be solved by some well-known algorithms such as Prim and Kruskal algorithm. By combining single valued neutrosophic sets theory [1, 2] with graph theory, references [6–9] introduced single valued neutrosophic graph theory (SVNGT for short). The SVNGT is generation of graph theory. In the literature some scholars have studied the minimum spanning tree problem in neutrosophic environment. In [4], Ye introduced a method for finding the minimum spanning tree of a single valued neutrosophic graph where the vertices are represented in the form of SVNS. Mandal and Basu [5] proposed an approach based on similarity measure for searching the optimum spanning tree problems in a neutrosophic environment considering the inconsistency, incompleteness and indeterminacy of the information. In their work, they applied the proposed approach to a network problem with multiple criteria. In another study, Mullai et al. [10] discussed about the minimum spanning tree problem in bipolar neutrosophic environment.

The main purpose of this paper is to propose a neutrosophic version of Kruskal algorithm based on the matrix approach for searching the cost minimum spanning tree in a network having trapezoidal fuzzy neutrosophic edge weight [12].

The rest of the paper is organized as follows. Section 2 briefly introduces the concepts of neutrosophic sets, single valued neutrosophic sets and the score function of trapezoidal neutrosophic number. Section 3 proposes a novel approach for searching the minimum spanning tree in a network having trapezoidal fuzzy neutrosophic edge length. In Sect. 4, a numerical example is presented to illustrate the proposed method. In Sect. 5, a comparative example with other method is provided. Finally, Sect. 6 presents the main conclusions.

2   Preliminaries and Definitions

In this section, the concept of neutrosophic sets single valued neutrosophic sets and trapezoidal fuzzy neutrosophic sets are presented to deal with indeterminate data, which can be defined as follows.

Definition 2.1 [1]. Let ξ be an universal set. The neutrosophic set A on the universal set ξ categorized in to three membership functions called the true $T_A(x)$, indeterminate $I_A(x)$ and false $F_A(x)$ contained in real standard or non-standard subset of $]^-0, 1^+[$ respectively.

$$^-0 \leq \sup T_A(x) + \sup I_A(x) + \sup F_A(x) \leq 3^+ \qquad (1)$$

Definition 2.2 [2]. Let ξ be a universal set. The single valued neutrosophic sets (SVNs) A on the universal ξ is denoted as following

$$A = \left\{ < x: T_A(x), I_A(x), F_A(x) > x \in \xi \right\} \tag{2}$$

The functions $T_A(x) \in [0. 1]$, $I_A(x) \in [0. 1]$ and $F_A(x) \in [0. 1]$ are named degree of truth, indeterminacy and falsity membership of x in A, satisfy the following condition:

$$0 \leq T_A(x) + I_A(x) + F_A(x) \leq 3 \tag{3}$$

Definition 2.3 [12]. Let ζ be a universal set and ψ [0, 1] be the sets of all trapezoidal fuzzy numbers on [0, 1]. The trapezoidal fuzzy neutrosophic sets (In short TrFNSs) $\breve{A}$ on the universal is denoted as following:

$$\breve{A} = \left\{ < x: \breve{T}_A(x), \breve{I}_A(x), \breve{F}_A(x) >, x \in \zeta \right\} \tag{4}$$

Where $\breve{T}_A(x): \zeta \to \psi[0, 1]$, $\breve{I}_A(x): \zeta \to \psi[0, 1]$ and $\breve{F}_A(x): \zeta \to \psi[0, 1]$. The trapezoidal fuzzy numbers

$$\breve{T}_A(x) = \left(T_A^1(x), T_A^2(x), T_A^3(x), T_A^4(x) \right) \tag{5}$$

$$\breve{I}_A(x) = \left(I_A^1(x), I_A^2(x), I_A^3(x), I_A^4(x) \right) \tag{6}$$

and

$$\breve{F}_A(x) = \left(F_A^1(x), F_A^2(x), F_A^3(x), F_A^4(x) \right), \text{ respectively denotes degree of truth, inde-}$$
terminacy and falsity membership of x in $\breve{A} \, \forall x \in \zeta$.

$$0 \leq T_A^4(x) + I_A^4(x) + F_A^4(x) \leq 3 \tag{7}$$

Definition   2.4.   [12]. Let $\breve{A}_1$ be   a   TrFNV   denoted   as $\breve{A}_1 = \langle (t_1, t_2, t_3, t_4), (i_1, i_2, i_3, i_4), (f_1, f_2, f_3, f_4) \rangle$ Hence, the score function and the accuracy function of TrFNV are denoted as below:

(i) $s(\breve{A}_1) = \dfrac{1}{12}[8 + (t_1 + t_2 + t_3 + t_4) - (i_1 + i_2 + i_3 + i_4) - (f_1 + f_2 + f_3 + f_4)] \tag{8}$

(ii) $H(\breve{A}_1) = \dfrac{1}{4}[(t_1 + t_2 + t_3 + t_4) - (f_1 + f_2 + f_3 + f_4)] \tag{9}$

In order to make a comparisons between two TrFNV, Ye [12], presented the order relations between two TrFNVs.

Definition 2.5 [12]. Let $\breve{A}_1$ and $\breve{A}_2$ be two TrFNV defined on the set of real numbers. Hence, the ranking method is defined as follows:

i. If $s(\tilde{A}_1) > s(\tilde{A}_2)$, then $\tilde{A}_1$ is greater than $\tilde{A}_2$, that is, $\tilde{A}_1$ is superior to $\tilde{A}_2$, denoted by $\tilde{A}_1 > \tilde{A}_2$

If $s(\tilde{A}_1) = s(\tilde{A}_2)$, and $H(\tilde{A}_1) > H(\tilde{A}_2)$ then $\tilde{A}_1$ is greater than $\tilde{A}_2$, that is, $\tilde{A}_1$ is superior to $\tilde{A}_2$, denoted by $\tilde{A}_1 > \tilde{A}_2$.

3 Minimum Spannig Tree Algorithm of TrFN- Undirected Graph

In this section, a neutrosophic version of Kruskal's algorithm is proposed to handle Minimum spanning tree in a neutrosophic environment and a trapezoidal fuzzy neutrosophic minimum spanning tree algorithm, whose steps are described below:

Algorithm:

Input: The weight matrix $M = \left[W_{ij}\right]_{n \times n}$ for which is constructed for undirected weighted neutrosophic graph (UWNG).

Step 1: Input trapezoidal fuzzy neutrosophic adjacency matrix A.

Step 2: Construct the TrFN-matrix into a score matrix $\left[S_{ij}\right]_{n \times n}$ by using the score function (8).

Step 3: Repeat step 4 and step 5 up to time that all nonzero elements are marked or in another saying all (n−1) entries matrix of S are either marked or set to zero.

Step 4: There are two ways to find out the weight matrix M that one is columns-wise and the other is row-wise in order to determine the unmarked minimum entries S_{ij}, besides it determines the weight of the corresponding edge e_{ij} in M.

Step 5: Set $S_{ij} = 0$ else mark S_{ij} provided that corresponding edge e_{ij} of selected S_{ij} generate a cycle with the preceding marked entries of the score matrix S.

Step 6: Construct the graph T including the only marked entries from the score matrix S which shall be the desired minimum cost spanning tree of G.

Step 7: Stop.

4 Numerical Example

In this section, a numerical example of TrFNMST is used to demonstrate of the proposed algorithm. Consider the following graph G = (V, E) shown Fig. 1, with fives nodes and fives edges. The various steps involved in the construction of the minimum cost spanning tree are described as follow:

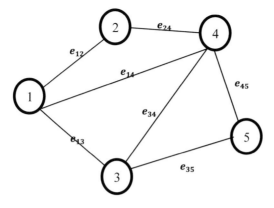

Fig. 1. A neutrosophic graph with TrFN edge weights

The TrFN- adjacency matrix A is written as follows:

$$= \begin{bmatrix} 0 & e_{12} & e_{13} & e_{14} & 0 \\ e_{12} & 0 & 0 & e_{24} & 0 \\ e_{13} & 0 & 0 & e_{34} & e_{35} \\ e_{14} & e_{24} & e_{34} & 0 & e_{45} \\ 0 & 0 & e_{35} & e_{45} & 0 \end{bmatrix}$$

Thus, using the score function, we get the score matrix:

$$S = \begin{bmatrix} 0 & 0.575 & 0.592 & 0.583 & 0 \\ 0.575 & 0 & 0 & 0.542 & 0 \\ 0.592 & 0 & 0 & 0.458 & 0.6 \\ 0.583 & 0.542 & 0.458 & 0 & 0.525 \\ 0 & 0 & 0.6 & 0.525 & 0 \end{bmatrix}$$

Fig. 2. Score matrix

We observe that the minimum record 0.458 according to Fig. 2 is selected and the corresponding edge (3, 4) is marked with red color. Repeat the procedure until the iteration will exist (Table 1).

Table 1. The values of edge weights

e_{ij}	Edge weights
e_{12}	< (0.2, 0.3, 0.5, 0.5), (0.1, 0.4, 0.4, 0.6), (0.1, 0.2, 0.3, 0.5) >
e_{13}	< (0.3, 0.4, 0.6, 0.7), (0.1, 0.3, 0.5, 0.6), (0.2, 0.3, 0.3, 0.6) >
e_{14}	< (0.4, 0.5, 0.7, 0.7), (0.1, 0.4, 0.4, 0.5), (0.3, 0.4, 0.5, 0.7) >
e_{24}	< (0.4, 0.5, 0.6, 0.7), (0.3, 0.4, 0.6, 0.7), (0.2, 0.4, 0.5, 0.6) >
e_{34}	< (0.1, 0.3, 0.5, 0.6), (0.4, 0.5, 0.6, 0.7), (0.3, 0.4, 0.4, 0.7) >
e_{35}	< (0.4, 0.4, 0.5, 0.6), (0.1, 0.3, 0.3, 0.6), (0.1, 0.3, 0.4, 0.6) >
e_{45}	< (0.3, 0.5, 0.6, 0.7), (0.1, 0.3, 0.4, 0.7), (0.3, 0.4, 0.8, 0.8) >

According to the Figs. 3 and 4, the next non zero minimum entries 0.525 is marked and corresponding edges (4, 5) are also colored.

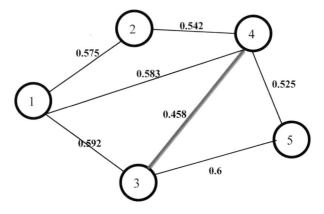

Fig. 3. An illustration of the marked edge

$$S = \begin{bmatrix} 0 & 0.575 & 0.592 & 0.583 & 0 \\ 0.575 & 0 & 0 & 0.542 & 0 \\ 0.592 & 0 & 0 & 0.458 & 0.6 \\ 0.583 & 0.542 & 0.458 & 0 & 0.525 \\ 0 & 0 & 0.6 & 0.525 & 0 \end{bmatrix}$$

Fig. 4. Score matrix

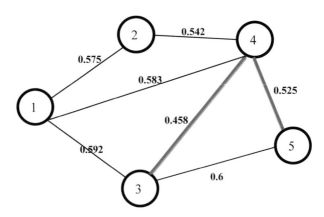

Fig. 5. An illustration of the marked edge (4, 5)

According to the Fig. 6, the next minimum non zero element 0.542 is marked (Figs. 5 and 7).

$$S = \begin{bmatrix} 0 & 0.575 & 0.592 & 0.583 & 0 \\ 0.575 & 0 & 0 & 0.542 & 0 \\ 0.592 & 0 & 0 & 0.458 & 0.6 \\ 0.583 & 0.542 & 0.458 & 0 & 0.525 \\ 0 & 0 & 0.6 & 0.525 & 0 \end{bmatrix}$$

Fig. 6. Score matrix

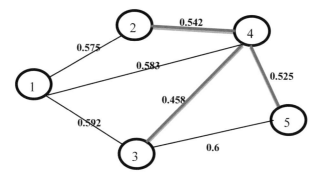

Fig. 7. An illustration of the marked edge (2, 4)

According to the Fig. 8. The next minimum non zero element 0.575 is marked, and corresponding edges (1, 2) are also colored (Fig. 9).

$$S = \begin{bmatrix} 0 & 0.575 & 0.592 & 0.583 & 0 \\ 0.575 & 0 & 0 & 0.542 & 0 \\ 0.592 & 0 & 0 & 0.458 & 0.6 \\ 0.583 & 0.542 & 0.458 & 0 & 0.525 \\ 0 & 0 & 0.6 & 0.525 & 0 \end{bmatrix}$$

Fig. 8. Score matrix

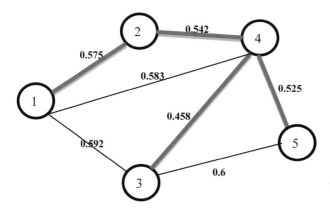

Fig. 9. An illustration of the marked edge (1, 2)

According to the Fig. 10. The next minimum non zero element 0.583 is marked. But while drawing the edges it produces the cycle. So we delete and mark it as 0 instead of 0.583.

$$S = \begin{bmatrix} 0 & \textbf{0.575} & 0.592 & 0.583 & 0 \\ 0.575 & 0 & 0 & \textbf{0.542} & 0 \\ 0.592 & 0 & 0 & \textbf{0.458} & 0.6 \\ \textbf{0.583}\ 0 & 0.542 & 0.458 & 0 & \textbf{0.525} \\ 0 & 0 & 0.6 & 0.525 & 0 \end{bmatrix}$$

Fig. 10. Score matrix

The next non zero minimum entries 0.592 is marked it is shown in the Fig. 11. But while drawing the edges it produces the cycle. So we delete and mark it as 0 instead of 0.592.

$$S = \begin{bmatrix} 0 & \textbf{0.575} & 0.592 & 0.583 & 0 \\ 0.575 & 0 & 0 & \textbf{0.542} & 0 \\ \textbf{0.592}\ 0 & 0 & 0 & \textbf{0.458} & 0.6 \\ \textbf{0.583}\ 0 & 0.542 & 0.458 & 0 & \textbf{0.525} \\ 0 & 0 & 0.6 & 0.525 & 0 \end{bmatrix}$$

Fig. 11. Score matrix

According to the Fig. 12. The next minimum non zero element 0.6 is marked. But while drawing the edges it produces the cycle so we delete and mark it as 0 instead of 0.6.

$$S = \begin{bmatrix} 0 & \textbf{0.575} & 0.592 & 0.583 & 0 \\ 0.575 & 0 & 0 & \textbf{0.542} & 0 \\ \textbf{0.592}\ 0 & 0 & 0 & \textbf{0.458} & \textbf{0.6}\ 0 \\ \textbf{0.583}\ 0 & 0.542 & 0.458 & 0 & \textbf{0.525} \\ 0 & 0 & 0.6 & 0.525 & 0 \end{bmatrix}$$

Fig. 12. Score matrix

After the above steps, the final path of minimum cost of spanning tree of G is portrayed in Fig. 13.

Based on the procedure of matrix approach applied to undirected neutrosophic graph. hence, the crisp minimum cost spanning tree is 2, 1 and the final path of minimum cost of spanning tree is $\{1, 2\}, \{2, 4\}, \{4, 3\}, \{4, 5\}$.

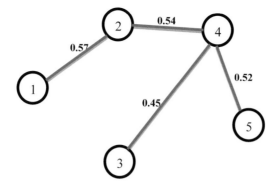

Fig. 13. Final path of minimum cost of spanning tree of G.

5   Comparative Example

To demonstrate the rationality and effectiveness of the proposed method, a comparative example with Mullai's algorithm [10] is provided. Following the step of Mullai's algorithm.

Iteration 1: Let $C_1 = \{1\}$ and $\overline{C_1} = \{2, 3, 4, 5\}$
Iteration 2: Let $C_2 = \{1, 2\}$ and $\overline{C_2} = \{3, 4, 5\}$
Iteration 3: Let $C_3 = \{1, 2, 4\}$ and $\overline{C_3} = \{3, 5\}$
Iteration 4: Let $C_4 = \{1, 2, 4, 3\}$ and $\overline{C_4} = \{5\}$

From the results of the iteration processes, the TrFN minimal spanning tree is:

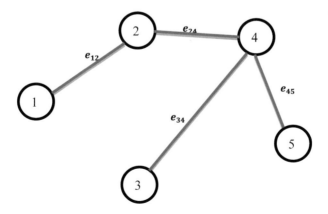

Fig. 14. TrFN minimal spanning tree obtained by Mullai's algorithm.

From the Fig. 14, it can be observed that the TrFN minimal spanning tree $\{1, 2\}, \{2, 4\}, \{4, 3\}, \{4, 5\}$ obtained by Mullai's algorithm, after deneutrosophication of edges' weight, is the same as the path obtained by the proposed algorithm.

The difference between the proposed algorithm and Mullai's algorithm is that Mullai's algorithm is based on the comparison of edges in each iteration of the algorithm and this leads to high computation whereas the proposed approach based on Matrix approach can be easily implemented in Matlab.

6   Conclusion

In this paper, a new approach for searching the minimum spanning tree in a network having trapezoidal fuzzy neutrosophic edge length is presented. The proposed algorithm use the score function of TrFN number, then a comparative example is worked out to illustrate the applicability of the proposed approach. In the next research paper, we can apply the proposed approach to the case of directed neutrosophic graphs and other kinds of neutrosophic graphs including bipolar neutrosophic graphs, and interval valued neutrosophic graphs.

References

1. Smarandache, F.: Neutrosophy. Neutrosophic probability, set, and logic. In: ProQuest Information & Learning, Ann Arbor, Michigan, USA (1998)
2. Wang, H., Smarandache, F., Zhang,Y., Sunderraman, R.: Single valued neutrosophic sets. In: Multisspace and Multistructure, vol. 4, pp. 410–413 (2010)
3. Kandasamy, I.: Double-valued neutrosophic sets, their minimum spanning trees, and clustering algorithm. J. Intell. Syst. 1–17 (2016). https://doi.org/10.1515/jisys-2016-0088
4. Ye, J.: Single valued neutrosophic minimum spanning tree and its clustering method. J. Intell. Syst. **23**, 311–324 (2014)
5. Mandal, K., Basu, K.: Improved similarity measure in neutrosophic environment and its application in finding minimum spanning tree. J. Intell. Fuzzy Syst. **31**, 1721–1730 (2016)
6. Broumi, S., Bakali, A., Talea, M., Smarandache, F., Kishore Kumar, P.K.: Shortest path problem on single valued neutrosophic graphs. In: 2017 International Symposium on Networks, Computers and Communications (ISNCC) (2017). (in press)
7. Broumi, S., Talea, M., Bakali, A., Smarandache, F.: Single valued neutrosophic graphs. J. New Theory **N 10**, 86–101 (2016)
8. Broumi, S., Talea, M., Smarandache, F., Bakali, A.: Single valued neutrosophic graphs: degree, order and size. In: IEEE International Conference on Fuzzy Systems (FUZZ), pp. 2444–2451 (2016)
9. Broumi, S., Smarandache, F., Talea, M., Bakali, A.: Decision-making method based on the interval valued neutrosophic graph. In: Future Technologie, pp. 44–50. IEEE (2016)
10. Mullai, M., Broumi, S., Stephen, A.: Shortest path problem by minimal spanning tree algorithm using bipolar neutrosophic numbers. Int. J. Math. Trends Technol. **46**(N2), 80–87 (2017)
11. http://fs.gallup.unm.edu/NSS/
12. Ye, J.: Trapezoidal fuzzy neutrosophic set and its application to multiple attribute decision making. In: Neural Computing and Applications (2014). https://doi.org/10.1007/s00521-014-1787-6

13. Zhang, C., Li, D., Sangaiah, A.K., Broumi, S.: Merger and acquisition target selection based on interval neutrosophic multigranulation rough sets over two universes. In: Symmetry, vol. 9, no. 7, p. 126 (2017). https://doi.org/10.3390/sym9070126
14. Abdel-Basset, M., Mohamed, M., Sangaiah, A.K.: Neutrosophic AHP-delphi group decision making model based on trapezoidal neutrosophic numbers. J. Ambient Intell. Hum. Comput. 1–17 (2017). https://doi.org/10.1007/s12652-017-0548-7
15. Abdel-Basset, M., Mohamed, M., Hussien, A.N., Sangaiah, A.K.: A novel group decision-making model based on triangular neutrosophic numbers. Soft Comput. 1–15 (2017). https://doi.org/10.1007/s00500-017-2758-5

Differential Evolution Assisted MUD for MC-CDMA Systems Using Non-orthogonal Spreading Codes

Atta-ur-Rahman[1(✉)], Kiran Sultan[3], Nahier Aldhafferi[2], and Abdullah Alqahtani[2]

[1] Department of Computer Science, College of Computer Science and Information Technology (CCSIT), Imam Abdulrahman Bin Faisal University, Dammam, Kingdom of Saudi Arabia
aaurrahman@iau.edu.sa
[2] Department of Computer Information Systems, College of Computer Science and Information Technology (CCSIT), Imam Abdulrahman Bin Faisal University, Dammam, Kingdom of Saudi Arabia
{naldhafeeri,aamqahtani}@iau.edu.sa
[3] Department of CIT, King Abdul Aziz University, Jeddah, Kingdom of Saudi Arabia
kkhan2@kau.edu.sa

Abstract. In this paper, receiver optimization techniques are being investigated into a Differential Evolution (DE) assisted Multiuser Detection scheme for a synchronous, MC-CDMA system. In multiuser detection, the induced multiple access interference (MAI) makes the detection very inefficient and critical. However, the proposed system is less vulnerable to this issue in MC-CDMA communication. In this proposed scheme, for sake of attaining frequency diversity gain, Orthogonal Frequency Division Multiplexing (OFDM) has been used. That is, same signal is transmitted over different sub-carrier frequencies and these sub-carrier frequencies being adequately separated in frequency domain, do not interfere with each other and hence end of the day capacity is added up. Moreover, the role of Walsh (orthogonal but less practical) and Gold spreading sequences (non-orthogonal) which are more practical in nature, is also investigated and the results are demonstrated for different number of users communicating at the same time. The proposed scheme can perform sufficiently well with very low computational complexity compared to the optimum maximum likelihood (ML) detection scheme with increasing users.

Keywords: MC-CDMA · OFDM · MUD · BER · Differential evolution

1 Introduction

Traditional wireless access techniques are consisted of Frequency Division Multiple Access (FDMA) [1], Time Division Multiple Access (TMDA) [1], Code Division Multiple Access (CDMA) [2, 3], and Space Division Multiple Access (SDMA) [4, 5]. In these techniques the users are distinguished (separated or uniquely identified by) by

© Springer International Publishing AG, part of Springer Nature 2018
A. Abraham et al. (Eds.): IBICA 2017, AISC 735, pp. 36–48, 2018.
https://doi.org/10.1007/978-3-319-76354-5_4

means of different frequencies, time slots, signature code and operating antinna beam resepectively. The MC-CDMA is a hybridization of Direct Sequence CDMA (DS-CDMA) and Orthogonal Frequency Division Multiplexing (OFDM). The terminology of hybrid CDMA comprises a group of techniques that combine two or more of the above-mention spread spectrum techniques. One of these hybrid techniques, known as Multicarrier CDMA (MC-CDMA), is of interest in recent many years. Parasad and Hara [1] have given a wonderful overview of MC-CDMA system. Briefly, MC-CDMA scheme may be classified into three major categories.

- Multi Tone CDMA (MT-CDMA) [1].
- MC-DS-CDMA [3].
- Frequency domain spreading MC-CDMA [4].

The common characteristic is these above mentioned techniques is that a spreading code is used for spreading user's signal either in time or in frequency domain and that more than one carrier frequency is used for transmission. In this way one is not obtaining the benefits of Spreading Spectrum but also the frequency diversity. In recent years, several excellent hybrid CDMA schemes were proposed, for example the one by Yang and Hanzo [9]. CDMA techniques have been standardized in the regime of several second generations (2G) [12] and third generation (3G) mobile systems [3]. In all the above-mentioned techniques, Multi-user detection (MUD) is critical process, where induced multiple access interference (MAI) limits the efficiency. One way to mitigate it, use of orthogonal spreading codes. This works well for less number of users, but as the number of users go beyond limit, two problems arise. First, long spreading codes are hard to find also they end up with high chip rate that may cause more bandwidth than ever. Second, the orthogonality does not survive due to a number channel hostilities. Evolutionary algorithms like Differential Evolution and Genetic Algorithms have been widely used for solving various problems in communication systems [13–24] over past many years. In this paper, a reduced complexity multiuser detection scheme is proposed using Differential Evolution (DE) and for sake of spreading the user data, non-orthogonal sequences are investigated (Gold sequences), which are more practical. Rest of the paper is structured as follow. Section 2 discusses the system model in detail, Sect. 3 presents the proposed dectection scheme, Sect. 4 presents the simulation results and Sect. 5 concludes the paper.

2 MC-CDMA System Model

In the proposed scheme the assumed MC-CDMA system model has M number of users being communicating over the same channel. Rayleigh flat fading channel with known Channel State Information (CSI) is considered for sake of experimentation. It is assumed that all these users are simultaneously transmitting the data in a bit synchronized fashion, that is a bit synchronous MC-CDMA. Each bit of each individual user is being spread using different spreading codes each of length L chips. Now this spread bit is modulated over a set of P frequencies which are assuming to be orthogonal. Moreover, the spreading code for each frequency is different that's they are not frequency specific, so the signals are not only orthogonal in time domain but in frequency domain as well. Hence that dual orthogonality helps in separation and demodulation of received composite signal. It also gives frequency diversity gain by using Maximal Ratio Combining (MRC). Hence the total coding gain along each frequency channel is added up and it will be a total of LP. So over here we need total of P spreading codes with each of length L. Here $\omega_1, \omega_2, \omega_3, \ldots \ldots, \omega_p$ are the modulating frequencies being used. With the fact that all these frequencies are separated by minimum $1/T$ interval so no overlap could occur. Hence the capacities will be added up end of the day. Both time and frequency domain orthogonality works as a double edge sword to fight with the impacts of noise and channel distortions. In this work, both orthogonal and non-orthogonal types of codes are investigated in time domain namely the Walsh Codes, and Gold Sequences, respectively. Since it is hard to find orthogonal codes for excessive number of users and we needed here a total of MP codes with length L. Figure 1 shows the transmitter model of MC-CDMA system.

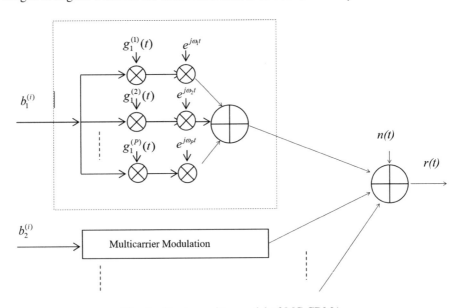

Fig. 1. The transmitter model of MC-CDMA

A. *Transmitter*

In Fig. 1, the assumed bit-synchronous MC-CDMA system is depicted. Here M numbers of simultaneous users are communicating over the same channel. Observe in the figure that the ith bit $b_1^{(i)}$ of the mth user is spread to P parallel subcarriers, each conveying one of the P number of L-chip spreading signature sequences $g_1^{(p)}(t)$, p = 1, . . ., P, each of which spans over (0, T_b) interval in time and we have $T_b/T_c = L$, where T_b and T_c are the bit duration and chip duration, respectively. Each of P spreading signatures is mapped onto a different frequency carrier. That makes the system orthogonal in both time and frequency domains. In other words, a single-carrier system occupying the same bandwidth as the multicarrier system considered would use a spreading signature having LP chips/bit, and both systems have a processing gain of LP.

Hence, the transmitted signal of mth user associated with the pth subcarrier can be expressed in an equivalent low pass representation is given in Eq. 1; composite signal of mth user over all subcarriers can be represented by Eq. 2; composite signal of all users over pth subcarrier can be represented by Eq. 3; and composite signal of all the users over all the subcarriers can be represented by Eq. 4.

$$s_m^p(t) = \sqrt{\frac{2\varepsilon_{b,m}}{P}} g_m^{(p)}(t) b_m^{(i)} e^{j\omega_p t} \tag{1}$$

$$s_m(t) = \sum_{p=1}^{P} s_m^p(t) = \sum_{p=1}^{P} \sqrt{\frac{2\varepsilon_{b,m}}{P}} g_m^{(p)}(t) b_m^{(i)} e^{j\omega_p t} \tag{2}$$

$$s^p(t) = \sum_{m=1}^{M} s_m^p(t) = \sum_{m=1}^{M} \sqrt{\frac{2\varepsilon_{b,m}}{P}} g_m^{(p)}(t) b_m^{(i)} e^{j\omega_p t} \tag{3}$$

$$s(t) = \sum_{m=1}^{M} s_m(t) = \sum_{m=1}^{M} \sum_{p=1}^{P} \sqrt{\frac{2\varepsilon_{b,m}}{P}} g_m^{(p)}(t) b_m^{(i)} e^{j\omega_p t} \tag{4}$$

where $\varepsilon_{b,m}$ is the mth user's signal energy per transmitted bit, $b_m^{(i)}$ belongs to $(1, -1)$ the antipodal signaling symbols, where total number of users are m = 1,...., M and 'i' denotes the ith transmitted bit of mth user, while the mth user's signature waveform is $g_m^{(p)}$, p = 1,...., P, m = 1, ..., M; on the pth subcarrier, which again has a length of L chips, and can be written as:

$$g_m^{(p)}(t) = \sum_{n=1}^{L} g_{m,n}^{(p)}(t) q(t - nT_c) \tag{5}$$

Where Tc is the chip duration, L is the number of chips per bit associated with each subcarrier and we have Tb/Tc = L as the coding gain. Again, the total processing gain in LP, while $q(t)$ is the rectangular chip waveform employed, can be expressed as:

$$q(t) = \begin{cases} 1, & 0 \le t < T_c \\ 0, & otherwise \end{cases} \tag{6}$$

Without loss of generality, we assume that the signature waveform $g_m^{(p)}(t)$ used for spreading the bits to a total of P subcarriers for all the M users has unit energy, which can be written as:

$$\int_0^{T_b} g_m^{(p)2}(t) dt = 1 \tag{7}$$

Where m = 1, 2... M, p = 1, 2,.., P.

B. *Channel*

It is assumed that signal of each user $s_m^p(t)$ transmitted on the pth subcarrier is propagated over an independent non-dispersive single-path Rayleigh Fading channel and where each user face a different amount of fading independent of each other.

Hence, the Channel Impulse Response (CIR) of the mth user on the p-subcarrier can be expressed as: $\alpha_m^p e^{j\theta_m^p}$, where the amplitude α_m^p is a Rayleigh distributed random variable, while the phase θ_m^p is uniformly distributed over $[0, 2\pi]$. That means it can take on any value in this range with equal probability.

C. *Receiver*

Having described the transmitter and the channel, the received signal on the pth subcarrier can be expressed as:

$$r_p(t) = \sum_{i=-\infty}^{\infty} \sum_{m=1}^{M} \sqrt{\frac{2\varepsilon_{b,m}}{P}} g_m^{(p)}(t - iT_b) \gamma_m^p b_m^{(i)} e^{(j\omega_p t + \phi_m^p)} + \eta(t) \tag{8}$$

Here M is the number of users supported and $\eta(t)$ is the additive white Gaussian noise (AWGN) with a variance of $N_0/2$. Fig. 2 depicts the receiver end of the proposed scheme.

The signal will be demodulated with the help of Matched Filters (MF) of matched to each of the M users and the outputs the match filters become the input to the proposed Differential Evolution based multi-user detector (DE-MUD). It is more convenient to express the associated signal in matrix and vector notation as:

$$r_p(t) = \mathbf{G_p H_p A b} + \boldsymbol{\eta} \tag{9}$$

$$\mathbf{G_p} = [g_1^P(t), \ldots, g_M^P(t)]$$
$$\mathbf{W_p} = diag[\alpha_1^p e^{j\theta_1^p}, \ldots, \alpha_M^p e^{j\theta_M^p}]$$
$$\mathbf{A} = diag[\sqrt{\frac{2\varepsilon_{b,1}}{P}}, \ldots, \sqrt{\frac{2\varepsilon_{b,M}}{P}}]$$
$$\mathbf{b} = [b_1, \ldots, b_M]^T$$
$$\boldsymbol{\eta} = [\eta_1, \ldots, \eta_M]^T$$

Based on Eq. 9, the output vector $\mathbf{U}_p$ of the bank of matched filters displayed in Fig. 2 can be formulated as:

$$\mathbf{U}_p = \mathbf{G}_p^T(r_p(t))$$
$$= \mathbf{G}_p^T \mathbf{G}_p \mathbf{W}_p \mathbf{A b} + \mathbf{G}_p^T \boldsymbol{\eta}$$
$$= \mathbf{R}_p \mathbf{W}_p \mathbf{A b} + \boldsymbol{\eta}$$

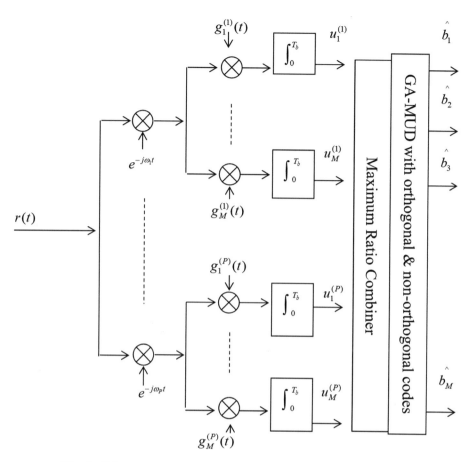

Fig. 2. Schematic of DE assisted MUD receiver of MC-CDMA system

Further that output of the matched filters is fed to a Maximum Ratio Combiner. Also in above equation, the correlation matrix R gives the possible correlation between the codes being used. It can be represented as;

$$\mathbf{R}_p = \begin{pmatrix} \rho_{11}^{(p)} & \rho_{12}^{(p)}\cdots & \cdots\rho_{1M}^{(p)} \\ \rho_{21}^{(p)} & \rho_{22}^{(p)}\cdots & \cdots\rho_{2K}^{(p)} \\ \vdots & \vdots & \ddots\vdots \\ \rho_{M1}^{(p)} & \rho_{M2}^{(p)}\cdots & \cdots\rho_{MM}^{(p)} \end{pmatrix} \tag{10}$$

Here the elements $\rho_{jk}^{(p)}$ of the matrix $\mathbf{R_p}$ are the auto and cross correlations of the spreading code being used, respectively, which can be expressed as:

$$\rho_{jk}^{(p)} = \int_0^{T_b} g_j^p(t)g_k^p(t)dt \qquad (11)$$

In case of Walsh codes that matrix is simply turned to Identity, since all cross correlations ends in zero, because of orthogonality impact. That can be shown by.

$$\rho_{jk}^{(p)} = \begin{bmatrix} \int_0^{T_b} g_j^p(t)g_k^p(t)dt = 1; j = k \\ 0; otherwise \end{bmatrix} \qquad (12)$$

However, this will not be possible in case of Gold sequences. Being non-orthogoanl, their cross correlation will not be zero but something between 0 and 1.

D. *Detection*

Per [25], the optimum multiuser detector of the pth subcarrier will maximize the following objective function:

$$J_p(\mathbf{b}) = 2\mathrm{Re}[\mathbf{b}^T\mathbf{A}\mathbf{W}_p^*\mathbf{U}_p] - \mathbf{b}^T\mathbf{A}\mathbf{W}_p\mathbf{R}_p\mathbf{W}_p^*\mathbf{A}\mathbf{b} \qquad (13)$$

Here the superscript * indicates the conjugate of complex version of matrix. Therefore, combining the contributions of a total of P parallel subcarriers, the objective function to be maximized in the context of an optimum multiuser detected MC-CDMA system can be expressed as the sum of all subcarriers outcome, as:

$$\begin{aligned} J(\mathbf{b}) &= \sum_{p=1}^{P} J_p(\mathbf{b}) \\ &= \sum_{p=1}^{P} \{2\mathrm{Re}[\mathbf{b}^T\mathbf{A}\mathbf{W}_p^*\mathbf{U}_p] - \mathbf{b}^T\mathbf{A}\mathbf{W}_p\mathbf{R}_p\mathbf{W}_p^*\mathbf{A}\mathbf{b}\} \end{aligned} \qquad (14)$$

Hence the decision rule for Verdu's optimum CDMA multiuser detection scheme based on the maximum likelihood (ML) criterion is to choose the specific M-user bit combination $\mathbf{b}$, which maximizes the metric of Eq. 14. Hence, we must find:

$$\hat{\mathbf{b}} = \arg\left\{ \max_{\mathbf{b}}[J(\mathbf{b})] \right\}. \qquad (15)$$

3   Differnetial Evolution Based Multi-user Detection

The maximization of Eq. 15 is a combinational optimization problem, which requires an exhaustive search for each of the 2^M combination of vector **b**, to find the one of that maximizes the metric of Equation. And in case of non-binary symbols this computational complexity is even high. Hence the complexity will increase exponentially with increasing number of users.

Hence, this situation suits Differential Evolution (DE) to find the optimum vector or a solution approximately near to that of optimum ML detector. Also since the orthogonal codes are not very practical so non-orthogonal codes are being utilized and effect of non-orthogonality is measured on the said scenario. Now to start with Differential Evolution algorithm, we need some initial points and considerations. In this case, if we consider M users' data as a single vector, then that can be designated as the initial vector as:

$$\tilde{\mathbf{b}}_n(y) = [\tilde{b}_{n,1}(y), \ldots, \tilde{b}_{n,M}(y)]$$

where y, $y = 1,\ldots,Y$ denotes the yth generation, and n, $n = 1,2,\ldots N$ denotes the nth individual of the population.

Here the received the signal from all subcarriers; summed them up using Maximum Ratio Combiner (MRC) and taken as initial chromosome. Then by mutating it in a special manner we get entire generation. The MRC-combined output vector $\hat{b}_{MRC}$ of the matched filter output can be expressed as: $\hat{\mathbf{b}}_{MRC} = [\hat{b}_{1,MRC}, \ldots, \hat{b}_{M,MRC}]$ where we have:

$$\hat{\mathbf{b}}_{m,MRC} = \sum_{p=1}^{P} u_m^p \gamma_m^p e^{-j\phi_m^p} \tag{16}$$

Having generated $\hat{\mathbf{b}}_{MRC}$, a 'mutated' version of the hard decision vector $\hat{\mathbf{b}}_{MRC}$ is taken for creating each individual in the initial population, where each bits of the MRC-vector is toggled according to the mutation probability used, in this case we utilized 0.1; means one of the 10 bits will be toggled. Hence, the first individual of the population namely $\tilde{b}_p(0)$ can be written as:

$$\tilde{\mathbf{b}}_p(0) = MUTATION[\hat{\mathbf{b}}_{MRC}] \tag{17}$$

So one can easily note that MUTATION is an operator, which when applied to a string of $(1, -1)$, will produce the toggled versions of initial vector.

4   Simulation Results

The basic parameters used for the simulation of the proposed DE assisted MUD for CDMA are considered in the following way.

The modulation scheme used is binary phase shift keying (BPSK), the CDMA spreading codes utilized are both orthogonal (Walsh codes) and non-orthogonal (Gold sequences). Number of subcarriers P is used as 8, the length of subcarrier spreading signature L is (8, 31) for (Walsh, Gold) respectively. So the according coding gain for Walsh is LP = 64 and for Gold is LP = 248. DE's selection was based upon fitness value returned by the cost function. Mutation was used as standard binary mutation. Bit flipping methodologies are used as standards multi-point. Bit mutation probability was 0.5 while crossover probability is 1. From Fig. 3 we can observe that the DE-assisted MUD's performance improves, when the population size P increases. The difference between P = 20 and 30 is more than an order while between 30 and 40 the difference in BER is exactly of an order for Walsh Code in higher SNRs.

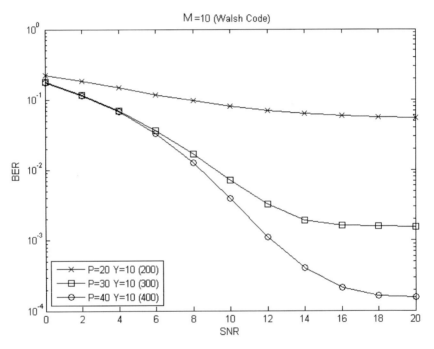

Fig. 3. BER performance of the DE assisted MUD using a 64-chip Walsh code

Similarly, in Fig. 4, in case of Gold Sequences population size plays a key role in reducing BER, especially when SNR > 13 dB. Number of users was 10 in Figs. 3 and 4.

For example, for Signal to Noise Ratio (SNR) values below 15 dB Bit Error Rate (BER) is significantly decreased for M = 10 users, when evaluating the objective function of Eq. 14, which imposes a complexity on order of O (P.Y) = O (40.10) = O (400) Furthermore, when the number of users M is increased to 20, the DE assisted MUD has a complexity of O (P.Y) = O (80.20) = O (1600), as seen in Fig. 5. Further results are obtained for M = 20 number of users using Gold sequences, which are more practical in nature due to their availability as well as relatively low chip rate compared to orthogonal codes.

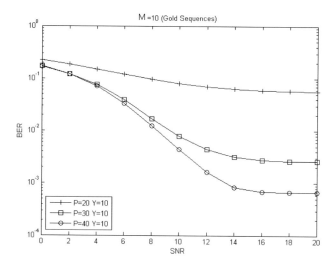

Fig. 4. BER performance of the DE assisted MUD using a 248-chip Gold code

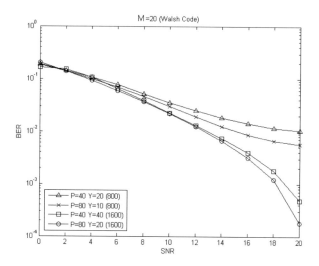

Fig. 5. BER performance of the DE assisted MUD using a 64-chip Walsh code

These results can be seen in Figs. 5 and 6. We can readily deduce that the population size P plays an important role to significantly reduce the bit error rate. Here almost same bit error rate is achievable as for M = 10 number of users, but definitely at the cost of complexity. Even in this scenario Walsh Codes perform better due to their inherent orthogonality. Furthermore, for very high SNR like SNR > 18 a bit error rate of 10-4 is achievable. Almost same effect of population size can be seen in Fig. 6 for 248-chip Gold sequences. In comparison to Walsh Code, the Gold sequences do not perform well but effect of DE can be seen for both codes. We can also observe that DE-assisted MUD is capable of significantly reducing the complexity of Verdu's

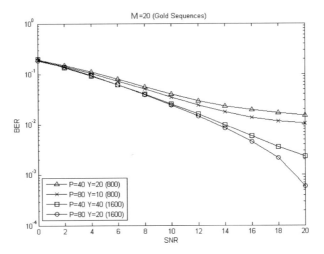

Fig. 6. BER performance of the DE assisted MUD for a bit-synchronous MC-CDMA

optimum MUD. For example, the complexity was reduced by a factor of 1300, when the number of users was M = 20. It is very interesting to note that the overall complexity in Figs. 4 and 6 is same for the cases O (40.40) and O (80.20) that is both are equal to O (1600) but graph of P = 80 in both figure shows a better performance. Hence, for the number of users, M > 14, population size P dominates the effect of number of generations Y.

Figure 7 demonstrates the complexity reduction factor versus number of users. This is because the increase in population size causes more crossovers and hence more parents are involved so the probability to find the optimum increases. An interesting fact can be seen here that with increase in complexity (though very small compare to

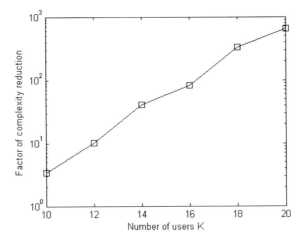

Fig. 7. The complexity reduction factor bounded at a BER of 10^{-3}

optimum) we can achieve the same results that of optimum detector. For example, in this figure below the complexity of 400 is giving good results and even good can be expected with 500 which is almost half of the optimum case which is 1024 for this case of 10 users.

5 Conclusion and Future Work

In this paper, a Differential Evolution (DE) assisted Multi-user detection (MUD) scheme for Multi-carrier CDMA (MC-CDMA) system is proposed. The proposed scheme significantly reduces the detection complexity compared to optimum Maximum Liklihood (ML) detector with high complexity, especially when the number of supported users is higher than 15. Population size plays a key role in decreasing the BER instead of number of generations in DE. Walsh code performs better in terms of detection compared to the Gold code being orthogonal in nature. However, practically it is hard to find the orthogonal codes, so Gold sequences are more practicle and also suitable for the systems with low chip rate where Walsh codes do not fit in.

References

1. Glisic, S.G., Leppannen, P.A.: Wireless Communications-TDMA Versus CDMA. Kluwer Academic Publishers, Dordrecht (1997)
2. Prasad, R., Hara, S.: Overview of multi-carrier CDMA. In: Proceeding of the IEEE International Symposium on Spread Spectrum Techniques and Applications (ISSSTA), (Mainz, Germany), pp. 107–114, 22–25 September 1996
3. Hanzo, L., Yang, L.L., Kuan, E.L., Yen, K.: Single and Multi-Carrier DS-CDMA, p. 1060. Wiley and IEEE Press, Hoboken (2003)
4. Hanzo, L., Munster, M., Choi, B.J., Keller, T.: OFDM and MC-CDMA. Wiley and IEEE Press, Hoboken (2003)
5. Bolgh, J., Hanzo, L.: Third-Generation Systems and Intelligent Networking. Wiley and IEEE Press, Hoboken (2002)
6. Steele, R., Hanzo, L.: Mobile Radio Communications, 2nd edn. IEEE Press-Wiley, Hoboken (1999)
7. Proakis, J.G.: Digital Communications, 3rd edn. Mc-Graw Hill International Editions, New York (1999)
8. Prasad, R., Hara, S.: Overview of Multicarrier CDMA. IEEE Commun. Mag. **35**, 126–133 (1997)
9. Yang, L.L., Hanzo, L.: Software-Defined-Radio-Assisted Adaptive Broadband Frequency Hopping Multicarrier DS-CDMA. IEEE Commun. Mag. **4**, 174–183 (2002)
10. Viterbi, A.: Principles of Spread Spectrum Communications. Addison-Wesley, Boston (1995)
11. Prasad, R.: CDMA for Wireless Personal Communications. Artech House Inc., London (1996)
12. Miller, L., Lee, J.S.: CDMA Systems Engineering Handbook. Artech House, London (1998)
13. Atta-ur-Rahman, Qureshi, I.M., Malik, A.N., Naseem, M.T.: QoS and rate enhancement in DVB-S2 using fuzzy rule base system. J. Intell. Fuzzy Syst. (JIFS) **30**(1), 801–810 (2016)

14. Atta-ur-Rahman, Qureshi, I.M., Malik, A.N., Naseem, M.T.: Dynamic resource allocation for OFDM systems using differential evolution and fuzzy rule base system. J. Intell. Fuzzy Syst. (JIFS) **26**(4), 2035–2046 (2014). https://doi.org/10.3233/ifs-130880

15. Atta-ur-Rahman, Qureshi, I.M., Malik, A.N., Naseem, M.T.: A real time adaptive resource allocation scheme for OFDM systems using GRBF-neural networks and fuzzy rule base system. Int. Arab J. Inf. Technol. (IAJIT) **11**(6), 593–601 (2014)

16. Atta-ur-Rahman, Naseem, M.T., Muzaffar, M.Z.: Reversible and robust watermarking using residue number system and product codes. J. Inf. Assur. Secur. (JIAS) **7**, 156–163 (2012)

17. Atta-ur-Rahman, Azam, M., Zaman, G.: Performance comparison of product codes and cubic product codes using FRBS for robust watermarking. Int. J. Comput. Inf. Syst. Ind. Manag. Appl. (IJCISIMA) **8**(1), 57–66 (2016)

18. Atta-ur-Rahman, Qureshi, I.M., Naseem, M.T.: Performance of modified iterative decoding algorithm for multilevel codes in adaptive OFDM system. Int. J. Comput. Inf. Syst. Ind. Manag. Appl. (IJCISIMA) **6**, 1–10 (2013)

19. Atta-ur-Rahman: Applications of softcomputing in adaptive communication. Int. J. Control Theory Appl. **10**(18), 81–93 (2017)

20. Atta-ur-Rahman, Qureshi, I.M., Malik, A.N.: Adaptive resource allocation in OFDM systems using GA and fuzzy rule base system. World Appl. Sci. J. (WASJ) **18**(6), 836–844 (2012)

21. Atta-ur-Rahman, Qureshi, I.M., Malik, A.N.: A fuzzy rule base assisted adaptive coding and modulation scheme for OFDM systems. J. Basic Appl. Sci. Res. **2**(5), 4843–4853 (2012)

22. Atta-ur-Rahman: Applications of evolutionary and neuro-fuzzy techniques in adaptive communications. In: Modeling, Analysis and Applications of Nature-Inspired Metaheuristic Algorithms, 1st edn., Chap. 10, pp. 183–217. IGI Global (2017)

23. Atta-ur-Rahman, Qureshi, I.M.: Effectiveness of modified iterative decoding algorithm for cubic product codes. In: International Conference on Hybrid Intelligent Systems (HIS 2014), pp. 260–265, Kuwait, 14–16 December (2014)

24. Atta-ur-Rahman, Qureshi, I.M.: Optimum resource allocation in OFDM systems using FRBS and particle swarm optimization. In: Nature and Biologically Inspired Computing (NaBIC 2013), North Dakota, USA, pp. 174–180 (2013)

25. Yen, K., Hanzo, L.: Genetic algorithm assisted joint multiuser symbol detection and fading channel estimation for synchronous CDMA systems. IEEE J. Sel. Areas Commun. **19**, 985–998 (2001)

Solving the Problem of Distribution of Fiscal Coupons by Using a Steady State Genetic Algorithm

Qëndresë Hyseni[1], Sule Yildirim Yayilgan[2], Bujar Krasniqi[1], and Kadri Sylejmani[1(✉)]

[1] Faculty of Electrical and Computer Engineering, University of Prishtina, Prishtina, Kosovo
{qendrese.hyseni,bujar.krasniqi,kadri.sylejmani}@uni-pr.edu
[2] Department of Information Security and Communication Technology,
Norwegian University of Science and Technology, Gjøvik, Norway
sule.yildirim@ntnu.no

Abstract. When customers buy goods or services from business entities they are usually given a receipt that is known with the name fiscal or tax coupon, which, among the others, contains details about the value of the transaction. In some countries, the fiscal coupons can be collected during a certain period of time and, at the end of the collection period, they can be handed over to the tax authorities in exchange for a reward, whose price depends on the number of collected coupons and the sum of their values. From the optimisation perspective, this incentive becomes interesting when, both the number of coupons and the sum of their value is large. Hence, in this paper, we model this problem in mathematical terms and devise a test set that can be used for benchmarking purposes. Furthermore, we propose a solution based on Genetic Algorithms, where we compare its results versus the results to the solution of the relaxed versions of the proposed problem. The computational experiments indicate that the proposed solution obtains promising results for complex problem instances, which show that the proposed algorithm can be used to solve realistic problems in a matter of few seconds by utilizing standard personal computers.

Keywords: Distribution of fiscal coupons · Mathematical modelling
Genetic algorithms

1 Introduction

The tax authorities of many countries try to find alternative ways to enforce business entities (e.g. shops, restaurants, travel agencies, etc.) to fully declare the profit they gain from their business activities, so that they have to pay taxes accordingly. The tax authorities from several countries, like for example Republic of Kosovo [1], utilize the strategy of encouraging the customers to collect the fiscal coupons when they do any kind of transaction with business entities. The collected coupons can be enveloped and submitted to the tax authorities in exchange of a reward that depends on the number and the total value of the fiscal coupons enclosed. In general, depending on the actual rules put in place by specific tax authorities, there can be different types of envelopes that can

© Springer International Publishing AG, part of Springer Nature 2018
A. Abraham et al. (Eds.): IBICA 2017, AISC 735, pp. 49–59, 2018.
https://doi.org/10.1007/978-3-319-76354-5_5

be submitted. Obviously, envelopes with more coupons and with higher total values, have higher rewards.

In more formal terms, in the case of the Distribution of Fiscal Coupons Problem (DFCP), each person has N number of coupons (see Fig. 1) collected for a period of time (e.g. a three month period). At the end of the collection period, the coupons will be distributed into a T number of envelopes by the person who possesses the coupons. The person has to make a decision related to which coupon is placed in which envelope. Each coupon has a value and consequently it should be placed in the envelope ultimately where the sum of values of the coupons in the envelope will lead to an overall higher reward. Each coupon can only be placed into a single envelope. The number of coupons in each envelope cannot be less than a minimum, whereas also the sum of all coupons cannot fall under minimum value. The achievable reward from each envelope type is predefined based on the number and values of the coupons placed inside it.

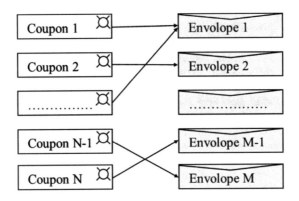

Fig. 1. The schematic view of the fiscal coupons problem

The significance of the work in this paper can be underlined by outlining its main contributions, which are: (1) introduction of a new optimization problem for the scientific community by presenting a mathematical formulation, as well as a test set that can be used for benchmarking purposes, (2) development of a metaheuristic based solution to the newly introduced problem, namely Genetic Algorithms, and (3) presentation of the systematic computation results that compare the performance of the proposed algorithm against solutions of the relaxed version of the envisioned problem, which can be used as benchmark results for future solutions.

The remainder of this paper is structured as is in the following. Section 2 presents a literature review of the related problems and their respective solution approaches. Next, in Sect. 3, we present the mathematical modelling of the DFCP problem as an Integer Linear Programming Problem. In Sect. 4, we present the proposed approach for solving the DFCP problem, while in Sect. 5, we show computation results of the proposed approach against a data set of 10 instances. Finally, in Sect. 6, we conclude the paper and present our view of the eventual future work.

2   Related Problems

In the classical Knapsack Problem (KP) there is a set of items and a container (knapsack) that has to be used for carrying a subset of items. Each item is characterized with two properties, namely value and weight, whereas the container has a single property, which is the maximum weight it can carry. The goal is to place a subset of items into the container such that the total value of the placed items is maximized subject to the capacity of the container. The Multiple Knapsack Problem (MKP) extends the KP problem by allowing multiple containers of the same capacity [2, 3], whereas the Distributed Multiple Knapsack Problem (DMKP) supports containers of varying capacities, which can be modeled as a general Distributed Constraint Optimization Problem (DCOP) [4, 5].

Another related problem is the Bin Packing Problem (BPP), where a set of items need to be placed into a set of containers (bins). Each item has a weight property, whereas each container has a maximum capacity property. The goal is to place each item into a container such that the number of containers used is minimised. In comparison to KP problem, where only a subset of items can be picked, in the BPP problem all items have to be picked up [6].

An additional related problem is the Rack Configuration Problem (RCP), where there is a set of items (electronic cards) that need to be placed (connected) into a set of containers (racks). Each item has a single property (i.e. power it requires), whereas each container has three properties, namely maximal power it can supply, number of connectors and the price. The goal is to plug in all the electronic cards into a set of racks with the smallest cost possible [7].

In Table 1, we compare the newly proposed DFCP problem against the above presented related problems, by outlining different characteristics of the individual problems, such as: type of objective function, capacity constraints (i.e. upper/lower limit), the number of containers, number of features per item/container, and whether all items need to be selected. By analysing the details given in the table, one can conclude that DFCP problem is closely related to RCP problem, in terms of number of containers and features per item/container, but differs in terms of the objective function, capacity constraints and in the aspect whether all items need to be selected.

Table 1. Comparison of features of various related problems

Problem	Objective	Capacity constraints		No. of. containers	No. of features per item	No. of features per container	All items to be selected
		Upper limit	Lower limit				
KP	Max	Yes	No	1	2	1	No
MKP	Max	Yes	Yes	>1	2	1	No
DMKP	Max	Yes	Yes	>1[a]	2	1	No
BPP	Min	Yes	No	>1	1	1	Yes
RCP	Min	Yes	No	>1	1	3	Yes
DFCP	Max	No	Yes	>1	1	3	No

[a]Containers might be of different capacities

To the best of authors' knowledge, there is no any problem in the literature that models or solves the problem of the distribution of fiscal coupons, hence in the following section, we present a mathematical modelling of this problem, along with the proposed solution.

3   Mathematical Modelling

The mathematical modelling for the problem of the optimal distribution of the fiscal coupons is formulated as an Integer Linear Programming (ILP) model that has a range of parameters and a couple of decision variables, as specified below:

Parameters:

N – Number of coupons
v_i – Value of coupon i, $\forall i = 1, \ldots, N$
T – Number of types of envelopes
C_k – Minimum number of coupons in envelope of type k, $\forall k = 1, \ldots, T$
S_k – Minimum sum of all coupons in envelope of type k, $\forall k = 1, \ldots, T$
R_k – Achievable reward from envelope of type k, $\forall k = 1, \ldots, T$

Decision variables:

M – Number of envelopes
x_{jk} – equals 1, if envelope j is of type k, otherwise it is 0, $\forall j = 1, \ldots, M$, $\forall k = 1, \ldots, T$
y_{ij} – equals 1, if coupon i is placed in envelope j, otherwise it is 0, $\forall i = 1, \ldots, N$, $\forall j = 1, \ldots, M$,

Objective function:

$$Max \sum_{j=1}^{M} \sum_{k=1}^{T} R_k x_{jk} \tag{1}$$

Constraints:

$$\sum_{i=1}^{N} y_{ij} x_{jk} \geq C_k, \forall j = 1, \ldots, M, \forall k = 1, \ldots, T \tag{2}$$

$$\sum_{i=1}^{N} v_i y_{ij} x_{jk} \geq S_k, \forall j = 1, \ldots, M, \forall k = 1, \ldots, T \tag{3}$$

$$\sum_{i=1}^{N} y_{ij} \leq 1, \forall j = 1, \ldots, M \tag{4}$$

$$\sum_{j=1}^{M} x_{jk} = 1, \forall k = 1, \ldots, T \tag{5}$$

In the mathematical formulation presented above, Eq. (1) denotes the objective function of the problem at hand, which is maximizing the total reward, by determining which combination of envelopes yields to the highest possible profit. Constraints (2) and (3) ensure the validity of envelopes in terms of the requirement for the minimum number of coupons and the minimum sum of their values, respectively. Constraint (4) guarantees that each single coupon is inserted into at most one envelope, whilst Constraint (5) makes sure that each single envelope can belong to only one particular type of envelope.

4 Genetic Algorithm Solution

In this section, we present the approach for solving the DFCP problem, where, we first present the basic algorithm ingredients, and then present the algorithmic solution in the form of pseudocode.

4.1 Basic Algorithm Ingredients

In this implementation, we use the Steady State approach of Genetic Algorithms, which was popularized by Whitley and Kauth [8], and its main idea, compared to the traditional generational approach, is to update the population in a slight manner rather than all at one time. The algorithm iteratively breeds a new child or two, assesses their fitness, and then restores them directly into the population itself, slaying off some preexisting individuals to make room for them. The Steady-State Genetic Algorithm has two essential features. First, it uses half the memory of a standard genetic algorithm, because there is only one population at a time. Second, it is more exploitative compared to a generational approach [9]. The particular details of the Steady-State Genetic Algorithm implemented here, can be summarized as in the following:

Representation of a given candidate solution is made as a list of lists, where the size of the main list corresponds to the number of assigned envelopes M, whereas each single member of the main list is also a list that corresponds to the number of coupons n_i placed inside a given envelope i. A sample representation of a given solution is: $S = \{E_1, E_2, \ldots, E_i, \ldots, E_M\}$, where $E_i = \{C_1, C_2, \ldots, C_{n_i}\}$.

Initialization of a given candidate solution begins by reading all the coupons values from the given problem instance, and then, randomly distributing them into a random number of envelopes, by considering the hard constraints about the minimum number/sum of coupons. The number of generated initial solutions is equal to the *population size (ps)* parameter.

Mutation mechanism of the algorithm consists of two operators, namely *swap* and *shift*, where the earlier swaps two coupons belonging to distinctive envelopes, while the later shifts a coupon from a given envelope to some other envelope, of the same candidate solution. In order to apply a number of swaps between different coupons of a given individual, the *swap* operator iterates through a loop for a number of iterations (as specified by *sw* parameter). During the course of a single iteration, initially, two distinct envelopes are selected randomly, and then, for swapping purposes, one random coupon is selected from each of these envelopes. The shift operator is also executed for several iterations, as specified by *sh* parameter. During the evolution of a given iteration, initially, two distinct envelopes are selected randomly, and then one random coupon is selected from the first envelope, and gets shifted to the second one.

Evaluation of a given candidate solution is done using Eq. (1), which, as described before, maximizes the total reward, by determining which combination of envelopes yields to the highest possible profit. Each member of the population has a certain number of envelopes of different types, and the sum of these envelope values denotes the fitness of the member.

Selection of the parent that will take part in breading the next population is completed by using the Tournament Selection algorithm. This algorithm is a simple and an effective one, as it returns the fittest individual of some *ts* individuals picked at random from the population [9].

Population update mechanism replaces the worst fit member of the current population (i.e. it replaces the current worst solution from the population with the best picked from tournament size individuals).

4.2 Pseudocode of the Algorithm

In abstract terms, as shown in Algorithm 1, the envisioned GA approach has 6 parameters, which can be used for fine tuning its performance for different problem complexities and sizes. Besides the default genetic algorithm parameters, such as population size (*ps*), maximum generations (*mg*) and tournament size (*ts*), the particular implementation at hand, uses three so called "intensity" parameters, namely swap mutate (*sw*) and shift mutate intensity (*sh*), for specifying the number of times a certain operator (i.e. swap or shift) will be applied when called upon. In addition, the algorithm uses a special parameter called the alternation frequency (*af*) to change the mutation operator from swapping to shifting and vice versa every *af* number of generations.

At the very start of the algorithm, a population P of n individuals is created by using the procedure for creating the initial solution explained above. Next, in the repetitive phase of the algorithm, at each iteration, the following steps are undertaken: (1) evaluation of all individuals, (2) selection of the parents based on tournament selection and mutation over the operators (i.e. swap and shift) used in the running iteration, and (3) formation of the new population by replacing the individual with the worst fitness, with the mutated new individual if the fitness of the second is better. The algorithm terminates when the maximum number of foreseen generations is achieved.

Algorithm 1 Steady State Genetic Algorithm

Require: coupons C (N, v_i, T, C_j, S_j, R_j), where $i=1\ldots N$ and $j=1\ldots T$; population size ps; maximum generations mg; tournament size ts; swap mutate intensity sw; shift mutate intensity sh; operator alternation frequency af.

  1:    $P = \{\}$; Best $= \emptyset$; Worst $= \emptyset$;
  2: **for** ps times **do**
  3:      $C_r =$ Random Individual (C)
  4:      AssessFitness(C_r)
  5:      **if** Worst $== \emptyset$ or Fitness$(C_r) <$ Fitness(Worst) **then**
  6:          Worst $= C_r$
  7:      $P = P \cup C_r$
  8: **for** each generation until mg **do**
  9:      $P_w =$ SelectWithReplacement(P),
10:      $P_b =$ TournamentSelection(P, ts),
11:      $C_b =$ Select best from P_b
12:      $C_b =$ Mutate(C_b, sw, sh, af)
13:      **if** Fitness$(C_b) <$ Fitness(Worst) **then**
14:          Worst $= C_b$
15:      SelectForDeath(P_w)
16:      $P = P - P_w$
17:      $P = P \cup C_b$
18:      **for** each individual $P_i \in P$ **do**
19:          AssessFitness(P_i)
20:          **if** Best $== \emptyset$ or Fitness$(P_i) >$ Fitness(Best) **then**
21:             Best $= P_i$
22: **return** Best

---

5   Computational Experiments

In this section, we initially present a test set of 10 instances that are used for conducting the evaluations of the presented solution. Further, we show the computational results for tuning the parameter values of the proposed approach. After that, we compare the obtained results against the lower bound values that are within reach when relaxing individual hard constraints of the problem at hand. The proposed GA based algorithm is developed by using the C# programming language through the developing environment of MS Visual Studio 2017. All experiments are done using a machine with an Intel Core processor i7-7500U CPU with the clock speed of 2.9 GHz and a RAM memory of 16 GB. The GA algorithm is tested under a MS Windows 10 Home 64 bit operating system.

5.1   Test Set

In order to test the algorithm for various scenarios of the distribution of fiscal coupons, we have set up a test set that consist of 10 different instances, where the values of individual coupons are generated randomly. Table 2 shows the characteristics of individual instances, which includes instance name, number of coupons and the total value of all coupons. The instance name, in addition to problem abbreviation DFCP, also encompasses the number of coupons and the total value present in a particular instance, e.g. Instance DFCP_2h_3k contains 200 (2hekta - 2*h*) coupons with a total value of 3000 (3kilo - 3*k*) currency units. In practice, the value of a fiscal coupon ranges from very small amounts (e.g. a chewing gum might cost less than a euro) to large amounts (e.g. a technological appliances might cost several, dozens, hundreds or even thousands of euros). However, during a certain period of time (e.g. a month or a year quartile), the number of large value transactions (i.e. fiscal coupons) made by a person is usually much lower than the number of transactions with small values. Hence, in order to make the test instances more realistic, 30% of coupons are set to have larger values, which range from several up to dozens of currency units (e.g. euros). Furthermore, based on the constraints enforced in practical situations, such as in the case of Tax Authorities of the Republic of Kosovo [1], three envelope types are defined throughout all test instances. In general, an envelope type is described with three properties, namely the minimum number of coupons, the minimum sum of the coupons and the foreseen reward. In particular, the types of envelopes used in the test set read as in the following: Type1 = {30, 250, 10}, Type2 = {40, 500, 15} and Type3 = {50, 800, 20}.

Table 2. Test set details and maximal reward when relaxing individual constraints

Instance name	Instance details		Envelope details					
			Number of coupons			Sum of coupons		
	Number of coupons	Total value	30	40	50	250	500	800
DFCP_2h_2k	200	2000	60	75	**80**	**80**	60	50
DFCP_2h_3k	200	3000	60	75	80	**120**	90	80
DFCP_5h_5k	500	5000	160	180	**200**	**200**	150	120
DFCP_5h_6k	500	6000	160	180	200	**240**	180	150
DFCP_1k_10k	1000	10000	330	375	**400**	**400**	300	250
DFCP_1k_11k	1000	11000	330	375	400	**440**	330	280
DFCP_2k_20k	2000	20000	660	750	**800**	**800**	600	500
DFCP_2k_22k	2000	22000	660	750	800	**880**	660	550
DFCP_5k_50k	5000	50000	1660	1875	**2000**	**2000**	1500	1250
DFCP_5k_55k	5000	55000	1660	1875	2000	**2200**	1650	1380

5.2 Upper Bound Limits

In addition, in Table 2, we present the maximal reward that can be achieved per instance if individual problem constraints are relaxed (i.e. either the constraint for the sum or number of coupons in the envelope is not enforced). In case the constraint for the sum of coupons is relaxed (i.e. it is not taken into account), the maximal reward that can be achieved, in all instances, is when the envelopes are all of Type3 (i.e. the number of coupons is 50). On the other hand, when the constraint for the minimum number of coupons is relaxed, the best scenario, in all instances, is when all the envelopes are of Type1 (i.e. the minimum sum of coupons is 250). If the constraint for the minimum sum of coupons is relaxed then the formula for calculation of upper bound values is $UB = [No.$ $coupons]/[Min.$ no. of coupons per envelope type] $* [Reward per envelope type]$, other-wise, if the constraint for the minimum number of coupons is relaxed the envisioned formula is $UB = [Total value]/[Min.$ sum of coupons per envelope type] $* [Reward per envelope type]$. In the case of relaxation of the minimum sum of coupon constraint, a sample calculation of the upper bound value for instance DFCP_2h_2k (the sixth column in Table 2) is $UB = 200/50 * 20 = 80$. Comparing the values in the sixth and the seventh column of Table 2, one can notice that the scenario of having envelopes of Type1 (i.e. the sum of coupons is 250) while relaxing the constraint for the minimum number of coupons, is the best scenario for all instances in the test set. Hence, in the following section, we use these values as Upper Bound (UB) limits (i.e. benchmark values) for evaluating the results that are obtained by the introduced solution in this paper.

5.3 Evaluation Results

In order to calibrate the values of the parameters of the GA approach, a systematic experimentation is performed by using the complete test set. Initially, based on some preliminary experimentation, for each parameter, a range of five best performing values is selected. Then, for each selected value, the algorithm is executed for each test instance 10 times. As a result, for each single parameter, the value that in average produces better results than the other four values, is adapted for the final round of the experimentation that is done with the aim of evaluating the performance of the proposed algorithm. The tuned values for the GA approach read as in the following: $mg = 10,000$; $ps = 5,000$; $ts = 20$; $sw = 15$; $sh = 20$ and $af = 10$.

In Table 3, we present the results of best, average and worst case scenario execution, for individual instances over ten unique executions of the algorithm, where the results are compared against the upper bound values (described in the previous section). In general, when the results are averaged over the whole test set, the gap of GA resulting

Table 3. Fitness results of GA versus upper bound limits

Instance name	Upper bound (UB)	Best	Average	Worst	$GA_{Avg.}$ vs. UB (%)
DFCP_2h_2k	80	65	61.82	60	22.73
DFCP_2h_3k	120	75	71.26	70	40.62
DFCP_5h_5k	200	170	159.57	145	20.21
DFCP_5h_6k	240	185	172.10	155	28.29
DFCP_1k_10k	400	330	297.20	280	25.70
DFCP_1k_11k	440	345	317.77	295	27.78
DFCP_2k_20k	800	600	574.25	545	28.22
DFCP_2k_22k	880	645	613.43	575	30.29
DFCP_5k_50k	2000	1365	1348.26	1340	32.59
DFCP_5k_55k	2200	1470	1427.17	1410	35.13
					29.16

values from the upper bound values is 29.16%. This gap remains below the average for instances with less than 2,000 coupons with total values of 22,000 currency units, except for instance DFCP_2h_3k (that has 200 coupons that have a total of 3000 currency units), which has a gap of 40.62%. These gaps should be considered as relative and only for comparison purposes, since the upper bound values do not represent actual solutions to the problem, but only solutions to the relaxed version of it. Hence, the obtained results can be considered as promising given that the computation time is short.

With regard to the computation time (see Table 4), the results show that the GA approach needs about 27.79 s, in average, to solve a DFCP problem from the test set. The results show that best and worst case computation time is relatively stable for the first four instances and the last instance (where the difference is at most 3.3 s), whereas, for the other instances, this difference is more than 10 s, especially for instance DFCP_1k_10k, where the difference goes up to 27.7 s. The worst case execution scenario, always remains under a computation time of less than 50 s, which shows the usability of the algorithm in practice, where generating good quality solutions would enable the user to gain more revenue from the practice of coupon collection that is applied in tens of countries around the globe (e.g. Republic of Kosovo [1]).

Table 4. Computation time of GA approach (in seconds)

Instance name	Best	Average	Worst
DFCP_2h_2k	20.16	21.44	23.44
DFCP_2h_3k	20.60	21.01	21.84
DFCP_5h_5k	21.02	21.53	22.22
DFCP_5h_6k	20.44	22.06	23.51
DFCP_1k_10k	21.39	26.37	49.07
DFCP_1k_11k	21.24	24.41	43.74
DFCP_2k_20k	25.33	30.21	37.77
DFCP_2k_22k	28.24	32.46	41.33
DFCP_5k_50k	37.68	39.95	50.02
DFCP_5k_55k	37.30	38.43	39.62
Avg.	25.34	27.79	35.26

6   Conclusion and Future Work

In this paper, we introduced a new problem for modelling the optimal distribution of fiscal coupons and devised an Integer Linear Programming (ILP) mathematical formulation. Further, we presented a metaheuristic approach based on Genetic Algorithms, which is able to solve the formulated problem at hand in matter of tens of seconds by using standard computing devices. In addition, a newly introduced test was used for benchmarking purposes, where it was shown that the proposed approach produces competitive results when compared to upper bound values. For additional comparison, as part of future work, we plan to develop exact methods from the field of dynamic programming and also investigate hybridization of the presented approach with other metaheuristics, as well as utilization of constraint satisfaction problem (CSP) techniques within the existing metaheuristic for the envisioned problem.

References

1. Tax Administration of Kosovo: Tax Administration of Kosovo, February 2015. http://www.atk-ks.org/
2. Chekuri, C., Khanna, S.: A polynomial time approximation scheme for the multiple knapsack problem. SIAM J. Comput. **35**(3), 713–728 (2005)
3. Chen, Y., Hao, J.-K.: Memetic search for the generalized quadratic multiple knapsack problem. IEEE Trans. Evol. Comput. **20**(6), 908–923 (2016)
4. Yeoh, W., Felner, A., Koenig, S.: BnB-ADOPT: an asynchronous branch-and-bound DCOP algorithm. In: Proceedings of the 7th International Joint Conference on Autonomous Agents and Multiagent Systems, vol. 2 (2008)
5. Mailler, R., Lesser, V.: Solving distributed constraint optimization problems using cooperative mediation. In: Proceedings of the Third International Joint Conference on Autonomous Agents and Multiagent Systems, vol. 1 (2004)
6. Karmarkar, N., Karp, R.M.: An efficient approximation scheme for the one-dimensional bin-packing problem. In: 23rd Annual Symposium on SFCS 2008. IEEE (1982)
7. Kızıltan, Z., Hnich, B.: Symmetry breaking in a rack configuration problem. In: IJCAI 2001 Workshop on Modelling and Solving Problems with Constraints (2001)
8. Whitley, D., Kauth, J.: GENITOR: a different genetic algorithm. In: 1988 Rocky Mountain Conference on Artificial Intelligence (1988)
9. Luke, S.: Essentials of Metaheuristics, 2nd edn. Lulu (2013)

A Survey of Cross-Layer Design for Wireless Visual Sensor Networks

Afaf Mosaif$^{(\boxtimes)}$ and Said Rakrak

LAMAI Laboratory, FSTG, Cadi Ayyad University, Marrakesh, Morocco
afaf.mosaif@edu.uca.ac.ma, s.rakrak@uca.ac.ma

Abstract. Wireless Visual Sensor Network is a collective network of nodes capable of collecting, processing, and transmitting a huge amount of image/video data from a region of interest to the base station. These nodes are equipped by cameras and are characterized by their limited resources in terms of computational capability, bandwidth and battery power. This type of Networks are more complicated and challenging compared to traditional wireless sensor networks. Therefore, a number of solutions have been recently proposed such as the cross-layer designs, which is an interesting research topic. It allows sharing information across all network layers even the nonadjacent ones, in order to improve the wireless network functionality and to obtain performance gains. In this article, we will present a survey of cross-layer design in Wireless Visual Sensor Networks where we will classify the recent proposals in this area in term of their architecture, interaction categories and theirs outcomes.

Keywords: Cross-layer design · Wireless visual sensor networks
Wireless multimedia sensor networks · Wireless video sensor networks

1 Introduction

Wireless visual sensor networks (WVSNs) are considered an extension of Wireless sensor Networks (WSNs) as shown in Fig. 1.

Recent developments of inexpensive CMOS (Complementary Metal Oxide Semiconductor) cameras brought the opportunity of imaging capabilities to sensor networks. In fact, sensor nodes can collect image/video data from an area of interest, process it collaboratively, and transmit the useful information to the Base Station (BS) for further analysis via multihops short range transmissions. These nodes are battery powered and equipped by cameras. Therefore, they operate with respect to the available and limited resources and they are called camera nodes.

Consisting of a large number of tiny low-power camera nodes, WVSNs will not only enhance existing sensor network applications such as tracking, home automation, and environmental monitoring, but they will also enable several new applications such as [1]:

- Storage of potentially relevant activities,
- Traffic avoidance,
- Enforcement and control systems;

© Springer International Publishing AG, part of Springer Nature 2018
A. Abraham et al. (Eds.): IBICA 2017, AISC 735, pp. 60–68, 2018.
https://doi.org/10.1007/978-3-319-76354-5_6

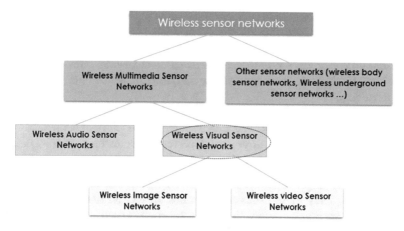

Fig. 1. Wireless sensor networks

- Advanced health care delivery;
- Automated assistance for the elderly and family monitors;
- Person locator services;

There are other applications mentioned in Fig. 2.

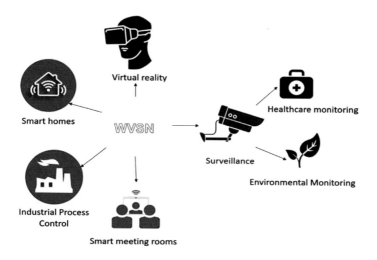

Fig. 2. Wireless visual sensor networks applications

However, WVSNs are more complicated than traditional WSNs and bring new research challenges and opportunities such as [2]:

- *Coverage problem*: the Field of View (FoV) coverage in WVSNs is determined by the camera's view angle, focal depth and occlusion caused by the obstacles instead of sensing range coverage.
- *Collecting/Processing/Transmitting visual data*: consumes much more resources (e.g. Bandwidth, Energy …) than scalar data due to huge amount and large size of image/video data.
- *Visual data reconstruction*: to reconstruct the whole picture of the interested objects at the sink node, multiple correlated data sources nodes have to cooperate.
- *Quality of Service (QoS) requirements*: QoS requirements for the visual data applications are more stringent than for scalar data applications.

These new challenges make the traditional WSNs algorithms or protocols inapplicable to WVSNs. Thus, more issues and solutions in WVSNs regarding QoS, security, mobility and energy efficiency have been recently studied such as cross-layer approach that seems a promising solution.

In this paper, we present a survey of the cross-layer design in WVSNs. Recent works and proposals in this topic are summarized and classified according to their architecture and interaction categories.

The remainder of this paper will be organized as follows. The Sect. 2 presents an overview of cross-layer design, then Sect. 3 discusses recent cross-layer design proposals in WVSN and finally conclusion is presented in Sect. 4.

2   Overview of Cross-Layer Design

For many years, the traditional seven-layer Open Systems Interconnect (OSI) have been used. It divides the overall networking task into layers and defines a hierarchy of services to be provided by the individual layers (Srivastava & Motani 2005). In this architecture, communication between adjacent layers is limited to procedure calls and responses and direct communication between nonadjacent layers is forbidden [3]. By using OSI model, the complex of network implementation could be reduced and it flexibility could be increased [4]. However, the new challenges of supporting multimedia applications and services over wireless networks, such as limited battery power, limited bandwidth, and stringent QoS requirements, cannot be solved via traditional layered architecture [5].

In [3], Srivastava & Motani explain that, in the framework of a reference layered architecture, protocols can be designed by respecting the rules of the reference architecture or by the violation of a reference layered communication architecture which is called cross-layer design.

Cross-layer design is a new research topic that refers to protocol design done by actively exploiting the dependence and interaction between protocol layers to obtain performance gains, for example by sharing variables between layers or allowing direct communication between protocols at nonadjacent layers. The special problems created by wireless links, the possibility of opportunistic communication on wireless links, and the new modalities of communication offered by the wireless medium, are the three main reasons that motivate designers to violate the layered architectures [3].

Authors in [6] declared that there are three issues that they can be viewed as goals of cross-layer designs, which are Security, Quality Of Service (QoS), and Mobility. The Security in a cross-layer design aims at providing a security communication by deploying encryption methods, such as SSH, Wi-Fi protected access. Improving the QoS could be achieved by enabling cross-layer communication between the upper layers (the application layer and the transport layer) and the lower layers (the physical layer and the data link layer). In wireless sensor networks, node movement would cause channel switch, route change, and other problems, thus, the mobility goal in cross-layer design aims at guaranteeing the uninterrupted communication in wireless networks.

In their survey of cross-layer Design [3], Srivastava and Motani discuss the basic types of cross-layer and categorize the initial proposals on how cross-layer interactions may be implemented. Figure 3 summarize the basic ways of violating the layered architecture and Fig. 4 summarize the cross-layer interactions.

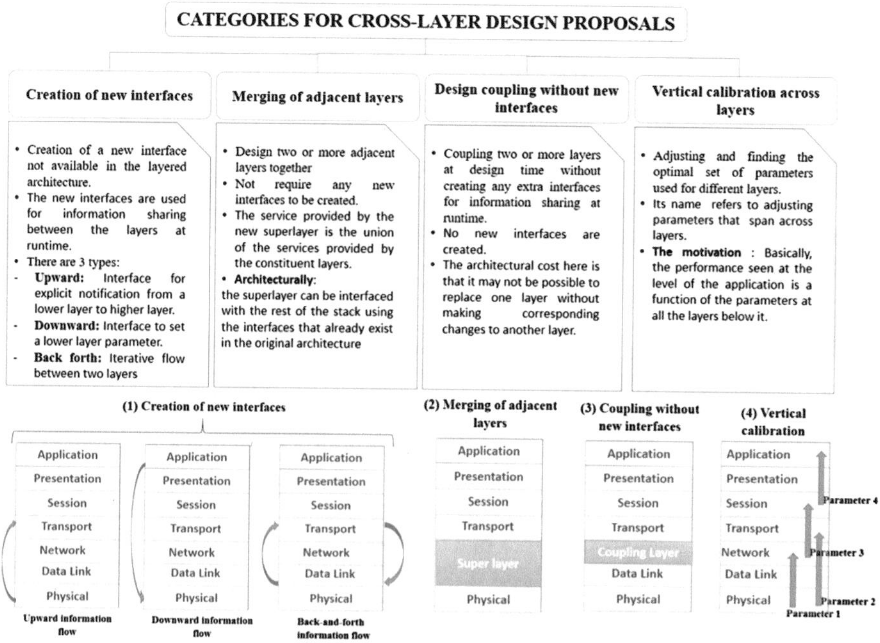

Fig. 3. Cross-layer design categories

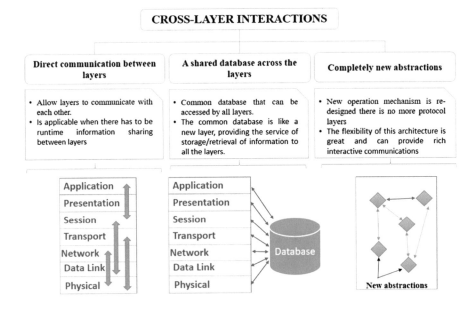

Fig. 4. Cross-layer interactions

More details and explanations of cross-layer categories and interactions are presented in [3, 4].

3   Recent Cross-Layer Design Proposals in WVSN

The cross-layer design topic has been investigated in many researches. Authors in [7] proposed a multichannel cross-layer architecture with a novel load balanced routing method (QS-LEERA-MS), where the next hop selection is done by considering the residual energy levels of the nodes in the coverage area. In this scheme, the original multimedia stream is segmented into multiple flows according to the number of paths constructed during the bandwidth reservation without exceeding the QoS constraint (the maximum number of hops that a packet must travel) defined in the request messages. As results, the network lifetime is significantly prolonged and possible congestions in a single channel-single path architecture is prevented due to the transmission of the packets over distinct paths, which increase the throughput of the system.

QoSMOS is a cross-layer QoS architecture presented in [8] that melts network and link layers with traffic classification, conditioning, forwarding (packet scheduling and buffer management) and service provisioning effects. Based on this architecture, the authors developed a cross- layer communication protocol (XLCP) that enables scalable service differentiation in wireless multimedia sensor network (WMSN), assumes geographic routing mechanism based on location awareness and optimized cost function for localized packet routing decision and provides soft QoS guarantees in latency, reliability and throughput domains without an explicit service level agreement.

Alaoui-Fdili et al. proposed in [9] a cross-layer approach for video delivery over WVSNs based on the video compression scheme H.264/AVC in its intra-only mode and the routing protocol MMSPEED taking into account the Avaible Buffer Size in the neighbour node's queue during the routing process. This approach is energy efficient and delivers good quality video streams.

A study on DS-CDMA VSN cross-layer resource allocation is presented in [10], two optimization criteria for the optimal allocation of the source and channel coding rates are studied, the first minimizes the average distortion of the video received by all nodes, and the second minimizes the maximum video distortion among all nodes. The resulting mixed integer optimization problems are tackled with the particle swarm optimization algorithm (PSO), as well as a hybrid scheme that combines PSO with the deterministic Active-Set optimization method. This cross-layer multi-node optimization design accounts for the overall system performance through all network layers. Particularly, at the application layer, data link layer, and physical layer, while the central server that can request from the nodes to properly adjust their transmission parameters depending on the amount of motion detected in each video sequence, lies at the network layer.

In order to optimize the multimedia transmission over WMSN, IEEE 802.11 g, IEEE 802.11e EDCA and H.264/SVC standards are utilized in a model of cross-layer design presented in [11] respectively in the Physical layer, the Data-link layer and the Application layer. Significant data from these three layers is gathered through parameters deliberation process where the data is enhanced to meet the prerequisites of QoS for video transmission. Then the output of the data optimization are delivered back into these layers. The outcomes of this model indicate that better results for throughput, packets end-to-end delay (latency), and packet drop rate can be attained compared with WSN standards.

An adaptive cross-layer framework for transmitting multimedia content over WSN (ACWSN) is presented in [12]. It combines cross-layer design with AOMDV as multi-path routing and AVQ (Adaptive Video Queue) as an algorithm for packer scheduling which optimizes both bandwidth and multimedia quality and solves multimedia transmission problems like limited bandwidth, wireless link failures and congested packets. ACWSN uses cross-layer communication between Physical, Network and Application layer where MPEG-4 encoder at application layer can adjust encoding parameters according to current wireless channel, which is communicated from physical layer. The limit of this framework is that do not consider the limited battery lifetime neither Qos parameters of video transmitted at time of evaluating and choosing optimum path.

Durdi, Kulkarni and Sudha [13] proposed an energy consumption reduction framework for transmission in wireless networks in order to maintain well- balanced QoS in multimedia network. In this framework, selective encryption using AES algorithm and H.264/AVC standard are used with a cross-layer approach in which the allotment of security and energy levels at both the application layer and physical layer takes place in accordance to the priority of video frames. Therefore, vital information, such as I frames that are given utmost prominence, is never lost.

An energy efficient cross-layer image transfer model for reliable image delivery across the network is proposed in [14] which employs cross-layer interaction between the application, transport, network and MAC layers. The Transport layer keep a close watch on the working of the MAC and physical layers and adjust its own parameters according to their parameter changes which ensures that the images are transported maintaining and in-fact improving the various image parameters.

According to the classifications in Sect. 2, we summarize the introduced literatures in Tables 1 and 2.

Table 1. Classifications of the introduced literatures

Referenced works	Crossed layers	Architecture category	Implementation category	Outcomes
(Çevik & Zaim 2013) [7]	Transport, Network, MAC and Physical layers	Back and forth	Direct communication	• Prolong network lifetime • Increase the throughput of the system • Prevent possible congestions in a single channel-single path architecture
(Demir & Demiray 2014) [8]	Network, MAC layers	Merging of adjacent layers	Direct communication	• Differentiate service classes in terms of soft delay, reliability and throughput • Eliminate local congestion • Distribute energy evenly along forwarding paths
(Alaoui-Fdili et al. 2015) [9]	Application, Network, MAC layers	Back and forth	Direct communication	• Extending the network lifetime • Video quality enhancement
(Pandremmenou et al. 2015) [10]	All layers	Vertical calibration	none	Optimal allocation of the source and channel coding rates and power consumption
(Emansa Hasri et al. 2015) [11]	Application, MAC, Physical layers	Vertical calibration	A shared database	Optimization of throughput, latency, and packet drop rate

The focus of the presented researches has been on the QoS topic in order to maximize the network lifetime or to enhance the quality of the transmitted image/video data. In their work [13] Durdi et al., proposed a framework that provides in addition of QoS the multimedia content protection and they presented more details of the security side of their framework in [15].

Table 2. Classifications of the introduced literatures

Referenced works	Crossed layers	Architecture category	Implementation category	Outcomes
(Youssif et al. 2015) [12]	Application, Network, Physical layers	Vertical calibration	Direct communication	• Optimizes video quality in different wireless channel conditions • Optimizes the bandwidth
(Durdi et al. 2016) [13]	Application and Physical layers	Downward	Direct communication	• Energy consumption reduction • Multimedia content protection
(Singh & Verma 2016) [14]	Application, Transport, Network and MAC layers	Coupling without new interfaces	Direct communication	Reliable transfer of images

4 Conclusion

Compared to the traditional wireless sensor networks that can only transmit scalar information such as temperature, wireless visual sensor networks are more complicated and raise new challenges such as the transmission of the huge amount of image/video data with the high bandwidth and QoS requirements over low-power visual sensor nodes. These challenges cannot be solved via traditional layered architecture. Therefore, cross-layer designs are used in order to achieve three goals (security, QoS, and mobility) by allowing one layer to exchange and share the information with other layers in the same node or in other nodes.

This paper presents a survey of the cross-layer design in WVSNs where recent existing works in this topic are overviewed and classified according to their architecture and interaction categories. By doing so, we created a platform over which new research can be built.

References

1. Akyildiz, I.F., Melodia, T., Chowdhury, K.R.: A survey on wireless multimedia sensor networks. Comput. Netw. **51**, 921–960 (2007). https://doi.org/10.1016/j.comnet.2006.10.002
2. Yap, F.G.H., Yen, H.H.: A survey on sensor coverage and visual data capturing/processing/transmission in wireless visual sensor networks. Sens. (Switzerland) **14**, 3506–3527 (2014). https://doi.org/10.3390/s140203506
3. Srivastava, V., Motani, M.: Cross-layer design: A survey and the road ahead. IEEE Commun. Mag. **43**, 112–119 (2005). https://doi.org/10.1109/MCOM.2005.1561928
4. Chao, H.-C., Chang, C.-Y., Chen, C.-Y., Chang, K.-D.: Survey of cross-layer optimization techniques for wireless networks. In: Fourth-Generation Wireless Networks: Applications and Innovations: Applications and Innovations, pp. 453–468 (2009)

5. Wang, H., Wang, W., Wu, S., Hua, K.: A survey on the cross-layer design for wireless multimedia sensor networks. In: Mobile Wireless Middleware, Operating Systems, and Applications, pp. 474–486 (2010)
6. Fu, B., Xiao, Y., Member, S., Deng, H.J., Zeng, H.: A survey of cross-layer designs in wireless networks. IEEE Commun. Surv. Tutorials **16**, 110–126 (2014)
7. Çevik, T., Zaim, A.H.: A multichannel cross-layer architecture for multimedia sensor networks. Int. J. Distrib. Sens. Netw. **2013**, 11 (2013)
8. Demir, A.K., Demiray, H.E.: QoSMOS: cross-layer QoS architecture for wireless multimedia sensor networks. Wirel. Netw. **20**, 655–670 (2014). https://doi.org/10.1007/s11276-013-0628-3
9. Alaoui-Fdili, O., Corlay, P., Fakhri, Y., Coudoux, F.-X., Aboutajdine, D.: A cross-layer approach for video delivery over wireless video sensor networks. arXiv preprint arXiv: 1501.07362 (2015)
10. Pandremmenou, K., Kondi, L.P., Parsopoulos, K.E.: A study on visual sensor network cross-layer resource allocation using quality-based criteria and metaheuristic optimization algorithms. Appl. Soft Comput. **26**, 149–165 (2015)
11. Emansa Hasri, P., Risanuri, H., Widyawan, I., Wayan, M.: Cross-layer design of wireless multimedia sensor network based on IEEE 802.11e EDCA and H.264/SVC. In: International Conference on Science in Information Technology (ICSITech), pp. 67–72 (2015)
12. Youssif, A.A.A., Ghalwash, A.Z., Abd El Kader, M.E.E.D.: ACWSN: an adaptive cross layer framework for video transmission over wireless sensor networks. Wirel. Netw. **21**, 2693–2710 (2015). https://doi.org/10.1007/s11276-015-0939-7
13. Durdi, V.B., Kulkarni, P.T., Sudha, K.L.: Cross layer approach energy efficient transmission of multimedia data over wireless sensor networks. In: Proceedings of the Second International Conference on Information and Communication Technology for Competitive Strategies, ICTCS 2016, p. 85 (2016)
14. Singh, R., Verma, A.K.: Efficient image transfer over WSN using cross layer architecture. Opt. Int. J. Light Electron Opt. (2016). https://doi.org/10.1016/j.ijleo.2016.10.143
15. Durdi, V.B., Kulkarni, P.T., Sudha, K.L.: Selective encryption framework for secure multimedia transmission over wireless multimedia sensor networks. In: Proceedings of the International Conference on Data Engineering and Communication Technology, pp. 469–480. Springer, Singapore (2017)

An IPv6 Flow Label Based Approach
for Mobile IPTV Quality of Service

Mohamed Matoui[✉], Noureddine Moumkine, and Abdellah Adib

Research Team: Networks, Telecommunications and Multimedia, Department of Computer
Sciences, Faculty of Science and Technology, Mohammedia, Morocco
matoui.mohamed@gmail.com

Abstract. The latest experiences hint that the QoS (Quality of Service)
approaches adopted by IMS (IP Multimedia Subsystem) technologies are still
suffering from a primary containment factor due to the nondifferentiation between
IPTV (Internet Protocol Television) video components. IMS system also cannot
ensure high IPTV data transfer due to the limitation of available cellular bandwidth. In this paper, we try to merge the advantage of high bandwidth assigned
to LTE (Long Term Evolution) and a new PHB (Per-Hop Behavior) that classify
and differentiate between IPTV sub traffics by using IPv6 Flow Label. This new
architecture permits high-quality IPTV video components with the capability to
prioritize the sub traffic according to the network administrator policy. The
proposed architecture is implemented using OPNET software. The results show
that IPTV users receive high-quality video data with a change in quantity
according to data priority.

Keywords: IPv6 · Flow Label · LTE · QoS · DiffServ · IPTV · IMS

1 Introduction

Digital video streaming has become widely spreading this days. IPTV services become
a wide demand as it provides the transfer of multimedia services over IP network to
provide the required QoS needed by the user as security, reliability, and interactivity. It
also requires carrying a video to a wide range of users with different screen sizes and
resolution as mobile phones and digital screen cinemas. So a continuous moving picture
and audio are transferred during transmission time. The existing IMS-based IPTV infrastructure doesn't take into account that the IPTV traffic consists of three sub-components
and the sensitivity of the linear television latency. In fact, classification of traffic uses
three classes: data, voice, and video. In the case of IPTV, we note that traffic can be
decomposed into three sub-traffics: BC (BroadCast), VoD (Video on Demand) and PVR
(Personal Video Recorder):

- BC: This is the broadcasting or multicasting of real- time video traffic in a network.
- VoD: Allows a user to select and view a video. It includes a library which will allow
  diffusing, in parallel with the video, a title, and music.
- PVR: Allows the end user to record the content of the received stream.

© Springer International Publishing AG, part of Springer Nature 2018
A. Abraham et al. (Eds.): IBICA 2017, AISC 735, pp. 69–80, 2018.
https://doi.org/10.1007/978-3-319-76354-5_7

DiffServ model makes the treatment of these three types of flows alike. The difference in sensitivity to QoS parameters requires a reclassification between them. Our contribution aims to remedy this problem. To achieve the required demand with an efficient QoS, the high bandwidth is required [1–4]. 4G cellular network has been assigned enormous bandwidth that ensures reliable delivery of IP traffic from smartphones that enables users to transfer high amount of data while moving inside the cell [5]. Many types of research had worked in improving IPTV services QoS. The poor-quality model has been merged in [3] to improve IPTV network accuracy and efficiency in case of pause or screen with less clarity. Li and Chen in [5] combined time slicing and discontinuous reception (DRX) schemes to build power saving technique for LTE network. The proposed mechanism decreases the UEs consumed power and saves the IPTV services quality. In [6], a new framework has been illustrated to measure the viewer's response and analyses TV content. This method leads to use IPTV network data according to users' opinion. A new proposed architecture with new coding has been proposed in [7] to mend robustness when the network capacity increase. To uphold IPTV in LTE network, Broadcast Multicast Service Centre had been designed in [8]. In [9], Chen and Liao succeeded to reduce the switching delay during video transfer using exact packet pairs that increase bandwidth. They also improved playback media stability by using buffers to store selected channel. IPTV network and QoS parameters had been explained and analyzed in details in [10]. Li and Chen support IPTV mobility over a wireless cellular network using spectrum allocation technique [11]. This offered better IPTV services besides it keeps voice service quality. In [12], IPTV data problems as dropping, blocking and bandwidth usage had been almost solved using new queue model that consider adaptive modulation and coding. IPTV seamless handover in wireless LAN had been achieved in [13] using Physical Constraint and Load-Aware. This technique allows the user to choose the next wireless LAN to access according to its strength, congestion and bit error rate.

To guarantee the best IPTV QoS, a new QoS-control paradigm based on adaptive control theory had been developed to enable this next-generation services to interact with the users. These new techniques will provide the user demand according to their QoS requirements. Knowing that the one implemented in the IMS-based IPTV is based on DiffServ, our approach is in the form of a new PHB (Per-Hop Behavior). As for the classification of IPTV sub-traffic, we propose a mechanism based on the use of the IPv6 FL (Flow Label) fields to enhance the QoS in the IMS network. The objective of this proposal is to differentiate between the IPTV packets and allow them to avoid the best effort treatment as they are part of the same traffic (Video).We also apply this proposed paradigm in LTE network to provide the best QoS. The outline of the paper is as follows: Sect. 2 shows briefly QoS issues, types of IP interworking networks and existing suggestions for improving QoS using IPv6 FL. It also explains our QoS optimization mechanism. That explanation is based on how to prioritize IPTV sub-traffic using the IPv6 FL field and how to generate new class of services. In Sect. 3, we discuss our implementation network and the studied scenario of the LTE-IMS-Based IPTV by using Opnet 17.5. Section 4 illustrates the performance analysis of the proposed mechanism. Finally, Sect. 5 concludes this paper and outlines the prospects.

2   Improving QoS Using Flow Label Proposed Technique

The capability of the network to provide the user requirements when using IPTV service taking into consideration the main parameters like delay, traffic losses, video jitter and quality is the heart of the definition of QoS in our network. Two main QoS models were proposed by IETF (Internet Engineering Task Force): IntServ and DiffServ [14–16]. The difference between these two models is explained in details in [17]. To improve QoS for IPTV services, during transmission, IPv6 FL had been used in addition to IMS system.

2.1   IPv6 Flow Label and Quality of Service

IPv6 FL is a 20-bits field just after the Traffic Class field of the IPv6 header. This field may be used to label packets of the same packet flow or an aggregation of flows [18]. Several approaches have been proposed to the IETF to use this field to improve QoS on the internet [19]. Some of them have suggested using it to send the bandwidth, delay, and buffer requirements. Others have recommended using this field to send the used port number and the transport protocol [20]. Other approaches have been proposed [21], but none of them have been standardized. However, there is an hybrid approach that takes into account the advanced approaches and applies them to DiffServ model [19]. This method has booked the first 3 bits of the IPv6 FL field to indicate the methods used and reserved the remaining 17-bit parameter relating to each particular approach. Table 1 summarizes this hybrid approach.

Table 1. The bit pattern for the first 3 bits of Flow Label

Value type of the used approach	
000	Default
001	A random number is used to define the Flow Label.
010	Int-Serv
011	Diff-Serv
100	A format that includes the port number and the protocol in the FL is used
101	A new definition explained in [22]
110	Reserved for future use
111	Reserved for future use

2.2   Optimization of IPTV Broadcasting Traffic

The IMS-Based IPTV was not limited to the provision of essential services of IPTV, but it opened the door to 'quadruple play' services and other more advanced ones. As FL allows the user to ask for a unique process for its real-time traffic flow [15], IMS provides a continuous connection as it allows users to get their data either with a fixed or mobile network. IMS had a choice of traditional or recent technologies for service management, especially using the experience inherited from the Internet model in QoS management, but the requirements to provide multimedia services compete to attract many customers and suppliers.

Video traffic is characterized by a variable data rate due to the dynamic nature of the captured scenes and also the encoding process [23, 24]. Assign EF (Expedited Forwarding) as PHB to video traffic in a DiffServ network can generate a control problem as video traffic has variable rates. Also, in the fact that several video streams are encapsulated in the same aggregate flows, it will be difficult to design the maximum inter-video traffic limit for his control at the entrance of the DiffServ domain (Ingres router). Large traffic with EF PHB may cause congestion of DiffServ core routers. Serving the EF packets continuously with high priority will increase the degree of this congestion. As the EF PHB uses only narrow queues, in addition to the growing delay after waiting in the queue, a significant delay is not desired by a real-time traffic. The succession of EF packets in the core network will make the refinement of the packets in the queue slower. Consequently, a significant number of packets will be dropped. In case the packets removal process is started, it will be no protection of the most sensitive packets. By adopting the same reasoning, EF packets at the edge of the DiffServ domain will be treated according to their importance in the GOP (GROUPE of Picture) video [25]. Using an AF (Assured Forwarding) class for IPTV traffic may cause several problems. The rejection priority for PHBs in AF is often implemented based on WRED (Weighted Random Early Detection). Because of static treatment experienced by packets AF, several large packets will be removed instead of other less important. The integration of the eTOM (enhanced Telecom Operation Map [26]) process in the IMS-based IPTV allowed classifying users by loyalty order. This approach gives a hand to the network administrator to differentiate between recipient-based packets.

In case we were in front of the same class of users "GOLD" for example, whose requesting the IPTV traffic, scoring inter users generates another factor that may affect the credibility of transmission, as these users have the same grade. In the case of congestion, routers will be forced to return to elimination process proposed by the DiffServ standard. Indeed, the IPTV stream is composed of three types of traffic noted BC, PVR and VoD. The PHBs EF (often attributed to the media stream) assumes that IPTV packets are within the same video type. It means that they will be treated together with the same priority, especially in the case of congestion, where routers carry the classification of these packets queued and disposal packets to alleviate the queues if the queue is full [27]. The treatment of all IPTV packets with the same priority occurs by using a First In First Out (FIFO) algorithm. All this will increase some latency packets that are sensitive to delay and loss rates. Knowing that the components of IPTV traffic differ in their sensitivity to latency and loss rate, seeking a reclassification mechanism between IPTV packets becomes a necessity. To overcome these limitations, we propose a new mechanism for identifying and subsequently classifying IPTV traffic with treatment and suppression priority which differs from one packet to another depending on the type of traffic. As the ToS (Type of Service) field of the IPV4 header is limited to a byte, we propose to make the mapping of DSCP (Differentiated Services Code Point) values to IPv6 FL fields that will give us more bits to differentiate IPTV traffic while remaining compatible with the DiffServ approach. Thus, Table 2 shows the new format of the IPv6 field:

Table 2. New IPv6 Flow Label values

0	1	2	3	4	5	6	7	8	9	0	1	2	3	4	5	6	7	8	9
0	1	1	DSCP							X	Y	Reserved for future use							

As the value of the DSCP field for the EF class is set to 101110, the IPv6 FL field can be written as shown in Table 3:

Table 3. New IPv6 Flow Label values

0	1	2	3	4	5	6	7	8	9	0	1	2	3	4	5	6	7	8	9
0	1	1	1	0	1	1	1	0	X	Y	Reserved for future use								

Where x and y are the bits used to differentiate the video Traffic intra-IPTV. The fact that IPTV packets take the same value of the DSCP field, exploiting only six bits, then we will use the following 10 and 11-bits in the IPv6 FL to a reclassification intra-IPTV. The remaining 9-bits will be reserved for future use. We give the name DSCP-FL to the first 11 bits of the IPv6 FL field. These new FL values are mapped to PHBs that are characterized by a high priority, low loss rate, jitter, and latency are similar to that of the current EF PHB. Indeed, three IPTV packets belonging successively under BC traffic, VoD, PVR will be subjected to a treatment illustrated through the algorithm in Fig. 1. When the DiffServ router becomes saturated, it will proceed with the removal of the packets with the highest priority level of dropping. In our case, it will be the one whose DSCP-FL field has a value close to 011 10111001 (Table 4).

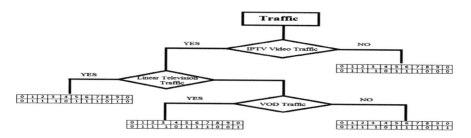

Fig. 1. The proposed algorithm to differentiate intra IPTV traffic

Table 4. IPv6 Flow Label values with highest priority level of suppression

0	1	2	3	4	5	6	7	8	9	0	1	2	3	4	5	6	7	8	9
0	1	1	1	0	1	1	1	0	0	1	Reserved for future use								

3   Implementation Scenario

In our considered network, we aim to increase the data received by BC, VoD and PVR user sequentially and decrease the delay and jitter that faces the BC traffic particularly. Using OPNET Modeler 17.5, we implemented our proposed technique in LTE cellular network. The main idea in the proposed framework architecture is to implement IMS-Based FL IPTV component in a 4G cellular system. A new modulated task application module had been developed as IMS-SIP server does not exist in the OPNET's modules. User registration in IMS network and session establishment of IPTV services are built in custom application in the proposed framework.

In this paper, we compare between performances of the network in two different ways. The first case without applying the FL QoS based system. The second scenario corresponds to when applying FL and WFQ (DSCP Based) QoS. Figure 2 presents three major components of the used architecture. The first element includes the IMS network, the second element contains three servers that represent the IPTV data center responsible for sending different types of multimedia contents (PVR, VoD, and BC). The final one is the personal receiver that receives data from the transmitter. As the IPTV users are not alone in a 4G network, there are 10 FTP, and 10 HTTP users that transfer data at the same time beside IPTV users. We must also mention that we compare the results of our proposed scenarios when using and disusing our proposed technique to measure the QoS parameters. The traffic sent is the same from the three different video servers (PVR, BC, and VoD), high-resolution video, and that after the user perform IMS authentication steps. In that scenario, the users move inside the cell with the same velocity 100 m/s.

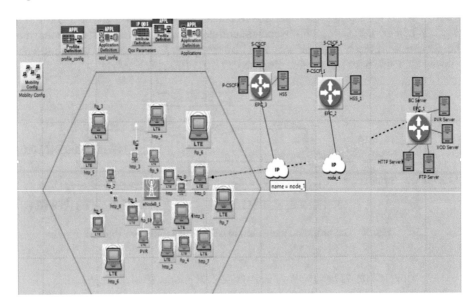

Fig. 2. Moving IPTV users

4   Performance Analysis

In this section, we gathered the collected results for our proposed scenarios; then we make overall performance analysis. The collected results are traffic dropped, packet end-to-end delay and delay variation. We compare the performance of the three users (BC,

VoD, and PVR) in case of using and disusing FL QoS. In this scenario, all the three users BC, VoD and PVR moves inside the cell with the same speed to make affair comparison between them when applying the proposed technique. Our proposed FL QoS shows a high performance for BC user.

4.1 Traffic Dropped

It can be defined as the data missing while sending from the server to the user. This missing data is due to the congestion of the network and imperfectly data links.

Figure 3 shows that all sources sent the same amount of data, although Fig. 4 shows that the amount of data received by BC user is higher than both VoD user and PVR user as BC user has the highest priority then VoD and PVR users.

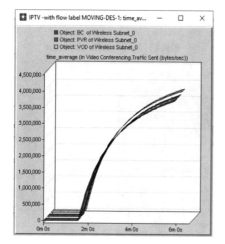

Fig. 3. Traffic sent (bytes/sec)   **Fig. 4.** Traffic received using Flow Label QoS (bytes/sec)

As shown in Figs. 5 and 6, BC user and VoD user received a higher amount of data when using FL QoS. In contrast, the amount of data received by the PVR user decreases when using FL QoS as shown in Fig. 7.

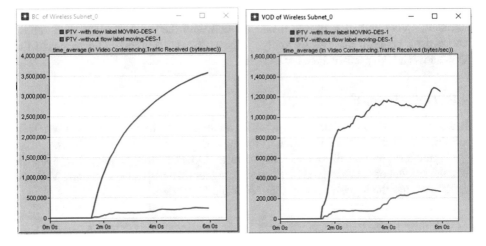

Fig. 5. BC traffic received   Fig. 6. VoD traffic received

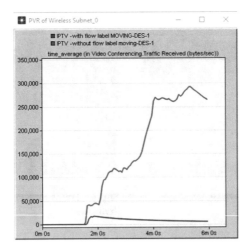

Fig. 7. PVR traffic received

4.2   End-to-End Delay

The time taken by the packets to travel from the server to the user can be called packet end-to-end delay. As shown in Fig. 8, end-to-end delay taken by the BC user is the lowest when using FL QoS. BC user delay as in Fig. 9 is lower than the delay when using FL QoS. In contrast, the delay by PVR users increases when using FL QoS as shown in Fig. 10, while the delay of VoD user remains almost the same as in Fig. 11.

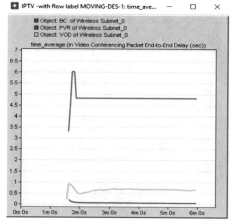

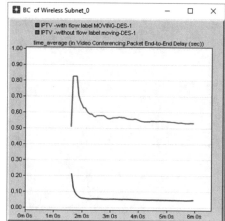

Fig. 8. End to End delay (sec)   **Fig. 9.** BC USER End to End delay (sec)

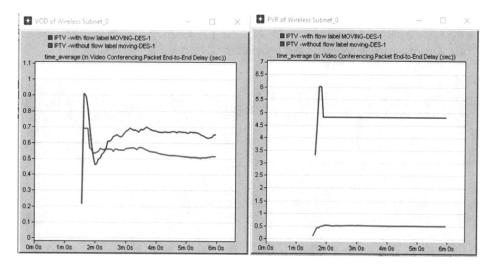

Fig. 10. VoD USER End to End delay (sec)   **Fig. 11.** PVR USER End to End delay (sec)

4.3   Packet Delay Variation

The playout buffers size is presented by Packet Delay Variation for regular delivery of packets. BC user delay variation is the lowest as shown in Fig. 12 when using FL QoS.

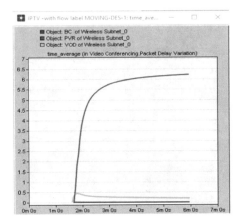

Fig. 12. Packet delay variation (sec)

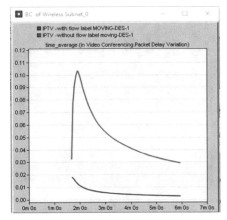

Fig. 13. BC USER Packet delay variation (sec)

Figures 13, 14 and 15 represent a comparison of packets delay variation by three users, in the case of using and disusing the FL QoS. BC user delay as shown in Fig. 13 is the lower amount of delay when using FL QoS. In contrast, the delay by the VoD and PVR users increases when using our approach as shown in Figs. 14 and 15.

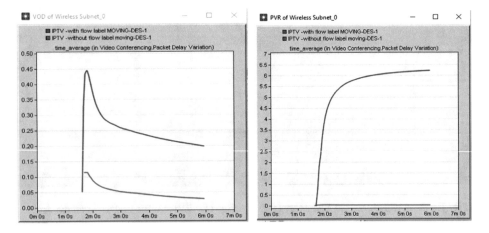

Fig. 14. VoD USER Packet delay variation (sec)

Fig. 15. PVR USER Packet delay variation (sec)

5   Conclusion and Perspective

In the recent years, many researchers try to improve the QoS of IPTV services that consider real-time traffic; mainly traffic losses and latency. None of them consider the problem of classification of IPTV sub traffic and the differentiation between the BC, VoD, and PVR packets. To fix this problem, we propose a new addressing algorithm that classifies between the packets using IPv6 FL field. This algorithm provides a reliable solution to increase QoS of IPTV sub traffic by increasing the priority of BC traffic over VoD and PVR traffics. We also improve the quality of IPTV services by applying that technique to a 4G cellular system. As LTE system provides the high bandwidth that helps in increasing the quality of sending data by increasing the amount of data sent. We study the performance of this algorithm using a practical lab as mentioned in Sect. 4. The performance results show that the amount of data received by BC user which had the highest priority is the highest in case of the moving user. Our results prove also that the packet losses, end-to-end delay and packet delay variation decreased for BC user, but increased for PVR which shows that our techniques work well. We are working on applying this technique to the next interworking heterogeneous network (LTE-WLAN-WiMAX). In the future, we will work on improving the security issues in IPTV IMS network and solve its related issues.

References

1. Grant, A.E., Meadows, J.H.: Communication Technology Update and Fundamentals, pp. 122–148. Twelfth Edition-Focal Press, Boston (2010)
2. Kumar, A.: Implementing Mobile TV: ATSC Mobile DTV, MediaFLO, DVB-H/SH, DMB, WiMAX, 3G Systems, and Rich Media Applications, pp. 261–290. Taylor & Francis (2010)
3. Li, R., Wu, Z., Huang, R., Wei, X., Qian, Y.: Evaluation for IPTV service based on poor-quality model. In: 19th International Symposium on Wireless Personal Multimedia Communications (WPMC) (2016)
4. Punchihewa, A., Malsha, A., Diao, Y.: Internet Protocol Television (IPTV), Multi-media Research Group (2011)
5. Li, M., Chen, H.-L.: Design of power saving mechanism for LTE-a networks supporting mobile IPTV services. In: IEEE International Conference on Communication Problem-Solving (ICCP) (2015)
6. Kren, M., Sedlar, U., Bešter, J., Kos, A.: Determination of user opinion based on IPTV data. In: 18th International Conference on Transparent Optical Networks (ICTON) (2016)
7. Kwon, M., Kwon, J., Park, B., Park, H.: An architecture of IPTV networks based on network coding. In: Ninth International Conference on Ubiquitous and Future Networks (ICUFN) (2017)
8. Bataa, O., Chuluun, O., Orosoo, T., Lamjav, E., Kim, Y., Gonchigsumlaa, K.: A functional design of BM-SC to support mobile IPTV in LTE network. In: 7th International Forum on Strategic Technology (IFOST) (2012)
9. Chen, Y.-C., Liao, C.-Y.: Improving quality of experience in P2P IPTV. In: 18th Asia-Pacific Network Operations and Management Symposium (APNOMS) (2016)
10. Elgeldawy, F., Salama, G., Abdel Fattah, M.: Performance of QOS parameters for IPTV through NGN. In: IEEE Student Conference on Research and Development (SCOReD) (2016)
11. Li, M., Chen, L.: Spectrum allocation algorithms for wireless cellular networks supporting mobile IPTV. Comput. Commun. **99**(1), 119–127 (2017)
12. Li, M.: Queueing analysis of unicast IPTV with adaptive modulation and coding in wireless cellular networks. IEEE Trans. Veh. Technol. **66**(10), 9241–9253 (2017)
13. Fard, H., Rahbar, A.: Physical constraint and load aware seamless handover for IPTV in wireless LANs. Comput. Electr. Eng. **56**, 222–242 (2016)

14. 3GPP: 3G TS 23.107 V5.0.0: Quality of Service, Concept, and Architecture. http://www.3gpp.org/ftp/Specs/archive/23_series/23.107/23107-a20.zip

15. Bhattarakosol, P.: Intelligent Quality of Service Technologies and Network Management, Models for Enhancing Communication. IGI Global, Hershey (2010)

16. Sambath, K., Abdulrahman, M., Suryani, V.: High Quality of Service video conferencing over IMS. Int. J. Inf. Educ. Technol. **6**(6), 470 (2016)

17. Jiang, S.: Future Wireless and Optical Networks: Networking Modes and Cross-Layer Design. Springer, London (2012)

18. Deering, S., Hinden, R.: Internet Protocol, Version 6 (IPv6) Specification. IETF Internet Draft, July 2017

19. Hu, Q., Carpenter, B.: Survey of proposed use cases for the IPv6 flow label. draft-hu-flow-label-cases-03, February 2011

20. Prakash, B.: Using the 20 bit flow label field in the IPv6 header to indicate desirable quality of service on the internet. Ph.D. dissertation, University of Colorado (2004)

21. Conta, A., Rajahalme, J.: A model for DiffServ use of the IPv6 flow label specification. IETF Internet Draft (2001)

22. Jee, R., Malhotra, S., Mahaveer, M.: A modified specification for use of the IPv6 flow label for providing an efficient quality of service using a hybrid approach. IPv6 Working Group Internet Draft, Draft-banerjee-flowlabel-ipv6-qos-03.txt, Technical report (2002–2004)

23. Lloret, J., Canovas, A., Tomas, J., Atenas, M.: A network management algorithm and protocol for improving QoE in mobile IPTV. Comput. Commun. **35**(15), 1855–1870 (2012)

24. Farmer, J., Lane, B., Bourg, K., Wang, W.: FTTx Networks: Technology Implementation and Operation. Morgan Kaufmann (2016)

25. Leghroudi, D., Belfkih, M. Moumkine, N., Ramdani, M.: Differentiation intra traffic in the IPTV over IMS context. In: e-Technologies and Networks for Development, pp. 329–336. Springer (2011)

26. Brahim, R., Bellafkih, M., Ranc, D., Errais, M.: WS-composite for management & monitoring IMS network. IJNGC **2**(3) (2011). http://perpetualinnovation.net/ojs/index.php/ijngc/article/view/143

27. Sabry, E.S., Ramadan, R.A., El-Azeem, M.A., ElGouz, H.: Evaluating IPTV network performance using OPNET. In: Communication, Management and Information Technology: Proceedings of the International Conference on Communication, Management and Information Technology (ICCMIT 2016), pp. 377–384. CRC Press (2016)

NWP Model Revisions Using Polynomial Similarity Solutions of the General Partial Differential Equation

Ladislav Zjavka[(✉)], Stanislav Mišák, and Lukáš Prokop

ENET Centre, VŠB-Technical University of Ostrava, Ostrava, Czech Republic
ladislav.zjavka@vsb.cz

Abstract. Global weather models solve systems of differential equations to forecast large-scale weather patterns, which do not perfectly represent atmospheric processes near the ground. Statistical corrections were developed to adapt numerical weather prognoses for specific local conditions. These techniques combine complex long-term forecasts, based on the physics of the atmosphere, with surface observations using regression in post-processing to clarify surface weather details. Differential polynomial neural network is a new neural network type, which generates series of relative derivative terms to substitute for the general linear partial differential equation, being able to describe the local weather dynamics. The general derivative formula is expanded by means of the network backward structure into a convergent sum combination of selected composite polynomial fraction terms. Their equality derivative changes can model actual relations of local weather data, which are too complex to be represented by standard computing techniques. The derivative models can process numerical forecasts of the trained data variables to refine the target 24-h prognosis of relative humidity or temperature and improve the statistical corrections. Overnight weather changes break the similarity of trained and forecast patterns so that the models are improper and fail in actual revisions but these intermittent days only follow a sort of settled longer periods.

Keywords: Polynomial neural network
Partial differential equation substitution · Relative sum derivative term
Similarity correction model

1 Introduction

Numerical Weather Prediction (NWP) models succeed in long-term forecasting upper large-scale patterns but are too crude to account for local variations in surface conditions. Pure statistical models using local measurements can forecast idiosyncrasies in local weather but are usually worthless beyond several hour horizon. Post-processing methods, called model output statistics (MOS), relate historical observations with the raw complex numerical forecasts, based on the physics of the atmosphere, using regression equations to reduce NWP model bias and systematic errors [6]. Adaptive intelligence methods try to clarify surface weather details and eliminate additionally random forecast errors of NWP models, induced due to uncertain initial conditions and

© Springer International Publishing AG, part of Springer Nature 2018
A. Abraham et al. (Eds.): IBICA 2017, AISC 735, pp. 81–91, 2018.
https://doi.org/10.1007/978-3-319-76354-5_8

data computational limitations. Extended Polynomial Neural Networks (PNN) can adopt some principles of the Similarity theory and procedures of the Operator calculus to decompose and substitute for the general linear Partial Differential Equation (PDE), which can describe any unknown dynamic system [10].

$$Y = a_0 + \sum_{i=1}^{n} a_i x_i + \sum_{i=1}^{n} \sum_{j=1}^{n} a_{ij} x_i x_j + \sum_{i=1}^{n} \sum_{j=1}^{n} \sum_{k=1}^{n} a_{ijk} x_i x_j x_k + \ldots \tag{1}$$

n - *number of input variables* x_i $a_i, a_{ij}, a_{ijk}, \ldots$ - *polynomial parameters*

Differential Polynomial Neural Network (D-PNN) is a new type of neural network, which extends the PNN structure to produce a series of relative derivative terms, whose convergent combinations can define and substitute for the general PDE in consideration of data samples. The GMDH (Group Method of Data Handling) algorithm, created by a Ukrainian scientist Aleksey Ivakhnenko in 1968, forms PNN in successive steps, adding layer by layer to decompose the Kolmogorov-Gabor polynomial (1). It expands the general connections between input and output variables into many simple relationships of low order polynomials (2) for every pair of input variables in each layer. The GMDH polynomials in PNN nodes can approximate any stationary random sequence of observations and can be computed in the last added layer by either adaptive methods or system of Gaussian normal equations. A typical PNN maps a vector input x to a scalar output Y, which is an estimate of the true function [4]. The number of layers of the PNN is not fixed in advance but becomes dynamically meaning that this self-organizing network grows over the trained period.

$$y = a_0 + a_1 x_i + a_2 x_j + a_3 x_i x_j + a_4 x_i^2 + a_5 x_j^2 \tag{2}$$

x_i, x_j - *input variables of polynomial neuron nodes*

The D-PNN decomposes the general PDE analogous to the PNN does the general connection polynomial (1). The derivative equality sum changes of selected substitution fraction terms, formed in the PNN nodes, can model an unknown target function in a PDE solution. The Operator calculus can convert the PDE into rational fractions, which represent the Laplace transforms of a searched function. The inverse transformation can be applied to them to obtain the originals and solve the PDE. In contrast with the Artificial Neural Network (ANN), each neuron (i.e. substitution PDE term) produced in any layer node, can be directly included in the total network output, which is the sum of selected (active) neurons outputs [10].

The general PDE can describe uncertain processes in specific local weather condition, which are too complex and require a mass of data to be solved by conventional regression or standard computing. These methods usually require a substantial reduction of input variables leading to model over-simplification [9]. The D-PNN models can process final NWP outcomes of the trained input variables to calculate output prediction series and improve MOS final corrections in the majority of cases. The derivative PDE models represent current fluctuant relations of local data over the training periods in more or less settled weather conditions. NWP model outputs can be adapted to them to obtain the successful revisions. If a front brings an overnight weather change, the predictions fail due to inconsistent trained and forecasted patterns.

Daily PDE models, formed for the estimated numbers of training days, can refine 24-h local forecasts of relative humidity (the lowest errors) or temperature.

2 NWP Model Forecasts Local Revisions

Meso-scale forecasts are produced by additional NWP models, based as well on the numerical integration of differential equations, which can describe circulation processes in the upper atmosphere on account of data observations. They are usually not equipped with their own data assimilation system, so their forecasts quality depends primarily on the global NWP systems providing the initial and boundary conditions. If the global forecast is questionable over the region of interest, the higher resolution will only magnify the problems. The physical down-scaling of the domain model can apply additional satellite atmospheric measurements from an unbiased observing system while the statistical down-scaling employs only independent surface observations. MOS typically derive a set of linear equations to relate NWP model outputs with the actual observations at a certain time. They attempt to minimize the systematic errors of the next forecast, which primarily result from the physical parameterization of weather events. The proposed forecast revisions (Sect. 5) are analogous to the Perfect-Prog (PP) approach, which applies forecasted variables to its analog or regression model, calculated with their corresponding observations in month periods. Other statistical algorithms can employ "Running-mean" and "Nearest neighborhood weighted mean" corrections to minimize the bias of the next forecast [2]. "Variational" method assumes the non-systematic component of forecast error is linearly dependent on some combination of the initial fields, end-time forecasts and the forecast tendency [5]. NWP model errors can be expressed in the form of the Lagrange interpolation polynomial, whose coefficients are determined by past model performance to solve the inverse problem [7]. The probabilities of certain local weather events can be estimated using a "germ–grain" model with non-negative least-squares approach to determine the local (rain-fall) intensities and a "semi-variogram" technique to find the grain-cell size [3]. Hybrid or ensemble methods can combine different techniques or models, which usually produce better results than single algorithms. Adaptive post-processing methods can equal or exceed traditional MOS and PP, starting from complex forecasts and the observations. The proposed daily revisions (Sect. 5) use PDE models that process final forecast of NWP systems, whose first outputs were corrected by several MOS and secondary data analyses.

3 The General PDE Decomposition and Substitution

D-PNN defines and substitutes for the general linear PDE, which exact form is not known in advance and which can describe any dynamic system, with a selected sum combination of relative polynomial terms (3). The unknown function u is calculated from the PDE sum of its derivative terms and bias. If u is a separable function, it can be expressed in the form of the sum of partial functions u_k (3) including convergent series arisen from their partial derivatives (4), formed in the PNN nodes.

$$a + bu + \sum_{i=1}^{n} c_i \frac{\partial u}{\partial x_i} + \sum_{i=1}^{n}\sum_{j=1}^{n} d_{ij} \frac{\partial^2 u}{\partial x_i \partial x_j} + \ldots = 0 \qquad u = \sum_{k=1}^{\infty} u_k$$

$$u(x_1, x_2, , \ldots, x_n) - unknown\ separable\ function\ of\ n - input\ variables$$
$$a, b, c_i, d_{ij}, \ldots - weights\ of\ terms \qquad u_i - partial\ functions$$

(3)

Considering 2-input variables of the PNN nodes, the 2nd order PDE (4) includes a series of 5 derivative terms, which corresponds to variables of the GMDH polynomial (2). This PDE is most often used to model physical or natural system non-linearities.

$$\left(\sum \frac{\partial u_k}{\partial x_1}, \sum \frac{\partial u_k}{\partial x_2}, \sum \frac{\partial^2 u_k}{\partial x_1^2}, \sum \frac{\partial^2 u_k}{\partial x_1 \partial x_2}, \sum \frac{\partial^2 u_k}{\partial x_2^2} \right)$$

(4)

u_k - node partial sum functions of an unknown separable function u

Fraction PDE terms are composed form the standard GMDH polynomials (2) in all PNN nodes (4) according to the adapted Similarity Dimensional Analysis (SDA). SDA applies various formal adaptations to PDEs according to the data dimensions to form dimensionless π_units, i.e. characteristic groups of variables, typically interpreted as the ratio of several original quantities [1].

$$L\{f^{(n)}(t)\} = p^n F(p) - \sum_{k=1}^{n} p^{n-i} f_{0+}^{(i-1)} \qquad L\{f(t)\} = F(p)$$

$$f(t), f'(t), \ldots, f^{(n)}(t) - originals\ continuous\ in\ <0+, \infty> \qquad p, t - complex\ and\ real\ variables$$

(5)

This SDA approach corresponds to a polynomial substitution (5) of the Operator calculus, which can convert differential equations into pure rational functions in consideration of the initial and boundary conditions. The reduced elementary fractions can be considered the Laplace transforms $F(p)$ of a searched separable function $f(t)$ of a real variable t, so that the inverse L-transformation can be applied to them to get the originals and solve the PDE in this way (6).

$$F(p) = \frac{P(p)}{Q(p)} = \sum_{k=1}^{n} \frac{P(\alpha_k)}{Q_k(\alpha_k)} \frac{1}{p - \alpha_k} \qquad f(t) = \sum_{k=1}^{n} \frac{P(\alpha_k)}{Q_k(\alpha_k)} e^{\alpha_k \cdot t}$$

(6)

a_k - simple real roots of the multinomial $Q(p)$ $F(P)$ - Laplace transform

The rational fractions, formed in each D-PNN node block (Fig. 1), fulfil the condition of the different multinomial degree in the numerator and denominator. The complete 2-variable GMDH polynomials replace the unknown partial functions u_k (3) of the PDE terms numerators, while reduced polynomials of the denominators represent the derivative variables (4). The inverse L-transform is applied to the selected polynomial fractions according to the Eq. (6) to substitute for the selected PDE terms (7) and get the originals, which sum represents an unknown separable output function (3) in a PDE solution. Each block contains a single output polynomial (2), without

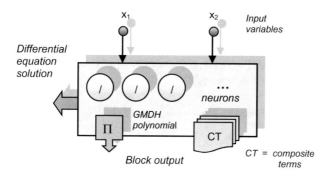

Fig. 1. D-PNN blocks forms simple (/) and composite (CT) substitution PDE terms (neurons)

derivative part. Neurons do not affect the block output but can be directly included in the total network output sum calculation of a DE solution. Each block has *1* and neuron *2* vectors of adjustable parameters *a*, and *a, b*.

$$F\left(x_1, x_2, u, \frac{\partial u}{\partial x_1}, \frac{\partial u}{\partial x_2}, \frac{\partial^2 u}{\partial x_1^2}, \frac{\partial^2 u}{\partial x_1 \partial x_2}, \frac{\partial^2 u}{\partial x_2^2}\right) = 0$$

$$where\ F(x_1, x_2, u, p, q, r, s, t)\ is\ a\ function\ of\ 8\ variables \qquad (7)$$

While using 2 input variables in the PNN nodes the 2nd order PDE can be expressed by the equality of 8 variables (7), including derivative terms formed in respect of all variables of the GMDH polynomial (2). Each D-PNN block can form *5* corresponding simple derivative neurons in respect of single x_1, x_2 (8) squared x_1^2, x_2^2 (9) and combination $x_1 x_2$ (10) derivative variables, whose selected combination sum can directly solve the 2nd order sub-PDEs in the PNN nodes (7).

$$y_1 = w_1 \frac{a_0 + a_1 x_1 + a_2 x_2 + a_3 x_1 x_2 + a_4 sig(x_1^2) + a_5 sig(x_2^2)}{b_0 + b_1 x_1} \cdot e^\varphi \approx \frac{\partial f(x_1, x_2)}{\partial x_1} \qquad (8)$$

$$y_3 = w_3 \frac{a_0 + a_1 x_1 + a_2 x_2 + a_3 x_1 x_2 + a_4 sig(x_1^2) + a_5 sig(x_2^2)}{b_0 + b_1 x_2 + b_2 sig(x_2^2)} \cdot e^\varphi \approx \frac{\partial^2 f(x_1, x_2)}{\partial x_2^2} \qquad (9)$$

$$y_5 = w_5 \frac{a_0 + a_1 x_1 + a_2 x_2 + a_3 x_1 x_2 + a_4 sig(x_1^2) + a_5 sig(x_2^2)}{b_0 + b_1 x_1 + b_2 x_{12} + b_3 x_1 x_2} \cdot e^\phi \approx \frac{\partial^2 f(x_1, x_2)}{\partial x_{12} \partial x_2}$$

$$\varphi = arctg(x_1 / x_2)\ \text{-}\ phase\ representation\ of\ input\ variables\ x_1, x_2 \qquad (10)$$

$$a_i, b_i\ \text{-}\ polynomial\ parameters \qquad\qquad sig\ \text{-}\ sigmoidal\ transformation$$

The Root Mean Squared Error (RMSE) is calculated for the simultaneous polynomial parameter optimization, neuron and node 2-input combination selection (11).

$$RMSE = \sqrt{\frac{\sum_{i=1}^{M}\left(Y_i^d - Y_i\right)^2}{M}} \to min \qquad (11)$$

$$Y_i\ \text{-}\ produced\ and\ Y_i^d\ \text{-}\ desired\ D - PNN\ output\ for\ i^{th}\ training\ vector\ of\ M - data\ samples$$

4 Backward Differential Polynomial Network

Multi-layer networks form composite functions (Fig. 2). Composite Terms (CT), which substitute for the derivatives with respect to variables of previous layers, are calculated according to the composite function (12) partial derivation rules (13).

$$F(x_1, x_2, \ldots, x_n) = f(z_1, zz_2, \ldots, z_m) = f(\phi_1(X), \phi_2(X), \ldots, \phi_m(X)) \qquad (12)$$

$$\frac{\partial F}{\partial x_k} = \sum_{i=1}^{m} \frac{\partial f(z_1, z_2, \ldots, z_m)}{\partial z_i} \cdot \frac{\partial \phi_i(X)}{\partial x_k} \quad k = 1, \ldots, n \qquad (13)$$

Each D-PNN block can form 5 simple neurons (8)–(10). The blocks of the 2nd and next hidden layers produce additional CTs, which substitute for the composite polynomial derivatives using output and input variables of the back connected previous layers blocks, e.g. the 3rd layer blocks can form CTs in respect of the 2nd (14) and 1st layer (15). The number of block neurons, i.e. CTs including composite derivatives, doubles with each previous back-connected layer. Thus the probability activations P_A of CTs, formed with respect to derivative block input variables of the previous layers, halves along with the increasing number of hidden layers, which are comprised backward in the network tree-like structure (Fig. 2) [10].

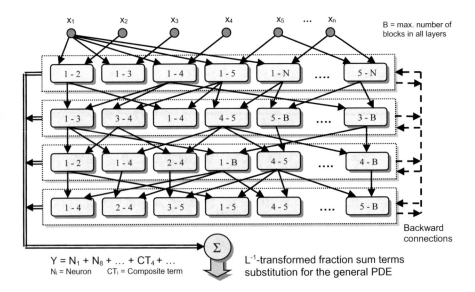

Fig. 2. N-variable D-PNN selects from 2-variable combination blocks in each hidden layer

$$y_i = w_2 \cdot \frac{a_0 + a_1 x_{21} + a_2 x_{22} + a_3 x_{21} x_{22} + a_4 sig(x_{21}^2) + a_5 sig(x_{22}^2)}{x_{22}} \cdot e^{\varphi_{31}}$$

$$\times \frac{x_{21}}{b_0 + b_1 x_{11} + b_2 sig(x_{11}^2)} \cdot e^{\varphi_{21}} \approx \frac{\partial^2 f(x_{21}, x_{22})}{\partial x_{11}^2} \tag{14}$$

$$y_i = w_3 \cdot \frac{a_0 + a_1 x_{21} + a_2 x_{22} + a_3 x_{21} x_{22} + a_4 sig(x_{21}^2) + a_5 sig(x_{22}^2)}{x_{22}} \cdot e^{\varphi_{31}} \cdot \frac{x_{21}}{x_{13}} \cdot e^{-\varphi_{21}} \cdot \frac{x_{11}}{b_0 + b_1 x_2} \cdot e^{\varphi_{11}} \approx \frac{\partial f(x_{21}, x_{22})}{\partial x_2}$$

$$y_i \; - \; i^{th} \, Composite \; Term \, (CT) \, output \qquad \varphi_{21} = arctg(x_{11}/x_{12}) \tag{15}$$

The square and combination derivative terms are formed analogously. The D-PNN using more than 3 input variables must select from the possible node blocks in each layer (i.e. select the optimal node 2-inputs analogous to the GMDH) as the number of the input combination couples grows exponentially in each next hidden layer. The complete D-PNN structure is fixed in advance and optimized as a whole, i.e. it is initialized with an estimated number of layers and the node blocks, which are not added one by one. A specific neurons combination, which can form a PDE solution, is not able to accept a disturbing effect of the rest of the neurons (able to form other local solutions) in the parameter optimization. The D-PNN total output Y is the arithmetic mean of active neurons output values so as to prevent their changeable number in a combination from influencing the total network output (16) [10].

$$Y = \frac{\sum_{i=1}^{k} y_i}{k} \quad k = the \; actual \; number \; of \; active \; neurons \tag{16}$$

2 random simultaneous processes of the optimal block 2-inputs and neurons combination selection are finished gradually in the initial D-PNN structure formation and primary general PDE definition. The iterations are done along with the continual polynomial parameters and term weights optimization using the Gradient Steepest Descent (GSD). The binary Particle Swarm Optimization (PSO), being able to solve large combinatorial problems, can optimize the neurons selection process.

5   NWP Outputs Refinements Using Daily PDE Correction Models

The National Oceanic and Atmospheric Administration (NOAA) operates free National Weather Service (NWS), which provides among other services forecasts of North American Meso-scale (NAM) system. NAM produces 4-day hourly forecasts of temperature, relative humidity, dew point, sky condition, sea level pressure, wind speed and direction at a selected locality [11]. Daily PDE correction models using 6-input - > 1-output variables at the same point-time process the above forecasts of the same trained input quantities to calculate the target predictions in Helena, Montana (Figs. 3, 4, 5 and 6). D-PNN was trained with hourly local data observations for the periods of 2 to 8 days

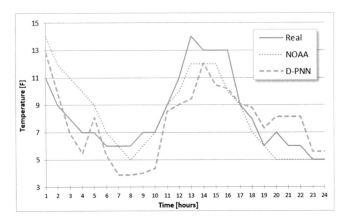

Fig. 3. 19.12.2013, Helena - RMSE: NOAA = *1.65*, D-PNN = *1.90*

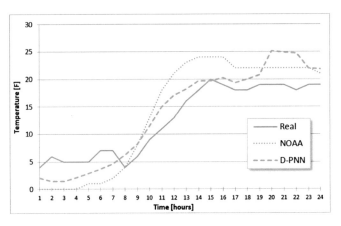

Fig. 4. 20.12.2013, Helena - Temperature RMSE: NOAA = *4.68*, D-PNN = *3.34*

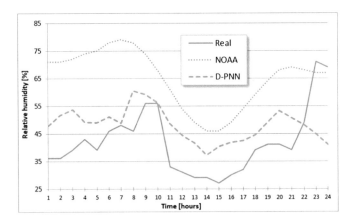

Fig. 5. 16.1.2013, Helena - RMSE: *NOAA = 25.35*, D-PNN = *12.75*

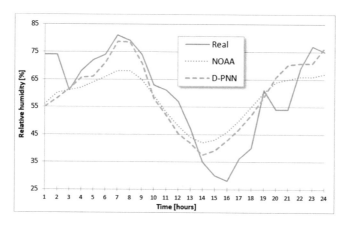

Fig. 6. 17.1.2014, Helena - RMSE: *NOAA = 10.42*, D-PNN = *9.08*

previously (i.e. 48–192 data samples). The optimal numbers of training days were esti-mated by additional supplementary models, trained analogously to the D-PNN correction models and tested with the previous day forecasts to compare the outputs with the latest observations. NOAA provides free historical weather data archives [12] and current hourly tabular observations [13] for many land-based surface stations. Weather Underground (WU) shares the historical data observations [14] and provides alternative forecasts, based on the NOAA forecasts and observations.

Relative humidity has typically a periodic daily behavior with increasing values at night hours, primarily inverse to the temperature drop. A steep or sudden increase in its values can indicate precipitation (Fig. 5), while slight or gradual changes in the slope curve mean variable cloudiness. The presented PDE models are rain-sensitive, they can indicate local storms. Weather conditions mostly do not change fundamentally within short time periods of several days, which are followed by overnight changes. The D-PNN models are not valid in the new situations, as they represent local data relations of the previous days and their corrections of NWP outputs are useless (Fig. 3). An appropriate NWP data analysis can indicate the sporadic days of frontal breaks to discard the revisions. NAM model forecasts can be also sufficient and accurate enough to allow for any valuable improvements using the PDE corrections models in same days (Fig. 6). The D-PNN daily revisions of relative humidity forecasts obtain the best results in comparison to other weather quantities, e.g. wind speed [8] or the most problematic temperature.

6   Conclusions

D-PNN local PDE solutions maintain the equality of derivative changes in selected polynomial relative terms to detail actual surface conditions. More or less settled weather periods allow to obtain successful revisions in daily forecast as the trained relations of current data observations do not change essentially in the forecasted days. Temporary frontal systems affect trained patterns character hence the PDE models do not represent changed local conditions and their corrections fail. The optimal training periods were initially estimated by secondary models, formed analogously to the prediction models but tested with the previous day input forecasts for the latest observations. The supplementary models can just as estimate the optimal training errors, which correspond to the minimal correction errors (in respect of the last data) and which can occasionally converge together with the prediction errors in some days. These 2 main training parameters are necessary to eliminate rapid dynamical changes in ground atmosphere and the inaccuracy of processed NWP outputs in operative predictions. The accuracy of the proposed method is naturally bound to the quality of processed input forecasts, which varies day to day. The D-PNN models using gridded data from several observation stations (and possibly atmospheric layers) would allow more precise daily predictions as the boundary (and sky) conditions in the object area will project themselves into the target central locality in forecasted daily horizon.

Acknowledgement. This paper is supported by the following projects: LO1404: Sustainable development of ENET Centre; CZ.1.05/2.1.00/19.0389 Development of the ENET Centre research infrastructure; SP2017/159 Students Grant Competition and TACR TH01020426, Czech Republic.

References

1. Chan, K., Chau, W.Y.: Mathematical theory of reduction of physical parameters and similarity analysis. Int. J. Theor. Phys. **18**, 835–844 (1979)
2. Durai, V.R., Bhradwaj, R.: Evaluation of statistical bias correction methods for numerical weather prediction model forecasts of maximum and minimum temperatures. Nat. Hazards **73**, 1229–1254 (2014)
3. Kriesche, B., Hess, R., Reichert, B.K., Schmidt, V.: A probabilistic approach to the prediction of area weather events, applied to precipitation. Spat. Stat. **12**, 15–30 (2015)
4. Nikolaev, N.Y., Iba, H.: Adaptive Learning of Polynomial Networks. Genetic and Evolutionary Computation. Springer, New York (2006)
5. Shao, A.M., Xi, S., Qiu, C.J.: A variational method for correcting non-systematic errors in numerical weather prediction. Earth Sci. **52**, 1650–1660 (2009)
6. Vannitsem, S.: Dynamical properties of MOS forecasts: analysis of the ECMWF operational forecasting system. Weather Forecast. **23**, 1032–1043 (2008)
7. Xue, H.-L., Shen, X.-S., Chou, J.-F.: A forecast error correction method in numerical weather prediction by using recent multiple-time evolution data. Adv. Atmos. Sci. **30**, 1249–1259 (2013)
8. Zjavka, L.: Wind speed forecast correction models using polynomial neural networks. Renew. Energy **83**, 998–1006 (2015)
9. Zjavka, L.: Numerical weather prediction revisions using the locally trained differential polynomial network. Expert Syst. Appl. **44**, 265–274 (2016)
10. Zjavka, L., Snášel, V.: Constructing ordinary sum differential equations using polynomial networks. Inf. Sci. **281**, 462–477 (2014)

11. NAM forecasts. http://forecast.weather.gov/MapClick.php?lat=46.5898&lon=-112.0212& lg=english&&FcstType=digital

12. NOAA National Climatic Data Center archives. www.ncdc.noaa.gov/orders/qclcd/

13. NOAA tabular weather observations. www.wrh.noaa.gov/mesowest/getobext.php?wfo= tfx&sid=KHLN

14. WU history. www.wunderground.com/history/airport/KHLN/2015/10/3/DailyHistory.html

Energy Consumption and Cost Analysis for Data Centers with Workload Control

Abdellah Ouammou[1(✉)], Mohamed Hanini[1], Said El Kafhali[2],
and Abdelghani Ben Tahar[1]

[1] Computer, Networks, Mobility and Modeling Laboratory,
Faculty of Sciences and Technology, Hassan 1st University, Settat, Morocco
abdellah.ouammou@gmail.com, haninimohamed@gmail.com, bentahara@yahoo.fr
[2] Computer, Networks, Mobility and Modeling Laboratory,
National School of Applied Sciences, Hassan 1st University, Settat, Morocco
said.elkafhali@uhp.ac.ma

Abstract. In the context of cloud computing, the energy consumed by
the data center is higher because it contains a large number of physical
machines, which in turn contain a number of virtual machines resulting
in high power consumption. In addition, the cloud provider must provide
a high quality of service (QoS) to its customers on the condition of not
consuming a large amount of energy. Among the techniques of minimizing
energy consumption is to turn down servers when the workload is low and
relocate its virtual machines to another server. In this paper, we propose
to combine this technique with another that uses a threshold ensuring
the condition of not crossing a given level of use capacity of each server.
We validate our model by numerical evaluation which demonstrates the
effectiveness of the proposition in terms of energy efficiency and QoS
improvement.

Keywords: Energy efficiency · Cloud computing
Energy consumption · Virtualization · Allocation of virtual machines
Live migration of virtual machines

1 Introduction

A cloud computing infrastructure consists of services that are offered and delivered through a data center, that can be accessed from a web browser anywhere
in the world [1]. A data center is a centralized repository for the storage, management, and dissemination of data and information. Typically, a data center is
a facility used to house computer systems and associated components, such as
telecommunications and storage systems. Cloud data centers (CDC) are made
up of a number of physical machines (PMs) consisting of groups and multiple virtual machines (VMs) running on each PM [2]. Many companies such as Amazon,
Microsoft, Google, and Apple offer cloud services more quickly and efficiently to
the client. Cloud market achieved up to 150 billion dollars in 2013 [3], but the
rising price of energy increases the total cost of ownership and reduces the return

© Springer International Publishing AG, part of Springer Nature 2018
A. Abraham et al. (Eds.): IBICA 2017, AISC 735, pp. 92–101, 2018.
https://doi.org/10.1007/978-3-319-76354-5_9

on Investment. According to [4], the energy consumption of CDC worldwide is estimated at 26GW corresponding to about 1.4% of the global electrical energy consumption with a growth rate of 12% per year. Another recent study estimates that a CDC can use 91 billion kilowatt-hours of electricity and its consumption is still growing rapidly [5]. This consumed energy can be classified into energy consumed by infrastructure facilities (e.g., cooling systems and power conditioning systems) and energy consumed by IT equipment (e.g., PMs, storage, networks, etc.).

Otherwise, the virtualization of computing resources is a basic technique in CDC used to improve the energy efficiency [6]. This technology allows multiple VMs to be consolidated on a minimum number of PMs to further minimize such power consumption in CDC that host Cloud Computing services. A significant amount of power is consumed even when the PM is idle (approximately 70% of the power consumed by the PM running at full CPU utilization) [7], thus opening an opportunity for VMs migration and turns off the PMs who are not used for reducing energy cost. In addition, cloud providers must provide reliable Quality of Service (QoS) for a customer that is negotiated in terms of Service Level Agreements (SLA), e.g. throughput, response time [8]. Therefore, to ensure better use of resources, cloud providers have to deal with power-performance trade-off, because aggressive consolidation of VMs can result in a significant loss of performance. To this end, in this work we exploit the dynamic migration of VMs based on current resource requirements while ensuring reliable QoS and minimizing energy consumption.

The rest of the paper is organized as follows: Sect. 2 summarizes the related work. The proposed model is presented in Sect. 3. Section 4 presents the analysis of the proposed model. Section 5 presents the performance analysis and the numerical results. Finally, Sect. 6 is devoted to the conclusion.

2   Related Work

Some studies about cloud energy consumption were proposed recently. For instance, Srikantaiah *et al.* [9] presented an energy-aware consolidation technique to decrease the total energy consumption of a cloud computing system. The authors empirically modeled the energy consumption of servers as a function of CPU and disk utilization. For that, they described a simple heuristic to consolidate the processing works in the cloud computing system. Performance of the solution is evaluated only for very small input size. Authors in [10], proposed a mathematical programming model for server consolidation, in which each service was implemented in a VM, to reduce number of used PM. Verma *et al.* [11], formulated a problem of Energy consumption for heterogeneous CDC with workload control dynamic placement of applications. They applied a heuristic for bin packing problem with variable bin sizes and costs and introduced the notion of cost of VM live migration, but the information about the cost calculation is not provided. The proposed algorithms, on the contrary to our approach, do not handle SLA requirements: SLA can be violated due to variability of the workload.

Beloglazov *et al.* [12], proposed a heuristic algorithm for dynamic adaption of VM allocation at run-time based on the current utilization of resources by applying live migration, switching idle nodes to sleep mode. The proposed algorithm evaluated by simulations significantly reduces the global power consumption of the simulated CDC infrastructure. Authors in [14], proposed a power performance problem for parallel computations in CDC. They adopt an optimal energy/time allocation among levels of tasks and equal power supply to tasks at the same level, and show significant performance improvement. Awada *et al.* [13], presented energy consumption formulas for calculating the total energy consumption in cloud environments to save energy. They described an energy consumption tools and empirical analysis approaches and provided generic energy consumption models for server idle and server active states. Boru *et al.* [15] proposed an energy-efficient data replication model in geographically distributed CDC. They introduced a unique replication solution which considers both energy efficiency and bandwidth consumption of the system. The proposed model improves communication delay and network bandwidth between geographically dispersed CDC as well as inside of each CDC.

In contrast to the discussed studies, we propose efficient heuristics for dynamic adaption of allocation of VMs in runtime applying live migration according to current utilization of resources and thus minimizing energy consumption. The proposed approach can effectively handle strict QoS requirements, heterogeneous infrastructure and heterogeneous VMs. The algorithms do not depend on a particular type of workload and do not require any knowledge about applications executed on VMs.

3   Proposed Model

3.1   Problem Statement

In this work, we consider a data center with m homogeneous PMs, each PM contains a number of heterogeneous VMs. In order to manage the CDC, there are many objectives that have been defined in the literature. Among them, minimizing the number of PMs required or minimizing the number of VMs executed per unit of time. All these objectives contribute to minimize energy consumption or to increase the level of performance. The proposed model in this paper aims to minimize the number of PMs in order to minimize the energy consumption of the server. However, consideration should be given to avoiding excessive consolidation which results in a negative impact on QoS.

The idea is to define two usage thresholds in each PM; namely, the upper threshold and the lower threshold. These thresholds are used to keep the total CPU usage of all VMs in the PM between these two values. If the total use of a PM is below the lower threshold, then all of the VMs running in this PM must be migrated from that host, and that host must be disabled to eliminate active power consumption, in the other case if total CPU usage exceeds the upper threshold, some VMs must be migrated from that host to reduce usage and to avoid potential contract violation at the service level (Fig. 1).

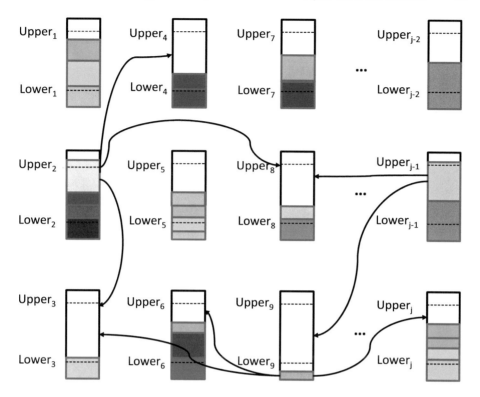

Fig. 1. Data center architecture and the proposed model

3.2   Mathematical Formulation

The proposed problem is studied as a mixed integer linear programming model. The notation used in this paper are given in the Table 1 below.

Objective Function: The objective used for the problem is about minimizing the number of PMs.

To model the objective function, intermediate variables are defined as below.

$$\forall j \in [1, m] \qquad y_j = \begin{cases} 1, \ Machine \ M_j \ is \ used \\ 0, \ otherwise \end{cases} \qquad (1)$$

Then, the objective function can be formulated as

$$\min \sum_{j=1}^{m} y_j \qquad (2)$$

Table 1. Notation and terminology

Notation	Description
m	Number of physical machines $(M_1, M_2, ..., M_m)$
n	Number of virtual machines for each physical machine j $(V_{1j}, V_{2j}, ..., V_{nj})$
C_{ij}	Capacity of each virtual machine i mapped in physical machine j
$Upper_j$	Threshold Up for each physical machine j
$Lower_j$	Threshold Down for each physical machine j
Cap_j	Capacity total the physical machines j
$Upper$	Threshold Up for total physical machines
Cap	Total capacity of the system

Note that value of the y_j variables will be calculated from decision variables defined in the following.

The decision variables in this model are defined as below:

$$x_{ij} = \begin{cases} 1, & i^{th} \text{ VM is mapped to } j^{th} \text{ PM} \\ 0, & \text{otherwise} \end{cases} , \quad \forall i \in [1, n], \forall j \in [1, m] \qquad (3)$$

Then, the y_j can be calculated by

$$y_j = \max_{i \in [1,n]} x_{ij} \qquad (4)$$

or $y_j \leq x_{ij}, \forall i \in [1, n]$ since they are in binary and the objective is relational to minimizing each value of y_j for all $j \in [1, m]$.

The constraints of the problem are:

– The total capacity does not exceed the capacity of system Cap
– Total capacity of virtual machines used for each physical machine M_j must be between the thresholds $Upper_j$ and $Lower_j$
– No virtual machine can exist in 2 physical machines at the same time

The constraints are formulated as below

$$\sum_{j=1}^{m} Cap_j \leq Upper, \qquad (5)$$

$$Lower_j \leq \sum_{i=1}^{n} C_{ij} \leq Upper_j, \quad \forall j \in [1, m] \qquad (6)$$

$$\sum_{j=1}^{m} x_{ij} \leq 1, \quad \forall i \in [1, n] \qquad (7)$$

4   Model Analysis

The proposed mechanism works in two phases: VM selection and VM placement.

VM Selection: The optimization of the current allocation of the VMs is done in two steps: in the first step, we select the VMs to be migrated, in the second step, the selected VMs are placed on the host using the modification of the Best Fit Decreasing (MBFD) algorithm [12]. We use a heuristic to choose the VMs to migrate. The Minimization of Migrations (MM) policy selects the minimum number of VMs needed to migrate from a host to reduce CPU utilization below the Upper threshold if the upper threshold is violated.

The MM policy finds a set R_j of VMs which must be migrated from the host j.

Let V_j the set of VMs currently allocated to the host j. For each VM_i in the host j, c_{ij} is its capacity.

The R_j as a subset of V_j is defined by the following equations

$$if \ \ Lower > Ucurr \ \ then \ \ R_j = V_j \tag{8}$$

$$if \ \ Upper < Ucurr \ \ then \ \ R_j = \{V_{ij} \in V_j | \min_{1 \leq k \leq n} \sum_{i=1}^{\binom{n}{k}} c_{ij} \geq Ucurr - Upper\} \tag{9}$$

VM Placement: Allocation of VMs can be divided into two steps, the first part is information about VMs, numbers and capacity; while the second part is to optimize the necessary allocation numbers of VMs. The first part can be considered a Bin Packing problem with variable bin sizes and prices. To solve it, we apply a MBFD algorithm such that we sort all the VMs in descending order of its utilization and allocate each VM to a host that provides the least increase of capacity due to this allocation. This makes it possible to take advantage of the heterogeneity of the nodes by choosing the most effective power.

The pseudo-code for the algorithm is presented in Algorithm 1.

5   Performance Analysis

In migration algorithm, the upper threshold is set to avoid the SLA violations. Each PM periodically executes an overload detection strategy to trigger migration. A PM is overload, when the resource utilization reaches the upper threshold. The upper threshold should be adjusted, depending on the specific system requirements, to avoid performance degradation and SLA violations. A PM is considered to be under loaded, when the resource utilization is under the lower threshold. The lower threshold significantly affects the energy efficiency and the amount of migrations.

Algorithm 1. Modified Best Fit Decreasing (MBFD)

Input: List of PMs, List of VMs
Output: Allocation of VMs
1. sortDecreasing all virtual machines
2. **for** each VM in List of VMs **do**
3.   **for** each PM in List of PMs **do**
4.   **if** Utilization of PM has not exceed upper threshold and has enough capacity for VM **then**
5.   Utilization of PM = Utilization of PM + Utilization of VM
6.   Remain=Upper-Utilization of PM
7.   **if** Remain is the minimum between the values provided by all VMs **then**
8.   Allocated VM in PM
9.   Remove VM in List of VM
10.   **endif**
11.   **endif**
12.   **endfor**
13. **endfor**
14. **return** VMs allocation

We perform modelization in Matlab to evaluate our model. Modelization has been chosen to evaluate the performance of the proposed algorithms and the number of PMs ranges from 10 to 100 by 10. Runs were performed to compare our method with the one where no policy is used.

5.1 Analysis of Energy Consumption

To evaluate the efficiency of the proposed model, we evaluate the amount energy consumed in CDC. To achieve this end, the following formula is used to calculate the energy

$$E_{Tot} = E + \sum_{activePM_j} (P_j + \sum_{VM_{ij} \in V_j} \alpha \times U(VM_{ij})) \tag{10}$$

where E is the power needed for monitoring the CDC in idle state, P_j is the power consumption in idle state for a PM_j, $U(VM_{ij})$ is the utilization of the VM_{ij} and α is a power weight coefficient.

The influence of high and low thresholds is evaluated on energy efficiency in data center which consists of 100 VMs.

In Fig. 2, we plot the amount of energy consumed under our proposed algorithm compared to the case where a random selecting algorithm is used. As the Fig. 2 shows, the system energy consumption vary with the value of number of PMs. The tendencies of the system energy consumption obtained from the two experiences are similar. We remark that, when the number of PMs increases from 30 to 100, there is obvious change of the system energy consumption. However, it increases faster from 70 to 90 and the total energy consumption is always lower for MBFD especially from 60 to 100 PMs. In average an amount of 20% of

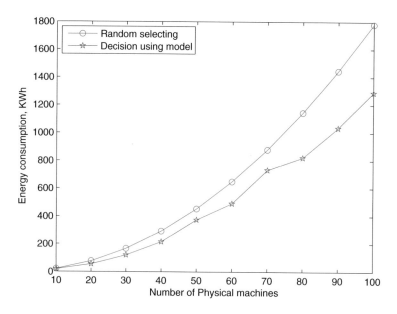

Fig. 2. Comparison of the system energy consumption

consumed energy is saved due to migration algorithm which permits us to turn off some PMs.

5.2   Cost of VM Migration

VMs migrate at the beginning of each monitor period. It has negative impacts on performance of the running tasks. We assume that each VM migration costs the same amount of resources. So it is crucial to minimize the number of VM migrations ensuring the QoS and energy conservation. In this work we are interested in evaluating the cost of migration with a view to study and minimize it in future work. Each migration has a cost in QoS. However, when the migration is monitored due to exceeding of the upper threshold, the QoS of the migrated machine and of remained VMs in the PM where the migration is performed from is improved. Then, the total cost of this operation can be computed using the following formula

$$Cost_{Mig} = \beta \times NM_{Tot} - \gamma \times NM_{up} \tag{11}$$

where NM_{Tot} is the total number of migrated VM, NM_{up} is the number of VM migrated due to exceeding of the upper threshold, β is the cost of a migration and γ is the gain in QoS when the migration is monitored due to exceeding of the upper threshold.

Moreover, we can suppose that β is very small compared to γ based to the fact that migration techniques used in CDC are developed and permits minimizing the cost due to this operation. In our experimentation, we use $\beta = 2$ and $\gamma = 4$.

Table 2. Cost of migration

Number of VM	Number of migration due to exceeding upper threshold	Cost
10	1	30
20	8	−32
30	12	−26
40	17	−50
50	14	−10
60	25	−80
70	29	−100
80	25	−58
90	38	−130
100	40	−122

Table 2 shows the cost of migration when the number of VMs is varying. We remark that when the number of VMs is increasing, the cost becomes not positive, which means that we gain at the QoS. This is due to the fact that there is more chance to have an exceeding of the upper threshold.

6   Conclusion

In cloud computing systems, managing the VM placement is a crucial task that can be used to save energy and to improve QoS. In this paper we have proposed a technique to manage the virtual machine migration between the PMs in a CDC. Our proposition is based on a set of thresholds that manages the selection and placement of VMs in physical machines, this management deal with power-performance trade-off. This model is presented as a mixed integer linear mathematical model. The numerical evaluation of the proposed technique, performed in Matlab, showed that our model enable saving an amount of energy consumed and improve the QoS. However, this work has to be detailed in terms of QoS performances.

References

1. Mell, P., Grance, T.: The NIST definition of cloud computing. Computer Security Division, Information Technology Laboratory, National Institute of Standards and Technology Gaithersburg (2011)
2. El Kafhali, S., Salah, K.: Stochastic modelling and analysis of cloud computing data center. In: Proceedings of 20th Conference Innovations in Clouds, Internet and Networks, pp. 122–126. IEEE (2017)

3. Arroba, P., Moya, J.M., Ayala, J.L., Buyya, R.: Dynamic voltage and frequency scaling-aware dynamic consolidation of virtual machines for energy efficient cloud data centers. In: Concurrency and Computation: Practice and Experience, vol. 29(10) (2017)

4. Koomey, J.: Estimating Total Power Consumption by Servers in the U.S. and the World, February (2007)

5. Research L.: Coal Computing: How Companies Misunderstand Their Dirty Data Centers. White paper (2016)

6. Dayarathna, M., Wen, Y., Fan, R.: Data center energy consumption modeling: a survey. IEEE Commun. Surv. Tutorials **18**(1), 732–794 (2016)

7. Kusic, D., Kephart, J.O., Hanson, J.E., Kandasamy, N., Jiang, G.: Power and performance management of virtualized computing environments via lookahead control. Cluster Comput. **12**(1), 1–15 (2009)

8. Chatterjee, T., Ojha, V.K., Adhikari, M., Banerjee, S., Biswas, U., Snášel, V.: Design and implementation of an improved datacenter broker policy to improve the QoS of a cloud. In: Proceedings of the Fifth International Conference on Innovations in Bio-Inspired Computing and Applications, pp. 281–290. Springer (2014)

9. Srikantaiah, S., Kansal, A., Zhao, F.: Energy aware consolidation for cloud computing. In: Proceedings of the 2008 Conference on Power Aware Computing and Systems, vol. 10, pp. 1–5, December 2008

10. Speitkamp, B., Bicher, M.: A mathematical programming approach for server consolidation problems in virtualized data centers. IEEE Trans. Serv. Comput. **3**(4), 266–278 (2010)

11. Verma, A., Ahuja, P., Neogi, A.: pMapper: power and migration cost aware application placement in virtualized systems. In: Proceedings of the 9th ACM/IFIP/USENIX International Conference on Middleware, pp. 243–264. Springer, New York (2008)

12. Beloglazov, A., Abawajy, J., Buyya, R.: Energy-aware resource allocation heuristics for efficient management of data centers for cloud computing. Future Gener. Comput. Syst. **28**(5), 755–768 (2012)

13. Awada, U., Li, K., Shen, Y.: Energy consumption in cloud computing data centers. Int. J. Cloud Comput. Serv. Sci. **3**(3), 145–162 (2014)

14. Li, K.: Power and performance management for parallel computations in clouds and data centers. J. Comput. Syst. Sci. **82**(2), 174–190 (2016)

15. Boru, D., Kliazovich, D., Granelli, F., Bouvry, P., Zomaya, A.Y.: Energy efficient data replication in cloud computing datacenters. Cluster Comput. **18**(1), 385–402 (2015)

A Stochastic Game Analysis of the Slotted ALOHA Mechanism Combined with ZigZag Decoding and Transmission Cost

Ahmed Boujnoui[1(✉)], Abdellah Zaaloul[1], and Abdelkrim Haqiq[1,2]

[1] Computer, Networks, Mobility and Modeling Laboratory, FST,
Hassan 1st University, Settat, Morocco
ahmed.boujnoui@gmail.com, zaaloul@gmail.com, ahaqiq@gmail.com
[2] e-NGN Research Group, Africa and Middle East, Rabat, Morocco

Abstract. This paper investigates the impact of transmission cost on stimulating the cooperation in decentralized Slotted ALOHA mechanism combined with ZigZag Decoding (SA-ZD). We model the system by a bi-dimensional Markov chain that integrates the effect of ZigZag Decoding (ZD), then we formulate the problem within a stochastic game framework. We evaluate and compare the performance parameters of our proposed approach with those of the same approach without transmission cost. We then show how to control the transmission cost to achieve the best performances. All found results show that our approach improves significantly the Quality of Service (QoS) of the system.

Keywords: Wireless networks · Performance evaluation
Game theory · Markov chain · Slotted ALOHA · Nash equilibrium
ZigZag Decoding · Transmission cost

1 Introduction

Medium access control is the way of access to common media and permission for transmission over them. Since wireless networks use a shared transmission medium, collision may occur because of simultaneous transmissions by two or several interfering nodes, hence the necessity to coordinate transmissions. The most popular multiple access protocols is probably the ALOHA family [1], Carrier Sense Multiple Access (CSMA) [3], and their corresponding variations. They have been widely studied as efficient methods to coordinate the medium access among competing users. To prevent contention issues between competitive nodes, various mechanisms have been implemented. For instance, to reduce contention under Slotted ALOHA (SA) mechanism, a user transmits a packet with certain probability during each time slot. However, we can see in CSMA mechanism, a user maintains a back off window and waits for a random amount of time bounded by the back off window before a transmission (or retransmission).

© Springer International Publishing AG, part of Springer Nature 2018
A. Abraham et al. (Eds.): IBICA 2017, AISC 735, pp. 102–112, 2018.
https://doi.org/10.1007/978-3-319-76354-5_10

The main motivation of this work is to provide an analytical framework to systematically study the behavior of selfish nodes in wireless network. We propose herein a game theoretic approach of the SA mechanism combined with a technique called ZigZag Decoding (ZD) and a Transmission Cost (TC). We evaluate our results using Matlab and compare them to the basic SA mechanism with a transmission cost (SA-TC) [2]. We also compare our results with those of the cooperative and the non-cooperative Slotted ALOHA mechanism combined with ZigZag Decoding (SA-ZD) without transmission cost [8,9].

The rest of the paper is organized as follows. In Sect. 2, we discuss some recent empirical studies in the context of selfish behavior in wireless networks. We present a brief overview of the proposed mechanism in Sect. 3. In Sect. 4, we formulate a non-cooperative model of the SA-ZD mechanism in which we integrate the concept of TC. Section 5 is dedicated to numerical results where we investigate the impact of adding the TC on the performance metrics of the system. Finally, Sect. 6 concludes the paper.

2   Related Work

Considerable work has been done for analyzing the behavior of selfish nodes in wireless networks. Below we highlight some of this literature.

Altman et al. [2] proposed to add a TC to every transmission attempt, the authors show that in non-cooperative game this pricing can be used to get the same throughput given by cooperative game. The same idea as [2] have been proposed by Karouit et al. in [5], where they studied the Binary Exponential Backoff (BEB) mechanism with Multiple Power Levels (MPLs). The resulting mechanism named MPL-BEB shows that under heavy traffic the Mobile Stations (MSs) act selfishly by attempting to get access to the channel using a retransmission probability close to 1, thus introducing the TC allows to control such a behavior and improve the performance of the system. In [6], the authors studied the influence of TCs on the behavior of selfish nodes in wireless Local Area Networks (LANs). Unlike [2], they assumed that active nodes know the current number of backlogged packets in the system. Therefore, nodes can decide whether or not to accept the new arriving packet, depending on the expected cost of successful transmission.

Zaaloul et al. studied the SA-ZD mechanism as a cooperative game in [9] and as a non-cooperative game in [8]. Our studies extend [8] by introducing the TC into the SA-ZD mechanism in order to mitigate the selfish behavior of competing nodes and hence improve the Quality of Service (QoS) of the system.

3   Proposed Mechanism

We propose a mechanism combining the SA, ZD and a TC. We name the resulting mechanism SA-ZDTC. ZigZag requires no changes to the MAC layer and introduces no overheard in the case of no collision [4]. If there is no collusion, ZigZag acts like a typical random access method. Furthermore, it achieves the same performance as if the colliding packets were a priori scheduled in separate time slots. Thus the SA-ZDTC mechanism works as follows:

- Time is divided into slots of one packet duration.
- The frame size is either one or two slots.
- If one node attempts transmission during a slot, the transmission is successful.
- If two nodes attempt transmission during a slot, the transmission is successful by ZigZag.
- Otherwise, a collision occurs and packets involved in a collision are lost.
- Collided packets are retransmitted after a random delay.
- If a new packet arrives during a slot, it will be transmitted in the next slot.
- If a transmission has failed, node becomes backlogged.
- For simplicity, we first assume that transmissions are cost free, but later costs are introduced in order to reduce selfishness.

4   A Stochastic Game Formulation

In many cases, Slotted ALOHA system is usually a decentralized entity, so the cooperative model is not efficient any more. We will develop a model for decentralized non-cooperative game which is more powerful and appropriate to SA mechanism. The Nash equilibrium concept replaces the concept of optimality in the team problem.

4.1   Non-cooperative Game

We consider $M + 1$ bufferless nodes that compete to get access to a shared medium. Let $\mathbf{q_r} = (q_r^1, q_r^2, ..., q_r^{M+1})$ be a vector of retransmission probabilities for all users. We define by $([\mathbf{q_r}]^{-i}, q_r^i)$ a retransmission policy where user i retransmits at any slot with probability q_r^i and any other node j retransmits with probability q_r^j for all $j \neq i$. We assume that all the M nodes retransmit with a given probability $[\mathbf{q_r}]^{-(M+1)} = (q_r, q_r, ..., q_r)$, whereas the node $M + 1$ retransmits with probability $q_r^{(M+1)}$.

In a non-cooperative game, each player attempts to optimize its utility (denoted by $objective_i(\mathbf{q_r})$), by either maximizing its own throughput or minimizing the expected delay of its packets. Our objective is to find a symmetric equilibrium policy $\mathbf{q_r^*} = (q_r, q_r, ..., q_r)$ such that for any user i and any retransmission probability $q_r^i \neq q_r$

$$objective_i(\mathbf{q_r^*}) \geq objective_i(q_r, ..., q_r^i, ..., q_r). \tag{1}$$

Due to symmetry, verifying (1) for a single player is a sufficient condition to have $\mathbf{q_r^*} = (q_r, q_r, ..., q_r)$ as an equilibrium. Hereafter, we choose the player $M + 1$ to be our tagged user.

We define the set of best response strategies of user $M + 1$ by

$$\mathcal{Q}^{M+1}(\mathbf{q_r}) = \underset{q_r^{(M+1)} \in [\epsilon, 1]}{\operatorname{argmax}} \left\{ objective_{(M+1)} \left([\mathbf{q_r}]^{-(M+1)}, q_r^{(M+1)} \right) \right\}, \tag{2}$$

where $\left([\mathbf{q_r}]^{-(M+1)}, q_r^{(M+1)}\right)$ denotes the retransmission policy, and the maximization is taken with respect to $q_r^{(M+1)}$. Then $\mathbf{q_r^*}$ is a symmetric equilibrium if

$$\mathbf{q_r^*} \in \mathcal{Q}^{M+1}\left(\mathbf{q_r^*}\right). \tag{3}$$

Let $Q_a(i, N)$ be the probability that i unbacklogged nodes transmit packets in a given slot. Then

$$Q_a(i, N) = \binom{M-N}{i}(1-p_a)^{(M-N-i)}p_a^i, \tag{4}$$

where p_a is the arrival probability. Let $Q_r(i, N)$ be the probability that i out of backlogged nodes retransmit packets in a given slot. Thus

$$Q_r(i, N) = \binom{N}{i}(1-q_r)^{(N-i)}q_r^i. \tag{5}$$

To find the performance metrics, we shall use a bi-dimensional Markov chain improved by ZD. The transition probability diagram is depicted in Fig. 1. Let the first state component be the number of backlogged packets among the users $1, ..., M$, and the second component be the number of backlogged packets of the user $M + 1$ (either 1 or 0). For any choice of values $q \in]0, 1]$, the state process is a Markov chain that contains a single ergodic sub-chain (and possibly transient states as well). Indeed, it is easy to check that the past and future are conditionally independent, given the present state (Markov property).

Since the state space is finite and all states communicate with each other, the Markovian chain is ergodic, and therefore the stationary distribution exists.

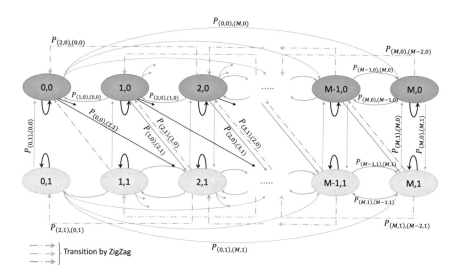

Fig. 1. Bi-dimensional Markov chain for the game problem

Let $\pi\left([\mathbf{q_r}]^{-(M+1)}, q_r^{(M+1)}\right)$ be the corresponding vector of steady state probabilities where its Nth entry $\pi_{N,a}\left([\mathbf{q_r}]^{-(M+1)}, q_r^{(M+1)}\right)$ denotes the probability that the state of the system is (N, a).

For simplicity purpose, we note $\bar{\mathbf{q}}_{(M+1)} = \left([\mathbf{q_r}]^{-(M+1)}, q_r^{(M+1)}\right)$. Then the steady state of the Markovian process is given by the following system

$$
\begin{cases}
\pi\left(\bar{\mathbf{q}}_{(M+1)}\right) = \pi\left(\bar{\mathbf{q}}_{(M+1)}\right) P\left(\bar{\mathbf{q}}_{(M+1)}\right), \\
\pi_{N,a}\left(\bar{\mathbf{q}}_{(M+1)}\right) \geq 0, \qquad N = 0, ..., M \quad \text{and} \quad a = 0, 1 \\
\sum_{N=0}^{M}\sum_{a=0}^{1} \pi_{N,a}\left(\bar{\mathbf{q}}_{(M+1)}\right) = 1.
\end{cases}
\tag{6}
$$

4.2   Frame Size

The frame size in a non-cooperative SA-ZD is either one or two slots and it is defined by

$$
T\left(\bar{\mathbf{q}}_{(M+1)}\right) = 2P_{ZigZag} + (1 - P_{ZigZag}),
\tag{7}
$$

where P_{ZigZag} is the probability that tow packets are transmitted by ZD, it is given by

$$
P_{ZigZag} = \sum_{N=0}^{M} P_1 \pi_{N,0}\left(\bar{\mathbf{q}}_{(M+1)}\right) + P_2 \pi_{N,1}\left(\bar{\mathbf{q}}_{(M+1)}\right),
\tag{8}
$$

where

$$
\begin{cases}
P_1 = p_a\left[Q_a(1, N)Q_r(0, N) + Q_a(0, N)Q_r(1, N)\right] \\
\quad + (1 - p_a)\left[Q_a(0, N)Q_r(2, N) + Q_a(2, N)Q_r(0, N) + Q_a(1, N)Q_r(1, N)\right], \\
P_2 = q_r^{(M+1)}\left[Q_a(1, N)Q_r(0, N) + Q_a(0, N)Q_r(1, N)\right] \\
\quad + (1 - q_r^{(M+1)})\left[Q_a(0, N)Q_r(2, N) + Q_a(2, N)Q_r(0, N) + Q_a(1, N)Q_r(1, N)\right].
\end{cases}
$$

4.3   Performance Metrics

We can now calculate the number of backlogged packets of user $M + 1$ as follows

$$
S_{(M+1)}\left(\bar{\mathbf{q}}_{(M+1)}\right) = \sum_{N=0}^{M} \pi_{N,1}\left(\bar{\mathbf{q}}_{(M+1)}\right).
\tag{9}
$$

The average throughput of user $M + 1$ is given by

$$
THp_{(M+1)}\left(\bar{\mathbf{q}}_{(M+1)}\right) = \frac{p_a}{T\left(\bar{\mathbf{q}}_{(M+1)}\right)} \sum_{N=0}^{M} \pi_{N,0}\left(\bar{\mathbf{q}}_{(M+1)}\right).
\tag{10}
$$

By Little's formula [7], the delay is given by

$$D_{(M+1)}\left(\bar{\mathbf{q}}_{(M+1)}\right) = 1 + \frac{S_{(M+1)}\left(\bar{\mathbf{q}}_{(M+1)}\right)}{THp_{(M+1)}\left(\bar{\mathbf{q}}_{(M+1)}\right)}. \tag{11}$$

The average throughput of backlogged packets of user $M + 1$ is

$$THp_{(M+1)}^{B}\left(\bar{\mathbf{q}}_{(M+1)}\right)$$
$$= \frac{1}{T\left(\bar{\mathbf{q}}_{(M+1)}\right)} \sum_{N=0}^{M} \sum_{N'=0}^{M} P_{(N,0)(N',1)}\left(\bar{\mathbf{q}}_{(M+1)}\right) \pi_{N,0}\left(\bar{\mathbf{q}}_{(M+1)}\right). \tag{12}$$

The delay of backlogged packets of user $M + 1$ is given by

$$D_{(M+1)}^{B} = 1 + \frac{S_{(M+1)}\left(\bar{\mathbf{q}}_{(M+1)}\right)}{THp_{(M+1)}^{B}\left(\bar{\mathbf{q}}_{(M+1)}\right)}. \tag{13}$$

4.4 Transmission Cost

In [5,8], we observe that as the arrival probability increases, the users tend to be more aggressive at equilibrium. This results in a dramatic decreases in the system's throughput. This is mainly due to the fact that the users act selfishly by attempting to access the channel with a retransmission probability q_r close to 1, which yields more energy consumption and more collisions. To avoid this collapse network, it is required to control the behavior of users. The main idea behind is to reduce the failure probability by limiting the aggressiveness of the competing nodes. Towards this end, we propose to associate a TC denoted by C (which can, in particular, represent the battery power cost) to each transmission attempt.

For illustrative purpose, we consider the example of one player J_1 versus a couple of players J_2 and J_3 together, as shown in Table 1. The first player can undertake two actions, either transmit or wait, while players $\{J_2, J_3\}$ can undertake three actions, either they both transmit (TT), they both wait (WW), or one transmits and the other waits (TW). If exactly one player decides to transmit while the others decide to wait, he receives $(1 - C)$ and the others receive 0. If exactly two players decide to transmit, each one receives $1 - C$ (due to ZD) while the remaining player (the one who decides to wait) receives 0. If all the three players decide to transmit, a collision occurs, and each one of them receives $-C$.

At the steady state, when the node $M + 1$ transmits successfully, it gains $(1-C)THp_{(M+1)}\left(\bar{\mathbf{q}}_{(M+1)}\right)$. Similarly, when the transmission fails, it pays a cost $Cq_r^{M+1} \sum_{N=0}^{M} \pi_{N,1}\left(\bar{\mathbf{q}}_{(M+1)}\right)$. Thus, the average utility of node $M + 1$ is given by

$$objective_{(M+1)}\left(\bar{\mathbf{q}}_{(M+1)}\right)$$
$$= (1 - C)THp_{(M+1)}\left(\bar{\mathbf{q}}_{(M+1)}\right) - Cq_r^{M+1} \sum_{N=0}^{M} \pi_{N,1}\left(\bar{\mathbf{q}}_{(M+1)}\right). \tag{14}$$

Table 1. Three nodes random access game with transmission cost

player J_1	players $\{J_2, J_3\}$		
	TT	TW	WW
Transmit (T)	$(-C, -2C)$	$(1 - C, 1 - C)$	$(1-C, 0)$
Wait (W)	$(0, 2(1 - C))$	$(0, 1-C)$	$(0,0)$

Therefore, in order to maximize its own profit, node $(M+1)$ is now faced with the following challenge

$$maximize_{q_r^{(M+1)} \in [\epsilon, 1]} \left\{ objective_{(M+1)} \left(\bar{\mathbf{q}}_{(M+1)} \right) \right\}. \tag{15}$$

We define as we did before, the set of best response strategies of user $M + 1$

$$\mathcal{Q}^{M+1} (\mathbf{q_r}) = \underset{q_r^{(M+1)} \in [\epsilon, 1]}{\operatorname{argmax}} \left\{ objective_{(M+1)} \left(\bar{\mathbf{q}}_{(M+1)} \right) \right\}, \tag{16}$$

then we seek the value $\mathbf{q_r^*}$ of retransmission probability that satisfies

$$\mathbf{q_r^*} \in \mathcal{Q}^{M+1} (\mathbf{q_r^*}), \tag{17}$$

which is the Nash equilibrium for the non-cooperative game.

5   Numerical Results

5.1   Fixed Transmission Cost

In this section, we compare the performance metrics of the non-cooperative SA-ZDTC mechanism, in one hand with the cooperative SA-ZD, and in the other hand with the non-cooperative SA-ZD. We show how the TC can affect the performance metrics of the system. We choose throughout this paper $\epsilon = 10^{-4}$.

We depict in Figs. 2 and 3 the global throughput at Nash equilibrium and Nash equilibrium retransmission probability, respectively, as a function of p_a for different values of C. We compare the non-cooperative SA-ZDTC (in which various TCs have been used $C = 0, 0.2, 0.5, 0.8$) with the cooperative SA-ZD. As shown in Fig. 2, the SA-ZDTC mechanism proves very effective compared with non-cooperative SA-ZD [8]. In Fig. 3, we see a decrease on the equilibrium retransmission probability as the cost increases, which means that the pricing mechanism strongly affect the behavior of nodes.

Figure 4 illustrates the expected delay of backlogged packets, as a function of arrival probability p_a for different values of C. We note that TC leads to a bounded delay. However, a large pricing could have a negative impact (e.g., a huge delay) since nodes will never transmit when the TC is greater than or equals to the gain obtained in a successful transmission. Furthermore, one may wonder why the pricing mechanism could looks very effective than cooperative

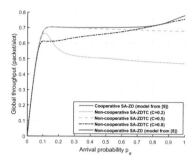

Fig. 2. Global throughput as a function of arrival probability p_a for $M + 1 = 10$, under different values of $C = [0\ 0.2\ 0.5\ 0.8]$

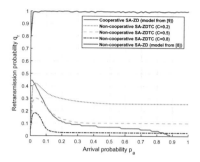

Fig. 3. Retransmission probabilities as a function of arrival probability for $M + 1 = 10$, under different values of $C = [0\ 0.2\ 0.5\ 0.8]$

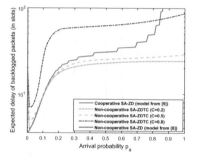

Fig. 4. Expected delay of backlogged packets as a function of the arrival probability p_a for $M + 1 = 10$, under different values of $C = [0\ 0.2\ 0.5\ 0.8]$

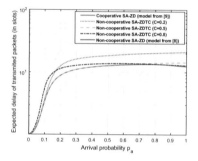

Fig. 5. Expected delay of transmitted packets as a function of the arrival probability p_a for $M + 1 = 10$, under different values of $C = [0\ 0.2\ 0.5\ 0.8]$

game when p_a is close to 1. This is mainly due to the fact that the cooperative system prioritizes the new arrival packets (since p_a is high, see Fig. 5) in order to maximize the global throughput. However, this priority mechanism does not appear in the non-cooperative game. Therefore, introducing the TC in the non-cooperative game makes the system more effective.

For all above, we note that the game equilibrium depends on the TC (C). Hence, this cost must be appropriately adjusted to achieve the best equilibrium.

5.2   Optimization on the Transmission Cost

In this section, we propose our approach to optimize the transmission cost.

To achieve the best average throughput, we propose the cost C_1 which is the solution of the following problem

$$\max_{C \in [0,1]} \left\{ THp_{(M+1)} \left(p_a, q_r^* \right) \right\}. \tag{18}$$

To maximize the ratio between the throughput and the delay of backlogged packets, we propose the cost C_2 which is the solution of the problem (19)

$$\max_{C \in [0,1]} \left\{ \frac{THp_{(M+1)} \left(p_a, q_r^* \right)}{D^B_{(M+1)} \left(p_a, q_r^* \right)} \right\}. \tag{19}$$

In order to compromise between the throughput and the delay of backlogged packets, we propose the cost C_3 given by

$$C_3 = \frac{1}{2} \left(C_1 + C_2 \right). \tag{20}$$

We present in Fig. 6, the optimal TC as a function of arrival probability for 10 nodes. We note that the cost C and C_1 reaches the maximum value when the arrival probability p_a is too large. In fact, the system prioritizes the new arriving packets in order to maximize the global throughput. Whereas C_2 and C_3 never reach their maximal values so that the backlogged packets are just as privileged as the new arriving packets.

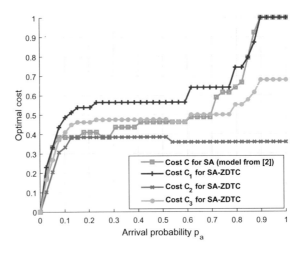

Fig. 6. Optimal cost as a function of arrival probability p_a for $M + 1 = 10$

We illustrate in Figs. 7 and 8, respectively, the average throughput and the delay of backlogged packets, as a function of arrival probability, for 10 nodes.

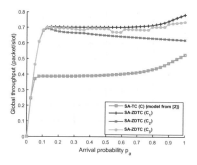
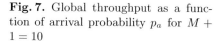

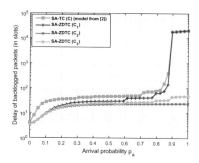

Fig. 7. Global throughput as a function of arrival probability p_a for $M + 1 = 10$

Fig. 8. Delay of backlogged packets as a function of arrival probability p_a for $M + 1 = 10$

We observe that a maximum global throughput is achieved when using the cost C_1. However, by using the cost C_2, we obtain a relatively acceptable throughput. Finally, a compromise is achieved by using C_3. In return, this compromise is compensated with a bounded delay of backlogged packets.

6   Conclusion

In this paper, we have studied the problem of stimulating the cooperation in the non-cooperative SA-ZD mechanism. Firstly, we formulated the problem as a non-cooperative game in which we proposed to add various fixed TCs. Then, in order to achieve the best performances, we have proposed three approaches to optimize the TC. We believe that our approach stimulates selfish nodes to be more tolerant in accessing the shared medium. We have shown through numerical results that our mechanism, named SA-ZDTC, is more efficient than the cooperative SA-TC mechanism and the non-cooperative SA-ZD mechanism.

References

1. Abramson, N.: The ALOHA system: another alternative for computer communications. In: Proceedings of the Fall Joint Computer Conference, 17–19 November 1970, pp. 281–285. ACM (1970)
2. Altman, E., Barman, D., El Azouzi, R., Jiménez, T.: A game theoretic approach for delay minimization in slotted aloha. In: 2004 IEEE International Conference on Communications, vol. 7, pp. 3999–4003. IEEE (2004)
3. Bianchi, G., Fratta, L., Oliveri, M.: Performance evaluation and enhancement of the CSMA/CA MAC protocol for 802.11 wireless LANs. In: Seventh IEEE International Symposium on Personal, Indoor and Mobile Radio Communications, PIMRC 1996, vol. 2, pp. 392–396. IEEE (1996)
4. Gollakota, S., Katabi, D.: Zigzag decoding: combating hidden terminals in wireless networks, vol. 38. ACM (2008)

5. Karouit, A., Sabir, E., Ramirez-Mireles, F., Barbosa, L.O., Haqiq, A.: A stochastic game analysis of the binary exponential backoff algorithm with multi-power diversity and transmission cost. J. Math. Modell. Algorithms Oper. Res. **12**(3), 291–309 (2013)
6. Marbach, P.: Transmission costs, selfish nodes, and protocol design. Wirel. Netw. **14**(5), 615–631 (2008)
7. Nelson, R.: Probability, Stochastic Processes, and Queueing Theory: The Mathematics of Computer Performance Modeling. Springer, New York (2013)
8. Zaaloul, A., Haqiq, A.: Analysis of performance parameters in wireless networks by using game theory for the non cooperative slotted ALOHA enhanced by ZigZag decoding mechanism. arXiv preprint arXiv:1501.00881 (2015)
9. Zaaloul, A., Haqiq, A.: Enhanced slotted aloha mechanism by introducing ZigZag decoding. arXiv preprint arXiv:1501.00976 (2015)

Analytic Approach Using Continuous Markov Chain to Improve the QoS of a Wireless Network

Adnane El Hanjri$^{(\boxtimes)}$, Abdellah Zaaloul, and Abdelkrim Haqiq

Computer, Networks, Mobility and Modeling Laboratory, FST,
Hassan 1st University, Settat, Morocco
adnane.elhanjri@gmail.com, zaaloul@gmail.com, ahaqiq@gmail.com

Abstract. QoS management is a crucial task in wireless networks. Continuous Markov chain is highly significant suitable model for resources management in wireless networks. In this paper, we propose a Continuous Markov chain, to evaluate the performance of Slotted ALOHA protocol implemented in wireless networks. The analysis has been done in two times, and we numerically evaluate system performance under different scenarios. In a first time, we used a Markov chain with two states and in a second time, the space of states has been extended at three states were considered in order to give a more appropriate model.

Keywords: Wireless networks · Performance evaluation
Continuous Markov chains · Slotted ALOHA cooperative

1 Introduction

Wireless networks [8] offer a widely accessible technology to connect a remote user terminal with its primary network. However, the limited resources in these networks cause a lot of problems that affect their performance parameters. To overcome these constraints, access management protocols in a multi-access environment are presented and studied.

Pure ALOHA (PA) [1] and Slotted ALOHA (SA) [5] are certainly the most studied protocols in the literature of telecommunication systems. These access methods are widely implemented in the satellite networks and mobile telephony and are considered resource reservation mechanisms.

In this paper, we are interested in SA that is a refinement over the PA, requires that time be segmented into slots of a fixed length exactly equal to the packet transmission time [2]. The transmission always occurs at the beginning of slot, i.e., transmission are synchronized with the system clock, and the ACK will be received at the end of the current slot. The collided packets, will be retransmitted after a random number of slots. In contrast, for PA, a packet transmission can begin at any time.

In this work, we consider a collision channel shared by M nodes unbuffered that send to a central receiver, i.e., the nodes do not generate new packets until

© Springer International Publishing AG, part of Springer Nature 2018
A. Abraham et al. (Eds.): IBICA 2017, AISC 735, pp. 113–124, 2018.
https://doi.org/10.1007/978-3-319-76354-5_11

the current is transmitted successfully. Afterwards, we study SA one of the most widely used random access protocols originally designed to enhance ALOHA throughput [3] an optimization problem, where all nodes are looking to optimize the same objective (maximize throughput and minimize the average delay). To do so, we model our system by a Continuous Markov chain with adjustable parameters.

The rest of the paper is organized as follows: We begin by an overview of some related work in Sect. 2, in Sect. 3 we construct a Continuous Markov Model with two states Busy state and idle or collision state and we evaluate the performance of the system. In Sect. 4 we extend the model to a Continuous Markov chain with 3 states, Busy, Idle and collision. Section 5 will be devoted to the conclusion.

2   Related Work

In the last years, the SA protocol is still a current topic in scientific research. Thus, Patet et al. [7] have analyzed the number of backlogged packets by using statistical approach and stabilize the expected number of backlogged packets minus mean number of packets which are successfully transmitted.

The authors in [6] proposed an adaptive SA Algorithm that can accelerate the adjusting speed and can acquire stable throughput on the conditions that there is large fluctuation of system load.

The authors in [4] propose a study based on the Markov chain model, to define optimal binomial distribution probabilities of retransmission and arrival packets, in the case of SA protocol; they defined the average packet delay and the throughput, and show the effect of the increase of the number of sources on these parameters.

The authors in [9] have argue why time-based fairness is desirable in some cases and they analyze the achieved throughput of competing nodes, possibly using different data rates and packet sizes in 802.16 cellular networks. They have interested the performance evaluation, in the case of state of the shared channel by M users. They provide a simple model that represents the time of occupation of the channel by a user once it transmits a packet with success.

In our work, we will study two Continuous Markov models for the cooperative SA protocol in IEEE 802.16 system where the users are competing for access to the system resources. Our objective is to evaluate the performances of the system in function of its different parameters. In these models we assumed that the users cooperate in order to optimize the performances of the system, especially the average delay and the average throughput. By doing an analytical analysis, close-form expressions of these parameters are derived.

3   Modeling Uplink Channel Utilization

3.1   Description of the Problem

In this section, we construct a Markov model from which we can analyze the throughput and we describe cooperative SA MAC protocol in which time is

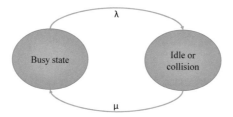

Fig. 1. Transition diagram of the Markov chain with two states

divided into units. Therefore the decision is actually Markovian for each node in SA. We will use a two system state as the following: one state is the busy state when only one of the users is transmitting; the other state is either when the system is idle or when collisions happen. We adopt the following notations for a generalized Markov model for SA type MAC protocol.

M: number of users in the system.

Th: Throughput function.

The transition time from the busy state to the idle or collision state is exponential with parameter.

The transition time from the idle or collision state to the busy state is exponential with parameter (Fig. 1). The infinitesimal generator is

$$Q = \begin{pmatrix} -\lambda & \lambda \\ \mu & -\mu \end{pmatrix}$$

So the transition matrix of Markov chain included is

- $p_{12} = \frac{q_{12}}{q_{12}} = 1,$
- $p_{21} = \frac{q_{21}}{q_{21}} = 1,$
- $p_{11} = 0,$
- $p_{22} = 0.$

$$\mathcal{P} = \begin{pmatrix} 0 & 1 \\ 1 & 0 \end{pmatrix}$$

The Markov chain included is irreducible, so the Continuous Markov chain is also irreducible, and the number of states is finite, so consequently it is ergodic and admits a stationary probability $\pi = (\pi_1, \pi_2)$

So the balance equations can be written as

$$\begin{cases} \lambda \pi_1 & = \mu \pi_2; \\ \lambda \pi_1 & = \mu \pi_2; \\ \pi_1 + \pi_2 = 1. \end{cases}$$

And the solution is

$$\begin{cases} \pi_1 = \frac{\mu}{\mu + \lambda}; \\ \pi_2 = \frac{\lambda}{\mu + \lambda}. \end{cases}$$

3.2   Performance Evaluation

We turn now to study some performance indicators.

The Time of Occupation of Channel. In general, after one user transmits successfully a packet, it obtains the channel. This user will continue to occupy the channel for random amount of time T. We assume that T is an exponential random variable with parameters $-q_{11} = \lambda$. Let $E[T]$ the average for T. If we put $E[T] = U$. We know that $E[T] = \frac{1}{\lambda}$. So $U = \frac{1}{\lambda}$.

Throughput. In each time slot, the system is either in the busy state or in the idle or collision state. The amount $\frac{\mu}{\mu + \lambda}$ represents the system utilization, using the above definitions, this amount equals the stationary probability that the system is in the busy state. So the throughput of the system is

$$Th = \pi_1 = \frac{\mu}{\mu + \lambda} \ .$$

Figure 2 shows the average throughput of the system depending on the parameter λ, for $\mu = 0$, 0.1, 0.4 et 0.8. Figure 2 shows that the throughput decreases when the parameter λ increases, cause the number of collisions increases. When μ increase, the system tends to be busy by a single user, so the throughput of the system increases.

An important note, to keep the throughput superior than 0.5, we must choose the parameter λ less than μ.

We can also note that for $\mu \neq 0$, the throughput will never equal to zero because there is atleast a user who transmitted successfully and when $\lambda = 0$, i.e., the system remains in the busy state, the throughput will be equal to 1. In Fig. 3 we can observe that when U increases (the occupation time of the channel by a user who transmits successfully) meaning that the number of users who transmit successfully increases, so the throughput also increases.

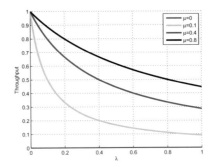

Fig. 2. Throughput depending on λ and μ

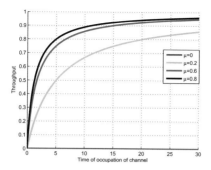

Fig. 3. Throughput depending on U and μ

Average Delay. A relevant quantity which must be taken into consideration is the average delay of packets which is defined as the average time, in slots that a packet takes to go from the SS to the BS.

We note by Q the number of users in the system.

$$Q = n_1 \pi_1 + n_2 \pi_2,$$

where $n_1 = 1$: represents the number of users in the cases where the system can be in busy state. $n_2 = \frac{(M+2)(M-1)}{2}$: represents the number of users in the cases where a collision is probable (the system can be in idle or collision state). So

$$Q = \pi_1 + \frac{(M+2)(M-1)}{2}\pi_2 .$$

Therefore by the formula of LITTLE [6], the delay is

$$D = \frac{Q}{\pi_1} .$$

We find

$$D = 1 + \frac{(M+2)(M-1)}{2}\frac{\lambda}{\mu} .$$

Figure 4 shows the total delay under different values of parameters λ and μ for M = 50. When λ increase, the delay increases because the system tends to the idle or collision state, intuitively, when μ increases the system will be in the busy state, so the delay decreases. Figure 5 shows the total delay under different values of λ, μ and M. Under low values of M, whatever the value of λ and μ, the delay is low. Therefore, if we want a low values of delay we have to choose λ less than μ (Fig. 6).

Note. For M = 1, i.e. we have one user who hold the channel, so the delay is one slot of time.

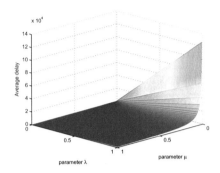

Fig. 4. The average delay depending on λ and μ for M $= 50$

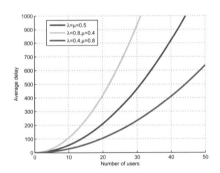

Fig. 5. The average delay depending on number of users, λ et μ

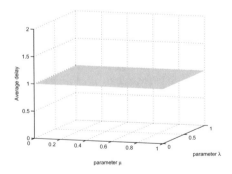

Fig. 6. The average delay for M $= 1$

4   Extended Model

In the previous section, we studied a mathematical model to evaluate the performances of a Wireless mobile networks. But for this kind of networks, we should also take into account the energy consumption problem and see how we can save it. For this reason and in order to enhance the performances of the network, we modified the number of states in the previous model and we considered a three state Markov chain. It is important for the operator to know when the network is in a collision state or in an idle state separately in order to optimize its exploitation.

4.1   Description of the Problem

We model the network by a three state Markov chain, where

- the first state describes the state of the network when it is busy.
- the second state describes the state of the network when it is idle.
- the third state describes the state of the network when there is a collision.

The transition time to the idle state is exponential with parameter μ.
The transition time to the busy state is exponential with parameter λ.
The transition time to the collision state is exponential with parameter σ (Fig. 7).

The infinitesimal generator is

$$\mathcal{Q} = \begin{pmatrix} -\mu - \sigma & \mu & \sigma \\ \lambda & -\lambda - \sigma & \sigma \\ \lambda & \mu & -\lambda - \mu \end{pmatrix}$$

So the transition matrix of Markov chain included is

- $p_{12} = \dfrac{q_{12}}{q_{12} + q_{13}} = \dfrac{\mu}{\mu + \sigma}$,

- $p_{21} = \dfrac{q_{21}}{q_{21} + q_{23}} = \dfrac{\lambda}{\lambda + \sigma}$,

- $p_{13} = \dfrac{q_{13}}{q_{12} + q_{13}} = \dfrac{\sigma}{\mu + \sigma}$,

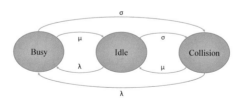

Fig. 7. Transition diagram of the Markov chain with three states

- $p_{23} = \dfrac{\sigma}{\lambda + \sigma}$,

- $p_{31} = \dfrac{\lambda}{\lambda + \mu}$,

- $p_{32} = \dfrac{\mu}{\lambda + \mu}$.

$$\mathcal{P} = \begin{pmatrix} 0 & \frac{\mu}{\mu+\sigma} & \frac{\sigma}{\mu+\sigma} \\ \frac{\lambda}{\lambda+\sigma} & 0 & \frac{\sigma}{\lambda+\sigma} \\ \frac{\lambda}{\lambda+\mu} & \frac{\mu}{\lambda+\mu} & 0 \end{pmatrix}$$

The Markov chain included is irreducible, so the Continuous Markov chain is also irreducible, and the number of states is finite, so consequently it is ergodic and admits a stationary probability: $\pi = (\pi_1, \pi_2, \pi_3)$. So the balance equations can be written as

$$\begin{cases} (\mu + \sigma)\pi_1 & = \lambda(\pi_2 + \pi_3); \\ (\lambda + \sigma)\pi_2 & = \mu(\pi_1 + \pi_3); \\ (\lambda + \mu)\pi_3 & = \sigma(\pi_1 + \pi_2); \\ \pi_1 + \pi_2 + \pi_3 = 1. \end{cases}$$

And the solution is

$$\begin{cases} \pi_1 = \frac{\lambda}{\mu+\lambda+\sigma} \; ; \\ \pi_2 = \frac{\mu}{\mu+\lambda+\sigma} \; ; \\ \pi_3 = \frac{\sigma}{\mu+\lambda+\sigma} \; . \end{cases}$$

4.2   Performance Evaluation

The Time of Occupation of Channel. As we have seen previously, when a user transmits successfully a packet, he continues to use the channel for a random period of time T. We assume that T follows an exponential distribution with parameter $-q_{11} = \mu + \sigma$. Let $E[T] = U$: the average occupation time of the system by a user. So we have: $U = \frac{1}{\mu+\sigma}$.

Average Throughput. As previously, the average throughput is the stationary probability that the system is in the busy state:

$$Th = \pi_1 = \frac{\lambda}{\mu + \lambda + \sigma} .$$

Figures 8, 9 and 10 shows the system throughput based on the parameters λ, μ and σ under different scenarios. In Fig. 8 we fix σ, we note that the throughput decreases when μ increases, on the other side when λ decreases the throughput increases. In Fig. 9 we fix μ, we can note that the throughput decreases when σ increases because of the increase of number of collision. And for the last curves Fig. 10, we fix λ, and this figure confirms the results obtained in the two previous

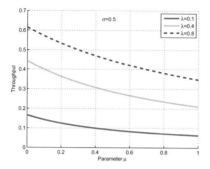

Fig. 8. Throughput in function of μ and λ for $\sigma = 0.5$

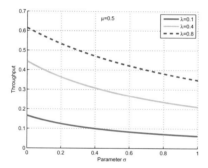

Fig. 9. Throughput in function de σ and λ for $\mu = 0.5$

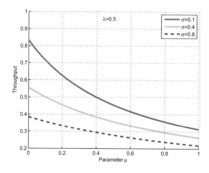

Fig. 10. Throughput in function of σ and μ for $\lambda = 0.5$

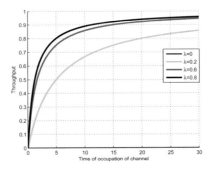

Fig. 11. Throughput in function of U and λ

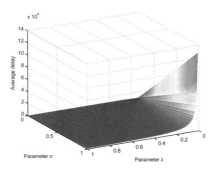

Fig. 12. Average delay in function of σ and λ for M = 50

figures. However, the throughput never drops to zero, i.e. a node still occupied the channel while the others are backlogged. For maximum throughput, we must increase λ and decrease μ and σ.

Figure 11 shows the average throughput depending to different values of U and λ, we can notice the same result of system with two states, i.e. when the occupation time of channel increases the throughput increases too.

Average Delay. As just defined in the previous section, the expected delay is the average time in slot that a packet takes to go from the SS to the BS.

Let Q to be the average number of users in the system, where

$n_1 = 1$: represents the number of users in the busy state of the Markov chain.
$n_2 = 0$: represents the number of users in the idle state of the Markov chain.
$n_3 = \frac{(M+2)(M-1)}{2}$: represents the number of users in the third state of the Markov chain.

So

$$Q = \pi_1 + \pi_3 \frac{(M+2)(M-1)}{2}.$$

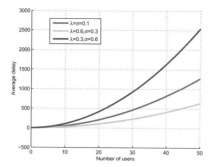

Fig. 13. Average delay in function of number of users under different scenarios of λ and σ

Therefore by the formula of LITTLE [6] the average delay is:

$$D = \frac{Q}{\pi_1}.$$

So:

$$D = 1 + \frac{\pi_3}{\pi_1} \frac{(M+2)(M-1)}{2} = 1 + \frac{\sigma}{\lambda} \frac{(M+2)(M-1)}{2}.$$

In Fig. 12, we plot the average delay of the transmitted packets in function of λ and σ for $M = 50$. This figure shows that when the transition parameter to the state of collision increases the average delay increases too and this is due to the increased number of collisions, on the other side, when the transition parameter to the busy state increases i.e. the system becomes more active so the average delay decreased.

In Fig. 13, we remark that the average delay is an increasing function of the number of users. An important result we can conclude is that to keep a minimum average delay and acceptable we have to select the value of λ superior than the value of σ.

5   Conclusion

In this work, we studied two Continuous Markov models for the cooperative Slotted ALOHA protocol. Our objective has been to evaluate the performances of the system in function of its different parameters.

In the first time, we modeled the system by a Continuous Markov chain with two states, a state where the system is busy and a state where the system is either idle or in collision. The analysis of the system showed that it is possible to increase the throughput and minimize the average delay by appropriately selecting the parameters λ and μ.

Secondly, we modeled the system by a three states Markov chain in order to manage it effectively and to take into account the interests of operators. The obtained results show that the choice of the parameters λ, μ and σ have a major impact on the system performances.

References

1. Abramson, N.: The ALOHA system-another alternative for computer communications. In: AFIPS Conference Proceedings, vol. 36, pp. 295–298 (1970)
2. Aboghsesa, S.M., Abdala, T.M., Daud, N.: Analysis and simulation of Slotted ALOHA-based RFID anti-collision protocol. In: International Conference on Network Security and Computer Science (ICNSCS 2015), Kuala Lumpur, Malaysia, 8–9 February 2015
3. Baek, H., Lim, J.: Performance analysis of Block ACK-Based Slotted ALOHA for wireless networks with long propagation delay. Ad Hoc Netw. **42**, 34–46 (2016)
4. Crowther, W., Rettberg, R., Walden, D., Ornstein, S., Heart, F.: A system for broadcast communication: reservation-ALOHA. In: Proceedings of the 6th Hawaii International Conference on System Sciences (1973)
5. Fang, F., Mao, Y., Leng, S.: An adaptive Slotted ALOHA algorithm. Elektronika, Energetyka, Elektrotechnika (Electrical Review), NR 5b (2012)
6. Belattar, M., Benatia, D., Benslama, M.: Analysis of Markov model of Slotted ALOHA protocol in satellite communication. Rev. Aerosp. Eng. **3**(3), 134 (2010)
7. Patel, S., Gupta, P.K.: Drift analysis of backlogged packets in Slotted ALOHA. Int. J. Mach. Learn. Comput. **2**(4), 415–418 (2012)
8. Rappaport, T.S.: Wireless Communications. Prentice-Hall, Englewood Cliffs (1996)
9. Zaaloul, A.: Amélioration de certains Mécanismes de Gestion des Collisions dans les Réseaux sans Fil: Approche Stochastique et Théorie des Jeux. Thèse de Doctorat, Faculté des Sciences et Techniques de Settat, 17 Juin 2015

Determining and Evaluating the Most Route Lifetime as the Most Stable Route Between Two Vehicles in VANET

Mohamed Nabil[1(✉)], Abdelmajid Hajami[2(✉)], and Abdelkrim Haqiq[1(✉)]

[1] Computer, Networks, Mobility and Modeling Laboratory,
FST, Hassan 1st University,
Settat, Morocco
nabilmed77@gmail.com, ahaqiq@gmail.com
[2] LAVETE Laboratory, FST, Hassan 1st University, Settat, Morocco
abdelmajidhajami@gmail.com

Abstract. One of the critical problems in VANET is the frequent breakdown of the path caused by the high mobility of vehicles. For this purpose, much searches has disclosed the route lifetime between source and destination vehicles as an important factor to improve the quality of service in the VANET network. Each solution propose his own technique to determine a stable route based on link lifetime and route lifetime; but they do not determine and select the most stable route between source and destination vehicles during the data packets transmission by selecting the longest route lifetime. For this reason, we propose two protocols that use vehicles' movement information to determine and evaluate the longest route lifetime as a most stable path for comfort applications in a highway environment. One uses the beacon message and the other does not use it. These new schemes increase the packets delivery ratio, the throughput and decrease the number of error messages generated during the transmission of data packets following of the vehicles density.

Keywords: Stable route · Link lifetime · Route lifetime · IDM_LC
Highway scenario · VANET · NS2

1 Introduction

Vehicular Ad Hoc Networks (VANETs) allow vehicles to communicate each other directly through the On Board Unit (OBU) device, forming vehicle to vehicle communications or with infrastructure via fixed equipment beside the road, referred to Road Side Unit (RSU) forming vehicle-to-infrastructure communications and they are a key component of intelligent transportation systems (ITS). VANETs support a wide range of safety and non-safety applications to make accurate decisions by drivers and to provide passengers comfort. They will play a vital role in road by providing and sharing information to the drivers or passengers such as traffic signals, location, speed of the neighboring vehicles, play online games, access the internet and check emails [1–3].

© Springer International Publishing AG, part of Springer Nature 2018
A. Abraham et al. (Eds.): IBICA 2017, AISC 735, pp. 125–132, 2018.
https://doi.org/10.1007/978-3-319-76354-5_12

The growth of route lifetime has been proposed by much research like [4, 5] to improve communication efficiency in vehicular ad hoc network. All these researches seek to grow the route lifetime by choosing vehicles that travel in de same direction, or by dividing the vehicles in groups like ROMSGP scheme in [4], or by building stable backbones on road using connected dominating sets (CDS) like SCRP scheme in [6], or by using an evolving graph from the source to the destination like in [7]. All these researches still suffer from a great number of route discovery messages for reactive schemes and suffer from vehicles density in case of proactive schemes. Furthermore, each one of them uses its method to calculate the link lifetime to determine the next link and then determine a stable route; but they not determine the most stable path based on the next link lifetime. As illustrated in Fig. 1, the route between source vehicle and destination one is S-I-K-D and its lifetime is 3 s based on the next link lifetime. However it is not the most stable route by comparing to the route S-A-K-D that is the most stable route and its lifetime is 5 s. Hence, the most stable route is not determined by the largest next link lifetime, but it is determined by previous links lifetimes and the next links lifetimes. To deal with these problems, we seek to determine the most stable route between source and destination by using route lifetime as a metric in a highway environment.

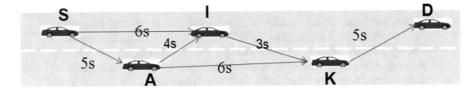

Fig. 1. Route lifetime

The idea behind this work is that each vehicle receives a route discovery message, will decides to forward this message or not in order to increase the route lifetime and to obtain the most route lifetime at the destination. This receiving vehicle will save a new route lifetime in its table and broadcasts the route discovery message, if it does not already received this route discovery message or, if the link lifetime and the route lifetime are greater than the route lifetime in the receiver's table. Where the link lifetime is the remaining time between the receiver and the previous forwarder for staying in direct communication; and the route lifetime is the delay between the source and the receiver.

The rest of the paper is organized as follows. Section 2 presents related works. Section 3 shows ours schemes. Section 4 presents simulation and results. Section 5 presents discussions and evaluation. Finally, we give a conclusion in Sect. 6.

2   Related Works and Motivations

The challenges of network routing protocols in VANETs have been attracting more research efforts, and a number of routing protocols have been proposed to determine the route based on the route lifetime. To determine a more stable route, authors in [4] proposed the scheme ROMSGP that group vehicles according to their movement directions. The most stable route is determined by selecting the path that has the longest link expiration time. The authors did not take into consideration the case where there is no vehicle travelling in the same direction of group movement. In [5], authors propose a cross-layer approach that estimates the remaining time for which a link's quality will remain above the specified threshold, called link residual time (LRT). The latter is defined as the time left of a given link that will be continued to be useful for satisfactory data transmission. The route construction is based on the selection of the link that has the largest link residual time. Authors not determine the most stable route and they assume that vehicles travel at a constant velocity for the duration of the link. In [8], authors propose a stable direction-based routing protocol (SDR) that combines direction broadcast and path duration prediction into AODV [9]. In SDR vehicles are grouped based on the position and the route selection is based on the link duration. In [7], authors propose an evolving graph-reliable ad hoc on-demand distance vector (EG-RAODV) that allows finding the most reliable route from the source to the destination. They proposed an extended version of the evolving graph model to model and formalize the VANET communication graph (VoEG) and they developed a new evolving graph Dijkstra's algorithm (EG-Dijkstra) to find the most reliable journey (MRJ) based on the journey reliability in VoEG. The problem of this protocol is in each period, it must be build a new evolving graph and it assumes that vehicles travel at a constant velocity along the same direction on the highway. Also, it does not take in to consideration the vehicles density. In [10], authors propose a method to select a reliable neighbor based on the residual lifetime of the corresponding communication link. They present an algorithm to predict the residual lifetime of links by making use of Kalman filter based prediction technique. The forwarding vehicle tries to predict the residual lifetime of one-hop links to all of its neighbor vehicles. The neighbor with maximum value for the link residual lifetime is chosen as the next forwarding vehicle. In [11] authors propose the scheme ARP-QD that is an QoS-based routing protocol in terms of hop count, link duration and connectivity, so as to cope with dynamic topology and keep the balance between stability and efficiency of the algorithm. However, it is not enough to only use a global distance to reflect the overall QoS of a routing path. All these schemes do not predict the most stable route at a given moment and they take determined directions. Therefore, the goal of this work is to predict the longest route lifetime as the most stable route whatever the direction of the vehicles on highway for non-safety applications.

3   Most Stable Route

We seek to determine the most stable route between source vehicles and destination ones to ameliorate the protocol performance using route lifetime as a factor in highway environment.

The network model consists of one road ended by two intersections in highway environment or in urban environment for roads segments. This road has the same characteristics such as length, width, number of lanes. Each lane has a distinctive traffic density. Each vehicle is equipped with a global positioning system (GPS) that provides information about its location, speed, and direction. Finally, each source vehicle knows the location of the destination by using a location service such as RLSMP [13] and ZGLS [14].

Given a directed graph G(V; E) that is defined by a finite set $V = \{v_1, v_2, v_3, ..., v_n\}$ of vertices where v_i is a vehicle, and by finite set $E = \{t_1, t_2, t_3, ..., t_m\}$ of edges where t_j is the remaining time between any two vehicles to stay in communication with each other.

We suppose that each vehicle that receives a discovery message of route saves the message identifier and the route lifetime (during a determined period) in a table, called Route Request Table (RRT).

The calculation of link lifetime between two vehicles is proposed in our works [12, 15]. The route lifetime is the minimum link lifetime between links that build the route between source and destination vehicles.

We seek to determine the most stable route between the source and the destination vehicles. As in Fig. 2, the most stable route is that built by vehicles S-A-I-K-D and the lifetime of this route is 4 s at instant t.

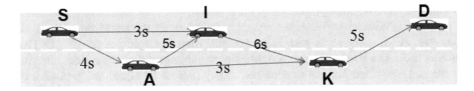

Fig. 2. Most route lifetime

The source vehicle broadcasts a new route discovery message if it wants to determine a new path between itself and the destination. To determine this route, we propose two schemes, one uses beacon message and the other does not use it.

3.1   Scheme Without Beacon Message

In this scheme, the source vehicle adds in the route request message its information (identifier, location, speed, direction, and route-lifetime) and broadcasts it in its communication range. The route lifetime value equals 0 s in the route request message at the source. Then, each vehicle receives this message, it calculates the time left between itself and the source; after that, it checks its table RRT if it has already received this message from the source. Otherwise, it broadcasts it in the half circle of its communication range in the side close to the destination. If it has received the message, it first checks whether the link lifetime (between itself and the source) and the route lifetime (in the route request message) are greater than the route lifetime in its table RRT; if it is the case, it checks again whether the link lifetime is greater than or equal to the route lifetime. If it is, it first updates the route lifetime value in its table RRT by the new calculated value, then it adds its information (identifier, location, speed, direction, route-lifetime) instead of those of the previous forwarder vehicle and finally broadcasts it in the half circle of its communication range in the side close to the destination. Otherwise, it deletes it. Each next receiving vehicle will do the same operations that have been done by the previous receiving vehicle until the route discovery message arrives to the destination.

3.2 Scheme with Beacon Message

It is assumed that each vehicle periodically sends its information in beacon message (location, speed, direction of movement, identifier, and current time) to its neighbors. Then, each vehicle constructs its neighboring list by information extracted from beacon messages. Whenever a new neighbor is discovered, a new entry is added and a timer is set. A vehicle waits two consecutive beacon intervals to hear from its neighbor. If no message was received, the neighbor's entry is deleted.

Each vehicle calculates periodically the time left (link lifetime) between each of its neighbor and itself. Then, it saves the link lifetime value and the identifier of its neighbor in its table, called Neighbors-Life-Time (NLT).

In this scheme, the source vehicle adds in the route request message its information (identifier, and route-lifetime) and broadcasts it in its communication range. The route lifetime value equals 0 s in the route request message at the source. Then, each vehicle receives this message; it checks its table RRT if it has already received this message from the source. If not, it broadcasts it in the half circle of its communication range in the side close to the destination. In case of receiving the message, it first checks whether the link lifetime in its table NLT and the route lifetime (in the route request message) are greater than the route lifetime in its table RRT; if it is, it checks again whether the link lifetime is greater than or equal to the route lifetime. If it is, it first updates the route lifetime value in its table RRT by the new value from the its table NLT, then it adds their information (identifier, route-lifetime) instead of those of the previous forwarder vehicle source) and finally broadcasts it in the half circle of its communication range in the side close to the destination. If not, it deletes it. Each next receiving vehicle will do the same operations that have been done by the first receiving vehicle until the route discovery message arrives to the destination.

4   Simulation and Results

We have used the pattern IDM-LC that is a microscopic mobility model in the tool Vehicular Ad Hoc Networks Mobility Simulator (VanetMobiSim) [16, 17] and we have used NS2 [18] to implement our protocol. Vehicles are deployed in a 4000 m × 100 m area. This area is a highway with four lanes bidirectional; its ends are set by traffic lights. Vehicles are able to communicate with each other using the IEEE 802.11 MAC layer. The vehicles' speed fluctuates between 0 m/s and 27 m/s. We have considered packet size of 512 bytes, Simulation Time of 400 s, hello interval of 1 s and packet rate of 4 packets per second. We setup ten multi-hop CBR flow vehicles over the network that start at different time instances and continue throughout the remaining time of the simulation. The transmission range is kept at 250 m. Simulation results are averaged over 20 simulation runs. Location-Aided Routing (LAR1) [19] is used to compare it with our schemes.

Figure 3 shows that our schemes have good packet delivery ratio and they outperform LAR1. This is because they forward data packets over roads by predicting the most stable route in contrary of LAR that selects the shortest path. And the selection of the most stable route allows the decrease of the number of route breaking. The packet delivery ratio of our schemes decreases during the growth of network density because the route lifetime decreases and the transmission time of data packets increases too.

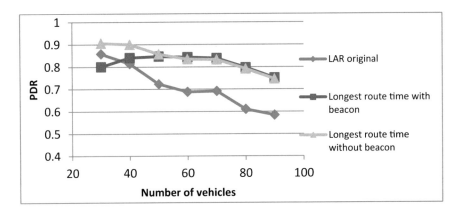

Fig. 3. PDR as a function of vehicle density

In Fig. 4 our scheme with beacon has the lowest throughput compared to our scheme without beacon and LAR1. This is explained by periodicity of beacon messages that charge the bandwidth.

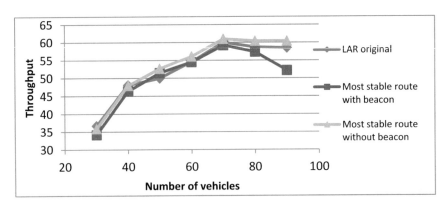

Fig. 4. Throughput as a function of vehicle density

Figure 5 shows that the number of errors increases during the increase of the network density. Our schemes have the lowest number of errors compared to LAR1. This is explained by predicting the route lifetime.

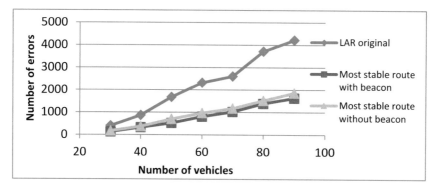

Fig. 5. Number of errors as a function of vehicle density

5   Evaluation

We observe that the number of errors remains high even if we have determined the most stable path and we have predicted the remaining time of the route, so that the source knows when it starts a new route request and when it stops the send of data. Also the prediction of the time that takes a data packet between the source and the destination vehicles remains a real problem in the VANET network. This number of errors causes an increase in network overhead, loss of data and decrease of the bandwidth capacity. Furthermore, we predict the most stable route for an instant t, but an instant t + T (where T is the route lifetime) may not be the most stable and the route that has been the least stable becomes the most stable according to the model of the mobility. For these reasons, we do not think that the protocols, which determine the path to be traversed beforehand send the data between the source and the destination vehicles, are suitable in the VANET network.

6   Conclusion

Our schemes are designed to enhance the communication in highway scenarios for the comfort applications. They strive to determine the most stable route by selecting the route that has the longest lifetime and to predict the remaining time of route to avoid the route disruption prior to its happening. They are based on the prediction of the link lifetime and the route lifetime. Our schemes increase the percentage of packets delivery, throughput and decrease the number of error messages generated during data packet transmission. They are evaluated as function of vehicle density and they are compared with LAR scheme1 in highway environment by using IDM_LC to generate realistic mobility files.

We will study and evaluate all the most known schemes that take the route lifetime as a factor and that determine the route to be taken beforehand sending the data to the destination vehicles. We will compare between them. Then, we will determine their limitations for non-safety applications in VANETs.

References

1. Sultan, S., Doori, M., Bayatti, A., Zedan, H.: A comprehensive survey on vehicular ad hoc networks. J. Netw. Comput. Appl. **37**, 380–392 (2013)
2. Dua, A., Kumar, N., Bawa, S.: A systematic review on routing protocols for vehicular ad hoc networks. Veh. Commun. **1**(1), 33–52 (2014)
3. Zeadally, S., Chen, Y., Hunt, R., Irwin, A., Hassan, A.: Vehicular ad hoc networks (VANETS): status, results, and challenges. Telecommun. Syst. **50**, 217–241 (2012)
4. Taleb, T., Sakhaee, E., Jamalipour, A., Hashimoto, K., Kato, N., Nemoto, Y.: A stable routing protocol to support ITS services in VANET networks. IEEE Trans. Veh. Technol. **56**(6), 3337–3347 (2007)
5. Sofra, N., Gkelias, A., Leung, K.: Route construction for long lifetime in VANETs. IEEE Trans. Veh. Technol. **60**(7), 3450–3461 (2011)
6. Togou, M.A., Hafid, A., Khoukhi, L.: SCRP: stable CDS-based routing protocol for urban vehicular ad hoc networks. IEEE Trans. Intell. Transp. Syst. **17**(5), 1298–1307 (2016)
7. Eiza, M.H., Ni, Q.: An evolving graph-based reliable routing scheme for VANETs. IEEE Trans. Veh. Technol. **62**(4), 1493–1504 (2013)
8. Liu, C., Shu, Y., Yang, O., et al.: SDR: a stable direction-based routing for vehicular ad hoc networks. Wirel. Pers. Commun. **73**(3), 1289–1308 (2013)
9. Perkins, C.E., Royer, E.M.: Ad-hoc on-demand distance vector routing. In: Proceedings of the 2nd IEEE WMCSA, pp. 90–100 (1999)
10. Shelly, S., Babu, A.V.: Link residual lifetime-based next hop selection scheme for vehicular ad hoc networks. J. Wirel. Com. Net. **2017**(1), 1–3 (2017)
11. Sun, Y., Luo, S., Dai, Q., Ji, Y.: An adaptive routing protocol based on QoS and vehicular density in urban VANETs. Int. J. Distrib. Sens. Netw. **1**(1), 1–14 (2015)
12. Nabil, M., Hajami, A., Haqiq, A.: A stable route and the remaining time prediction to send a data packet in highway environment. In: Abraham, A., Haqiq, A., Alimi, A.M., Mezzour, G., Rokbani, N., Muda, A.K. (eds.) HIS 2016. AISC, vol. 552, pp. 11–20. Springer, Cham (2017)
13. Saleet, H., Basir, O., Langar, R., Boutaba, R.: Region-based location-service-management protocol for VANETs. IEEE Trans. Veh. Technol. **59**(2), 917–931 (2010)
14. Rehan, M., Hasbullah, H., Faye, I., et al.: ZGLS: a novel flat quorum-based and reliable location management protocol for VANETs. Wirel. Netw. 1–19 (2017)
15. Nabil, M., Hajami, A., Haqiq, A.: Improvement of route lifetime of LAR protocol for VANET in highway scenario. In: 2015 IEEE/ACS 12th International Conference of Computer Systems and Applications (AICCSA), Marrakech, Morocco, pp. 1–8 (2015)
16. Harri, J., Fiore, M.: VanetMobiSim- vehicular ad hoc network mobility extension to the CanuMobiSim framework. Manual, Institut Eurecom/Politecnico di Torino, Italy (2006)
17. Fiore, M., Harri, J., Filali, F., Bonnet, C.: Vehicular mobility simulation for VANETs. In: Proceedings of the 40th Annual Simulation Symposium (ANSS 2007), pp. 301–309. IEEE Computer Society, Washington, DC (2007)
18. NS-2 Mannual. http://www.isi.edu/nsnam/ns/ns-documentation.html
19. Ko, Y.-B., Vaidya, N.H.: Location aided routing in mobile ad hoc networks. ACM J. Wirel. Netw. **6**(4), 307–321 (2000)

Reverse Extraction of Early-Age Hydration Kinetic Equation of Portland Cement Using Gene Expression Programming with Similarity Weight Tournament Selection

Mengfan Zhi[1], Ziqiang Yu[1(✉)], Bo Yang[1,2(✉)], Lin Wang[1,2],
Liangliang Zhang[1], Jifeng Guo[1], and Xuehui Zhu[1]

[1] Shandong Provincial Key Laboratory of Network Based Intelligent Computing,
University of Jinan, Jinan 250022, China
{ise_yuzq,yangbo}@ujn.edu.cn
[2] School of Informatics, Linyi University, Linyi 276000, China

Abstract. The early stages of portland cement hydration directly affect the physical properties of cement. Therefore, it is necessary to research the hydration process in the early stages of portland cement. Owning to the cement hydration process includes a large number of chemical and physical changes, researching the cement hydration process faces many difficulties. In this paper, early-age hydration kinetic equation is reverse extracted from cement hydration heat data using gene expression programming (GEP) with similarity weight tournament (SWT) selection operator. The method clever use the cement hydration heat data and the powerful performance of genetic expression programming. In addition, the effectiveness of the proposed method is improved using SWT selection operator. The result shows that the performance of GEP method with SWT selection operator is better than traditional GEP.

1 Introduction

Cement is a important building material and its performance including strength, heat resistance, corrosion resistance, etc., is directly affected by the cement hydration process [1–3]. Thus the study of hydration process of cement is helpful to understand the formation mechanism of cement performance. Even that it can helps material scientists to design higher quality cement. Therefore, it is important to research the hydration process of cement.

There are many models for the simulation and modeling of cement hydration process. Jennings and Johnson [4] proposes the microstructure simulation model, and Bullard [5] develops stochastic simulation model called HydratiCA. But these models are based on simulation, which means these models can not express the real hydration process. In another perspective, the cement hydration process is an exothermic process. Real hydration heat data can be obtained in the five stages of cement hydration: pre-induced phase, induction phase, accelerated

© Springer International Publishing AG, part of Springer Nature 2018
A. Abraham et al. (Eds.): IBICA 2017, AISC 735, pp. 133–142, 2018.
https://doi.org/10.1007/978-3-319-76354-5_13

phase, deceleration phase and stable phase. Hydration heat data can indirectly reflect the degree of hydration of cement. Time series of cement hydration degree can be obtained by using method of microcalorimetry. Moreover, the real hydration process of cement can be simulated by the above time series data.

Recent learning algorithms [6] and neural networks are widely used in data processing. Based on real data, recent studies have shown that the target equation can be obtained from evolutionary computations. Nitsure [7] uses genetic programming (GP) algorithm to estimate an important ocean parameter: significant Wave Height (SWH) using the wind information. It is proved that the parameter estimation generic programming algorithm has reasonable accuracy. Therefore, the hydration kinetic equation of cement can be reverse extracted from data by same idea.

In this work, early-age hydration kinetic equation is reverse extracted from cement hydration degree data using the GEP [8] algorithm with SWT selection operator. GEP can obtain complex mathematical equations according to data. In addition, SWT is used to increase the diversity of individual populations and it improves the performance of GEP.

In this paper, the Sect. 2 introduces the related works of this paper. The Sect. 3 introduces the method used in this article. The Sect. 4 introduces the relevant content of the experiment. Section 5 is the conclusion.

2   Related Works

There are many methods to research the cement hydration process. Jennings and Johnson developed a mathematical model describe the hydration of tricalcium silicate (C3S). This is the first simulation of the hydration process of cement. But the research of the cement hydration process through the simulation of the cement microstructure has some limitations, this model does not express the real hydration process. On the other hand, the cement releases a lot of heat in the process of hydration. The amount of heat released reflects the degree of hydration. Therefore, the important issue is uses hydration heat data to study cement hydration.

In the field of marine science, S.P. Nitsure uses GP algorithm to estimate an important ocean parameter. It means that intelligent algorithms can be used to research cement hydration processes. There are a lot of intelligent algorithms: intelligent neural network, genetic algorithm, simulated annealing algorithm etc. Many scholars do study on improving neural network classifiers [9]. The particle swarm optimization (PSO) [10] is widely used in the field of neural network. Genetic algorithm draw lessons from the concept of life science. Genetic algorithm has the principle of survival of the fittest. The individuals with higher adaptability are more likely to survive. GEP algorithm is a new type of self-evolution algorithm in combination with the advantages of genetic algorithm. Lin Wang used the method of GEP with PSO [11] to extract the kinetic equation of cement hydration. It prove that GEP has powerful ability to generate equations.

3   Methodology

3.1   Gene Expression Programming

GEP algorithm is a new member of the intelligent computing family. GEP algorithm combines the advantages of genetic algorithm and genetic programming which borrows the concept of genetics and chromosomes in the life sciences and the concept of population multiplication in biological sciences. First, establishing an initial population. Then the evolution of population is achieved by selection and genetic operators. Repeat this operation until find the optimal solution.

Genetic operators in GEP include insertion sequence, mutation, root insertion sequence, two-point recombination operators and one-point recombination. Each individual in the population is called an chromosome. An chromosome is a fixed-length symbol sequence consisting of one or more genes. The symbol sequence is represented by an expression tree. The chromosome sequence is divided into coding and non-coding regions. Figure 1 is an instance of a chromosome. The symbol sequence takes from the function set F and the terminator set T. The function set F contains functions. For instance, $+$, $()^2$, $()^3$ etc. The T set includes constants and variables. For example, a, b, c etc. Genes are divided into head and tail. The head part take the value from the F and T sets. The tail part take the value from the T set. The chromosome length is the sum of the length of the head and tail. First, determine the length of the head h, then the tail length t is obtained by the following formula.

$$t = 1 + (n - 1) \times h \tag{1}$$

where n is the maximum value of the operand of the function in the F set.

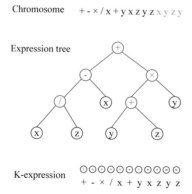

Fig. 1. Chromosomes and its corresponding expression trees and k-expressions. The black part is the coding region of the chromosome, and the gray part is the non-coding region. The red part is a function of the chromosome

3.2   SWT Selection

SWT selection is a new selection operator proposed by Wang Lin which introduces the concept of possibility $p[i]$. The possibility $p[i]$ refers to the probability that the individual is selected to participate in the competition. *similarity* is used to indicates the similarity degree between two chromosomes. Calculate the *similarity* between the individual i and the historical optimal solution and then converting it to $p[i]$. In the SWT selection, the K individuals are selected according to the probability p. Individual with higher fitness is inherited into the next generation. Repeating this operation until the number of individuals are equal to the size of the population.

Calculate the *similarity* between individual i and the history of the best individual, it is only necessary to compare the coding region sequences of the two chromosomes and the measure of calculate *similarity* starts with each symbol of the two chromosomes. First calculating the overlap, overlap refers to the number of the same symbol of the two sub-string. The overlap between the two chromosomes is represented by *Same*. Calculate the *Same* formula as shown below.

$$f_{(c_1,c_2)}(i,j) = \begin{cases} 1, c_1(i) = c_2(i) \\ \\ 0, else \end{cases} \tag{2}$$

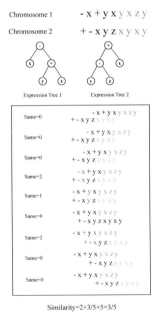

Similarity=2×3/5+5=3/5

Fig. 2. Example of calculating the value of similarity. The black part in chromosomes represent the coding region while the gray part is the noncoding region.

$$Same(start_1, start_2, lenth) = \sum_{i=1}^{lenth} f_{c_1, c_2}(start_1 + i, start_2 + i) \tag{3}$$

Two substrings come from the coding region of chromosome c_1 and chromosome c_2. $c_1(i)$ and $c_2(j)$ represent the i-th and j-th symbols in the two chromosome. And lenth is the total number of symbols. $start_1$ and $start_2$ represent the starting positions in c_1, c_2. Then the *similarity* of the two chromosomes are calculated. The formula as shown below.

$$Similarity(c_1, c_2) = \frac{2 \times MaxSame}{len(c_1) + len(c_2)} \tag{4}$$

MaxSame represents the maximum *Same* of chromosome c_1 and chromosome c_2, and *len* represents the length of the coding region of the chromosome c_1 and chromosome c_2. The value of *similarity* is calculated as shown in Fig. 2. The formula for probability P is shown below.

$$p[i] = \frac{1 - Similarity(c_i, c_{best})}{\sum_{j=1}^{PopulationSize} (1 - Similarity(c_j, c_{best}))} \tag{5}$$

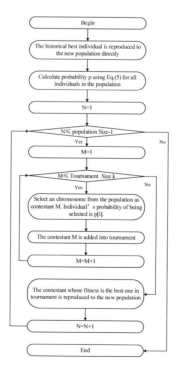

Fig. 3. Algorithm of SWT selection

where c_{best} represents the history of the best individual. The algorithm of SWT selection is shown in Fig. 3. Thus, the higher the similarity with the best individual in history, the less likely to be chosen to compete. The lower the similarity to the best individual in history, it is easier to be chosen to compete. But the individuals with low fitness in the competition will still be eliminated.

3.3   The Reverse Extraction Method for Hydration Kinetic Equation

According to the cement hydration heat data, the early cement hydration kinetics equation [12] is obtained. It is forecasted to get the differential equation as follows

$$\frac{d\alpha}{dt} = f(t, \alpha) \tag{6}$$

where α is the hydration value of a certain moment, t represents the time. There are thirteen function operators in function set F, and eighteen variables and constants in the end point set.

$F = \{+, -, \times, \div, -1 \times (), \frac{1}{()}, \sqrt{}, \sqrt[3]{}, ()^2, ()^3, ()^0, e^{()}, \ln()\}$

$T = \{C_1, C_2, C_3, C_4, C_5, C_6, C_7, C_8, C_9, C_{10}, \alpha^3, \alpha, t, 1 - \alpha, \sqrt[3]{1 - \alpha}, \sqrt[3]{(1 - \alpha)^2}, \ln(1 - \alpha),$ $1 - \sqrt[3]{1 - \alpha}\}$

GEP algorithm needs to use fitness to update the individual in the population and achieves the purpose of finding the optimal solution. The numerical solution

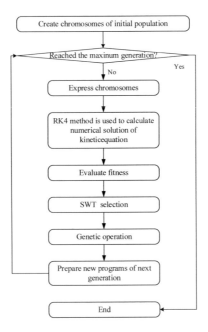

Fig. 4. Algorithm of GEP-SWT.

of equations are obtained by fourth-order Runge-Kutta method (RK4). The relative error between the numerical solution and the true hydration value are calculated.

$$\Delta = |\frac{\alpha - \hat{\alpha}}{\alpha}| \tag{7}$$

where $\hat{\alpha}$ is the degree of hydration of the kinetic equation obtained by the RK4 method. The fitness function is defined as

$$fitness = \frac{1}{1 + \sum\limits_{i=1}^{N} \Delta_{ij}} \tag{8}$$

where N represents the number of time points which are selected from the time series data. Figure 4 indicates the flow chart of the proposed algorithm.

4   Experiments

In this section, the hydration process from the rapid reaction period to the stable period during the hydration was carried out 24 h. The cement hydration reaction was observed for 24 h. The sampling interval of the hydration heat data [13] was set to 5 min. Eight hydration heat data were used. Each set of data contains 288 time points. The GEP-SWT and the traditional GEP were carried out ten experiment with each time series of data and the average of the ten experimental results is calculated. The performance of the two algorithms was compared in the case of small population and small iteration.

In this experiment, the comparison between GEP-SWT and GEP are organized. A small population size and number of iterations are set. The population size of the two algorithms is set to 1000, number of iterations is 1000. The number of genes is 1. Gene head length is defined as 20. The maximum operand in the function set F is 2, get the gene tail is 21 by Eq. (1) and chromosome's length is the sum of head part and tail part. Insertion sequence probability is ten percent. One-point recombination Probability is twenty percent. Two-point recombination probability is twenty percent. Mutation probability is thirty percent. Root insertion sequence probability is twenty percent. $C_1 - C_{10}$ in the T set represent constant 3, constant 9, constant 4 constant 17, constant 20, constant 2, constant 8, constant 19, constant 13, constant 15.

The traditional GEP and GEP-SWT are used to do the experiment and record the 1000 generation optimal individual and its fitness value according to the above parameter setting. The population size and the number of iterations of the two methods are exactly the same. Compare the optimal fitness value obtained from the experiments of the traditional GEP and GEP-SWT. The performance of the two algorithms can be got by the experiment and analysis.

According to the eight hydration heat data, eight comparison between GEP and GEP-SWT are obtained. As shown in Fig. 5, eight comparison show that the 1000th generation optimal individual fitness is higher than GEP. The trend of fitness growth of GEP-SWT is significantly faster than GEP. In the series 1–3 and

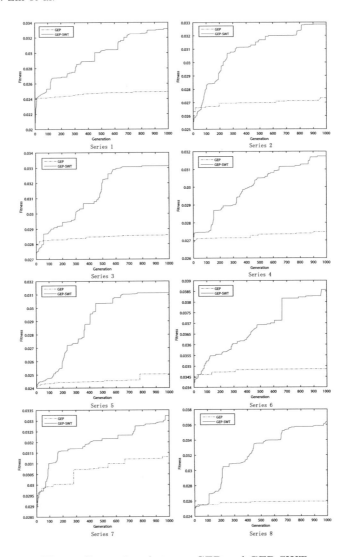

Fig. 5. Comparison between GEP and GEP-SWT.

series 6, the first-generation optimal individual fitness obtained by the GEP is higher than the optimal individual fitness obtained by the GEP-SWT. However, with the population alternation, the 1000th-generation optimal individual fitness obtained by GEP-SWT is higher than the 1000th-generation optimal individual fitness obtained by GEP. For each data, the best equation with the highest fitness is obtained from ten experiments. Eight kinetic equations are shown as below.

$$\frac{d\alpha}{dt} = \frac{(1-\alpha)^{\sqrt{13\alpha^3}}}{340\sqrt{3^{1-\alpha}}} \tag{9}$$

$$\frac{d\alpha}{dt} = \frac{\varrho^{\frac{-1}{\sqrt{1-t\alpha^3-(t-19)^{1-\alpha}}}}}{400} \tag{10}$$

$$\frac{d\alpha}{dt} = \frac{\left(64 \cdot (1-\alpha)^{\frac{2}{3}}\right)}{\left(9 \cdot (1444^{3\alpha} \cdot \alpha^3 + 64)^2 - 960\right)} \tag{11}$$

$$\frac{d\alpha}{dt} = \sqrt{\frac{(1-\alpha)^2 \cdot \sqrt[3]{1-\alpha}}{\log\left[(t-15)^2 - 8000\right]^5 + t^2 + 2t}} \tag{12}$$

$$\frac{d\alpha}{dt} = \left(\frac{\alpha}{8 \cdot \left(\alpha \cdot 1959319 \cdot \alpha^9 + \alpha\right) + 8} + \frac{(1-\alpha)^{\frac{2}{3}}}{8}\right)^3 \tag{13}$$

$$\frac{d\alpha}{dt} = 1 \div 243 \cdot \left[\alpha^5 \left(t - \alpha^6 + 18\alpha^3 - 81\right)^3\right]^{5\alpha^3} \tag{14}$$

$$\frac{d\alpha}{dt} = \left[24 - (1-\alpha)^{2\alpha+2t-80\sqrt[3]{1-\alpha}-46}\right]^{1-\alpha} \div 8000 \tag{15}$$

$$\frac{d\alpha}{dt} = 9 \div \left[\varrho^{\alpha\left(\alpha^2+3\alpha^6+17t+\alpha^6t+51\alpha^5\right)+\sqrt[3]{1-\alpha}+8} - t + 20\right] \tag{16}$$

The following reasons led to this phenomenon. In the traditional GEP, individuals are selected according to individual fitness. In a random case, population may be premature convergence. The diversity of individuals of the population is reduced, which represents a decrease in the diversity of chromosomal genes. In the method of GEP-SWT, if the coding region of an individual is exactly the same as the optimal individual the probability of the individual participating in the competition is 0. This method can maintain the diversity of offspring population and avoid premature convergence. The above is the reason why GEP-SWT performance is better than GEP performance on most datasets.

5   Conclusions

In this paper, early-age hydration kinetic equation is reverse extracted from cement hydration heat data using GEP-SWT algorithm and the performance of GEP-SWT and GEP are compared. The performance of GEP is improved by using SWT selection operator. Experimental results show that GEP-SWT can faster extract the hydration kinetic equation of cement from the data than tradition GEP.

Acknowledgment. This work was supported by National Natural Science Foundation of China under Grant No. 61573166, No. 61572230, No. 81671785, No. 61373054, No. 61472164, No. 61472163, No. 61672262, No. 61640218. Shandong Provincial Natural Science Foundation, China, under Grant ZR2015JL025, ZR2014JL042. Science and technology project of Shandong Province under Grant No. 2015GGX101025. Project of Shandong Province Higher Educational Science and Technology Program under Grant no. J16LN07. Shandong Provincial Key R&D Program under Grant No. 2016ZDJS01A12, No. 2016GGX101001.

References

1. Bullard, J.W., Jennings, H.M., Livingston, R.A., Nonat, A., Scherer, G.W., Schweitzer, J.S., Scrivener, K.L., Thomas, J.J.: Mechanisms of cement hydration. Cem. Concr. Res. **41**(12), 1208–1223 (2011)
2. Krstulović, R., Dabić, P.: A conceptual model of the cement hydration process. Cem. Concr. Res. **30**(5), 693–698 (2000)
3. Thomas, J.J., Biernacki, J.J., Bullard, J.W., Bishnoi, S., Dolado, J.S., Scherer, G.W., Luttge, A.: Modeling and simulation of cement hydration kinetics and microstructure development. Cem. Concr. Res. **41**(12), 1257–1278 (2011)
4. Jennings, H.M., Johnson, S.K.: Simulation of microstructure development during the hydration of a cement compound. J. Am. Ceram. Soc. **69**(11), 790–795 (1986)
5. Bullard, J.W.: Approximate rate constants for nonideal diffusion and their application in a stochastic model. J. Phys. Chem. A **111**(11), 2084–2092 (2007)
6. Chen, C.L.P., Liu, Z.: Broad learning system: an effective and efficient incremental learning system without the need for deep architecture. IEEE Trans. Neural Netw. Learn. Syst. (2017). https://doi.org/10.1109/TNNLS.2017.2716952
7. Khare, K.C., Nitsure, S.P., Londhe, S.N.: Application of genetic programming for estimation of ocean wave heights. IEEE (2009)
8. Ferreira, C., Gepsoft, U.: What is gene expression programming (2008)
9. Wang, L., Yang, B., Chen, Y., Zhang, X., Orchard, J.: Improving neural-network classifiers using nearest neighbor partitioning. IEEE Trans. Neural Netw. Learn. Syst. **28**(10), 2255–2267 (2017)
10. Kennedy, J., Eberhart, R.: Particle swarm optimization. In: Proceedings of the IEEE International Conference on Neural Networks, 1995, vol. 4, pp. 1942–1948 (2002)
11. Wang, L., Yang, B., Orchard, J.: Particle swarm optimization using dynamic tournament topology. Appl. Soft Comput. **48**, 584–596 (2016)
12. Li, T., Rogovchenko, Y.V.: Oscillation of second-order neutral differential equations. Math. Nachr. **288**, 1150–1162 (2015)
13. Wang, L., Yang, B., Zhao, X., Chen, Y., Chang, J.: Reverse extraction of early-age hydration kinetic equation from observed data of portland cement. Sci. China Technol. Sci. **40**(5), 582–595 (2010)

Performance Improvement of Bio-Inspired Strategies Through Feedback Laws

Lairenjam Obiroy Singh[1(✉)] and R. Devanathan[2]

[1] Department of Electronics and Communication Engineering,
Hindustan Institute of Technology Science, Chennai 603103, India
obi3925@gmail.com
[2] Department of Electrical and Electronics Engineering,
Hindustan Institute of Technology Science, Chennai 603103, India
devanathanr@hindustanuniv.ac.in

Abstract. Pursuit evasion game is prevalent in nature and also has many civilian and defense applications. A feedback law has been applied to pursuer assuming a constant behavior of evader. This paper proposes that PI/PID feedback control laws can improve the performance of the pursuer strategy compared to the existing strategy of P alone. Specifically it is shown that using PI/PID control, the time for the pursuer to reach a camouflage manifold as well as to capture the evader is shortened. The results are applicable both for the case where the evader follows a prescribed or controlled motion. The results are shown through a computer simulation.

Keywords: Bio-inspired · Pursuit evasion game · Feedback laws
Close loop control

1 Introduction

Pursuit evasion behavior is widely seen in nature. It is observed when animal species search for food, fight for survival, and battle for territory and while mating. Pursuit evasion game has been originally studied from a game-theoretic perspective [1]. In terms of applications, pursuit evasion game has been studied in the context of missile guidance and avoidance, aircraft pursuit and evasion, protection of maritime assets both civilian and military [2] etc.

Geometric pattern associated with pursuit has been studied [3]. The pursuit has been modeled as an interaction between two particles moving at constant speeds. The geometric patterns are studied in terms of pursuit manifolds, which contain states that satisfy certain criteria relating to the relative distance and the relative velocity between the pursuer and the evader. Three types of pursuit are studied in particular: classical pursuit, motion camouflage and constant bearing.

The trajectory leading from an initial point to the pursuit manifold is governed by feedback laws that are chosen to minimize an associated cost function. Also, the application of feedback law ensures that the speed of the pursuit particle is not altered. The focus is usually on the pursuer with the evader being assumed moving at a constant speed. Both open loop and closed loop steering strategies are used by the pursuer.

© Springer International Publishing AG, part of Springer Nature 2018
A. Abraham et al. (Eds.): IBICA 2017, AISC 735, pp. 143–156, 2018.
https://doi.org/10.1007/978-3-319-76354-5_14

The author in [4] formulates the confinement – escape problem of a defender and evader to provide general characteristics of the system with bio-inspired control laws for both defender and evader. The author in [5] try to solve two-player pursuer –evader game problem based on bio-inspired method.

The motivation of the paper is that though the basic feedback law has been derived in [3], one would like to investigate how the performance of the pursuer can be improved through additional feedback control techniques. In particular, one is interested in finding out how the adoption of proportional-integral and proportional-integral-derivative laws will improve the pursuer performance. The performance in question is time to reach the manifold from an arbitrary starting point and time it takes pursuer to intercept the evader for the first time.

The main contribution of the paper is (i) to reformulate the pursuit evasion problem as a feedback control problem (ii) derive expressions for Proportional (P), Proportional-Integral (PI) and Proportional, Integral and Derivative (PID) controllers in the context of pursuit-evasion problem and (iii) show improvement in the performance of the pursuit strategy using PI and PID controllers through simulation over that of P controller alone as in the existing results.

To summarize the rest of the paper, Sect. 2 gives the background necessary for the paper. Section 3 discusses the development P, PI and PID control laws. Section 4 describes the simulation results and discusses the proposed improvements in the pursuer performance. Section 5 concludes the paper and discusses future work.

2   Background

We model planar pursuit interactions using gyroscopically interacting particles, as in Wei et al. [6]. The (unit speed) motion of the pursuer is described by

$$\dot{r}_p = x_p, \ \dot{x}_p = y_p u_p, \ \dot{y}_p = -x_p u_p \tag{2.1}$$

where r_p is the position of the pursuer, x_p its velocity and y_p is the acceleration of the pursuer. The motion of the evader (with speed v) is given by

$$\dot{r}_e = v x_e, \ \dot{x}_e = v y_e u_e, \ \dot{y}_e = -v x_e u_e \tag{2.2}$$

where r_e is the position, x_e is the velocity and y_e is the acceleration of the evader. The steering control of the evader, u_e, is prescribed or controlled, and the steering control of the pursuer, u_p, is given by a feedback law. We also define

$$r = r_p - r_e \tag{2.3}$$

which is referred to as the 'baseline' between pursuer and the evader.

In three dimensions, Wei et al. [7] use the concept of natural frame to formulate interacting particles models. In that setting, each particle has two steering (natural curvature) controls.

2.1   Pursuit Manifolds and Cost Functions

The state space for the two-particle pursuer-evader system is a subset of the Euclidean plane of two dimensions. Following [3], we define the cost functions F: $G \times G \to R$ associated with different pursuit strategies as follows:

$$\Gamma = \left(\frac{r}{|r|} \cdot \frac{\dot{r}}{|\dot{r}|}\right) = \frac{\frac{d}{dt}|r|}{\left|\frac{dr}{dt}\right|} \text{ (motion camouflage)} \tag{2.4}$$

and

$$\Lambda = \left(\frac{r}{|r|} \cdot R \, x_p\right) \text{ (constant bearing)}, \tag{2.5}$$

where

$$R = \begin{bmatrix} cos\theta & -sin\theta \\ sin\theta & cos\theta \end{bmatrix}, \tag{2.6}$$

for $\theta \in (-\pi/2, \pi/2)$.

For $R = I_2$ the identity matrix of order 2, we define

$$\Lambda_0 = \left(\frac{r}{|r|} \cdot x_p\right) \tag{2.7}$$

to be the cost function associated with classical pursuit.

All three cost functions Γ, Λ and Λ_0 are well defined for $|r| > 0$, and that they take values on the interval $[-1, 1]$.

The cost functions Γ, Λ and Λ_0 define the respective pursuit manifolds. Γ is seen to correspond to the cosine of the angle between r and $\dot{r}$. Camouflage pursuit manifold is

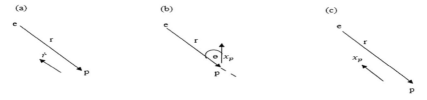

Fig. 1. Geometric representation of pursuit manifolds: (a) motion camouflage pursuit, (b) constant bearing pursuit and (c) classical pursuit.

defined by the condition $\Gamma = -1$, which corresponds to the case of the angle between r and $\dot{r}$ being π. (see Fig. 1(a)).

The constant bearing pursuit manifold is represented by the condition $\Lambda = -1$. This condition is satisfied when the heading of the pursuer makes an angle θ with the baseline vector (see Fig. 1(b)). Similarly, the classical pursuit manifold is defined by the condition $\Lambda_0 = -1$. This condition is satisfied when the heading of the pursuer is aligned with the baseline (see Fig. 1(c)).

In [3], considering deterministic or random nonreactive evasive strategies, the authors have shown, through Monte Carlo simulation, that three strategies of classical, constant bearing, and camouflage converge to pure motion camouflage in an evolutionary game. However, Pais and Leonad [8] generalizing the work of [3] considered the reactive control strategies for the evader and conclude that the evolutionary game between the three strategies does not converge to motion camouflage.

3   System Modelling

3.1   Pursuit – Evasion System

Representing

$$r_p = \begin{bmatrix} r_{px} & r_{py} \end{bmatrix}^t$$
$$r_e = \begin{bmatrix} r_{ex} & r_{ey} \end{bmatrix}^t$$

we write (2.1) and (2.2) in terms of state equations as follows:

$$
\begin{aligned}
x_1 &= r_{px} \\
x_2 &= r_{py} \\
x_3 &= x_{px} \\
x_4 &= x_{py} \\
x_5 &= y_{px} \\
x_6 &= y_{py} \\
x_7 &= r_{ex} \\
x_8 &= r_{ey} \\
x_9 &= x_{ex} \\
x_{10} &= x_{ey} \\
x_{11} &= y_{ex} \\
x_{12} &= y_{ey}
\end{aligned}
\qquad (3.1)
$$

$$\begin{aligned}
\dot{x}_1 &= \dot{r}_{px} = x_{px} = x_3 \\
\dot{x}_2 &= \dot{r}_{py} = x_{py} = x_4 \\
\dot{x}_3 &= \dot{x}_{px} = y_{px}u_p = x_5 u_p \\
\dot{x}_4 &= \dot{x}_{py} = y_{py}u_p = x_6 u_p \\
\dot{x}_5 &= -\dot{y}_{px} = -x_{px}u_p = -x_3 u_p \\
\dot{x}_6 &= -\dot{y}_{py} = -x_{py}u_p = -x_4 u_p \\
\dot{x}_7 &= v\dot{r}_{ex} = vx_{ex} = vx_9 \\
\dot{x}_8 &= v\dot{r}_{ey} = vx_{ey} = vx_{10} \\
\dot{x}_9 &= v\dot{x}_{ex} = vy_{ex}u_e = vx_{11}u_e \\
\dot{x}_{10} &= v\dot{x}_{ey} = vy_{ey}u_e = vx_{12}u_e \\
\dot{x}_{11} &= -v_{ex} = -vx_{ex}u_e = -vx_9 u_e \\
\dot{x}_{12} &= -v\dot{y}_{ey} = -vx_{ey}u_e = -vx_{10}u_e
\end{aligned} \tag{3.2}$$

$$r = r_p - r_e = \begin{bmatrix} x_1 \\ x_2 \end{bmatrix} - \begin{bmatrix} x_7 \\ x_8 \end{bmatrix} = \begin{bmatrix} x_1 - x_7 \\ x_2 - x_8 \end{bmatrix} \tag{3.3}$$

$$\dot{r} = \dot{r}_p - \dot{r}_e = \begin{bmatrix} \dot{x}_1 \\ \dot{x}_2 \end{bmatrix} - \begin{bmatrix} \dot{x}_7 \\ \dot{x}_8 \end{bmatrix} = \begin{bmatrix} x_3 \\ x_4 \end{bmatrix} - v \begin{bmatrix} x_9 \\ x_{10} \end{bmatrix} \tag{3.4}$$

3.2   Feedback Laws

In this section, we formulate the pursuit strategies in terms of feedback control laws. The maintenance of the Γ-manifold represents $\cos\phi = -1$ where ϕ is the angle between r and $\dot{r}$. This is represented in the form of a feedback control system as shown in Fig. 2.

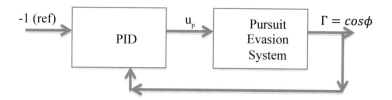

Fig. 2. Γ Manifold control

If now an angle ψ is defined between r and $\dot{r}^{\perp}$ where $\dot{r}^{\perp}$ is $\dot{r}$ rotated counter clockwise by $\frac{\pi}{2}$ radian, $\phi = \pi \Rightarrow \psi = \pi/2$, then a control measurement

$$\Gamma' = \left\langle \frac{r}{|r|} \cdot \dot{r}^{\perp} \right\rangle \tag{3.5}$$

corresponds to an angle ψ such that $\cos \psi \to \cos \frac{\pi}{2} = 0$ on the Γ' manifold. We can rewrite Fig. 2 as

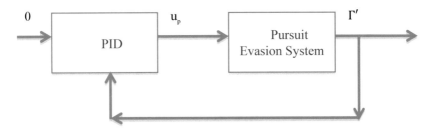

Fig. 3. Γ' Manifold control.

Feedback laws are derived as follows
P control: $u_p = -\mu_1 \Gamma'$

$$u_p = -\mu_1 \left\langle \frac{r}{|r|} \cdot \dot{r}^{\perp} \right\rangle$$

$$= -\left\langle \frac{(x_1 - x_7, x_2 - x_8)}{\sqrt{(x_1 - x_7)^2 + (x_2 - x_8)^2}} \cdot (\upsilon x_{10} - x_4)(x_3 - \upsilon x_9) \right\rangle \tag{3.6}$$

PI control: : $u_p = -\mu_1 \Gamma' - \mu_2 \int \Gamma' dt$

$$u_P = -\mu_1 \left\langle \frac{r}{|r|} \cdot \dot{r}^{\perp} \right\rangle - \mu_2 \int \left\langle \frac{r}{|r|} \cdot \dot{r}^{\perp} \right\rangle dt$$

$$= -\mu_1 \left\langle \frac{(x_1 - x_7, x_2 - x_8)}{\sqrt{(x_1 - x_7)^2 + (x_2 - x_8)^2}} \cdot (\upsilon x_{10} - x_4)(x_3 - \upsilon x_9) \right\rangle \tag{3.7}$$

$$- \mu_2 \int \left\langle \frac{(x_1 - x_7, x_2 - x_8)}{\sqrt{(x_1 - x_7)^2 + (x_2 - x_8)^2}} \cdot (\upsilon x_{10} - x_4)(x_3 - \vartheta \upsilon) \right\rangle dt$$

PID Control:
Using the identity

$$\frac{d}{dt} \langle \alpha, \beta \rangle = \langle \dot{\alpha}, \beta \rangle + \left\langle \alpha, \dot{\beta} \right\rangle$$

where $\alpha = \frac{r}{|r|}$ and $\beta = \dot{r}^{\perp}$, we have

$$\dot{\beta} = \left(v^2 x_{12} u_e - x_6 u_p, x_5 u_p - v^2 x_{11} u_e \right)^t.$$

Dropping u_p, we approximate $\dot{\beta}$ as

$$\hat{\dot{\beta}} = \left(v^2 x_{12} u_e, -v^2 x_{11} u_e \right)^t$$

Further simplification for implementation yields the following.

$$\frac{d}{dt} \left\langle \frac{r}{|r|} \cdot \dot{r}^\perp \right\rangle = \langle x \cdot \hat{y} \rangle \langle [(x_1 - x_7), (x_2 - x_8)], \left[\left(v^2 x_{12} u_e, -v^2 x_{11} u_e \right) \right] \rangle$$

Hence we have

$$
\begin{aligned}
u_P = & -\mu_1 \left\langle \frac{r}{|r|} \cdot \dot{r}^\perp \right\rangle - \mu_2 \int \left\langle \frac{r}{|r|} \cdot \dot{r}^\perp \right\rangle dt - \mu_3 \left\langle \frac{r}{|r|} \cdot \dot{r}^\perp \right\rangle \\
= & -\mu_1 \left\langle \frac{(x_1 - x_7, x_2 - x_8)}{\sqrt{(x_1 - x_7)^2 + (x_2 - x_8)^2}} \cdot (v x_{10} - x_4)(x_3 - v x_9) \right\rangle \\
& -\mu_2 \int \left\langle \frac{(x_1 - x_7, x_2 - x_8)}{\sqrt{(x_1 - x_7)^2 + (x_2 - x_8)^2}} \cdot (v x_{10} - x_4)(x_3 - v x_9) \right\rangle dt \\
& -\mu_3 \left\langle [(x_1 - x_7), (x_2 - x_8)], \left[\left(v^2 x_{12} u_e, -v^2 x_{11} u_e \right) \right] \right\rangle
\end{aligned}
\tag{3.8}
$$

For the evader the steering control can be as follows

$$u_e = \cos t \quad \text{(prescribed)}$$

$$u_e = \left\langle \frac{r}{|r|} \cdot \dot{r}_e^\perp \right\rangle \quad \text{(reactive)}$$

$$= \left\langle \frac{(x_1 - x_7, x_2 - x_8)}{\sqrt{(x_1 - x_7)^2 + (x_2 - x_8)^2}} \cdot (-x_{10}, x_9) \right\rangle$$

4 Simulation Results

4.1 Evader Motion Presented

In this section, we provide the results of simulation of the dynamic equations given Sect. 3 and discuss the same. We assume u_e corresponds to (3.9).

Table 1 provides data on the results obtained using the computer simulation of the pursuit evasion game and its control as explained in the previous sections. The first

Table 1. Time to reach $|r| = 0$ and $\Gamma = -1$ for different combination of μ_1, μ_2, μ_3 for different initial values of r_p and x_p.

μ_1, μ_2, μ_3	$(r_{px}, r_{py}), (x_{px}, x_{py})$ (4, −6), (0.735, 0.678) (1)		$(r_{px}, r_{py}), (x_{px}, x_{py})$ (−4, 9), (0.327, 0.945) (2)		$(r_{px}, r_{py}), (x_{px}, x_{py})$ (0, 8), (0.530, 0.848) (3)		$(r_{px}, r_{py}), (x_{px}, x_{py})$ (−1, 1), (0.632, 0.775) (4)		$(r_{px}, r_{py}), (x_{px}, x_{py})$ (−2, 10), (0.335, 0.942) (5)		$(r_{px}, r_{py}), (x_{px}, x_{py})$ (10, −1), (0.9, 0.436) (6)	
	Time at $r = 0$	Time at $\Gamma = -1$	Time at $r = 0$	Time at $\Gamma = -1$	Time at $r = 0$	Time at $\Gamma = -1$	Time at $r = 0$	Time at $\Gamma = -1$	Time at $r = 0$	Time at $\Gamma = -1$	Time at $r = 0$	Time at $\Gamma = -1$
1, 0, 0	6.7	0.8	21.2	NA	25	24.9	21.8	0.9	27	NA	9.1	5.82
1, 1, 0	6.4	0.5	19	2	14.8	16	3.2	3.1	19.7	24	8.49	1.9
1, 2, 0	6.4	0.5	18.2	1.5	13.7	16.4	2.9	5	18.3	21	7.8	1.3
1, 3, 0	6.3	0.4	17.6	1.2	13.1	15	2.83	8.8	17.3	18	8.7	2.3
1, 1, 1	6.5	0.4	19.5	2	13.9	2.7	3.4	0.9	19.7	20	9	3.1
1, 1, 2	6.5	0.4	20.1	2.2	13	2.3	3.7	1	19.8	20	8.4	3.6
1, 1, 3	6.6	0.4	22.2	25	12.5	2.1	4.1	1	19.8	20	8.1	1.7
1, 2, 1	6.4	0.3	18.6	1.6	14	2.63	3	4.6	18.3	19.1	8.7	1.7
1, 2, 2	6.4	0.3	19.2	1.8	12.6	1.9	3.2	0.8	18.3	27	8.7	2.8
1, 2, 3	6.4	0.3	20	2.31	12	1.73	3.5	0.9	18	NA	8.1	3.2
1, 3, 1	6.3	0.2	18	1.3	14	20.9	2.9	4.1	17	17.8	7.8	1.3
1, 3, 2	6.3	0.3	18.5	1.5	12.4	1.7	3	4.1	17.3	18	9	2.1
1, 3, 3	6.3	0.3	19.7	1.9	11.9	1.5	3.2	0.8	17.3	23	8	2.7

column of Table 1 provides the P, I and D gains used corresponding to μ_1, μ_2, μ_3 respectively. The next six columns correspond to different initial starting coordinates for r_p and x_p. r_e and x_e are assumed to be $(0, 0)^t$ and $(1, 0)^t$ respectively. The two sub-columns of each of the six columns in Table 1 correspond to time to reach zero for the magnitude of r and the time to reach $\Gamma = -1$ manifold in that order.

It is seen in Table 1 that PI and PID laws tend to improve on the performance of the pursuer compared to using P alone, which is the existing method. For example, comparing the rows corresponding to (1, 3, 0) and (1, 3, 3) against (1, 0, 0) in the first column, it is seen that the time for the magnitude of r to reach zero and the time to reach the value of $\Gamma = -1$ both for the first time is much reduced in the case of PI and PID settings compared to P alone. This holds for all possible starting points considered. The cells in Table 1 which are blank correspond to the case when $\Gamma = -1$ could not be reached within the simulation period of 30 s.

Figure 4 presents the plot of $|r|$, the magnitude of the baseline, for different initial points and P, PI, PID control laws. The rows correspond to P, PI, PID control strategy of the pursuer yielding the control input u_p. The vertical columns in Fig. 4 correspond to the different initial conditions of r_p and x_p which corresponds to the position and heading of pursuer.

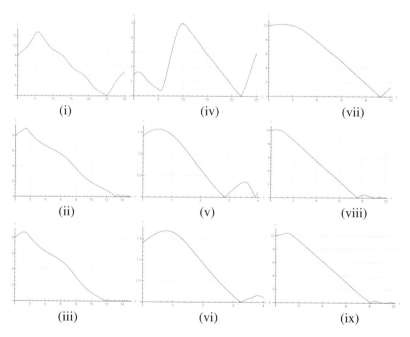

(i)	(iv)	(vii)
(ii)	(v)	(viii)
(iii)	(vi)	(ix)

Fig. 4. Simulated plot of magnitude of 'r'. The row corresponds to control laws (P, with $\mu_1 = 1, \mu_2 = 0, \mu_3 = 0$), (PI, with $\mu_1 = 1, \mu_2 = 3, \mu_3 = 0$), (PID, $\mu_1 = 1, \mu_2 = 3, \mu_3 = 3$) respectively and columns corresponds to initial conditions for r_p and x_p {(0, 8), (0.530, 0.847)}, {(-1, 1), (0.631, 0.775)}, {(10, -1), (.9, 0.417)} respectively.

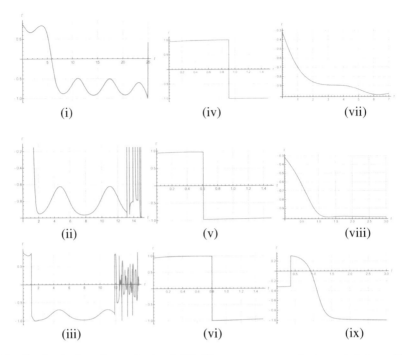

Fig. 5. Simulated plot of magnitude of 'Γ'. The row corresponds to control laws (P, with $\mu_1 = 1, \mu_2 = 0, \mu_3 = 0$), (PI, with $\mu_1 = 1, \mu_2 = 3, \mu_3 = 0$), (PID, $\mu_1 = 1, \mu_2 = 3, \mu_3 = 3$) respectively and columns corresponds to initial conditions for r_p and x_p {(0, 8), (0.530, 0.847)}, {(−1, 1), (0.631, 0.775)}, {(10, −1), (.9, 0.417)} respectively.

Considering the first column of Fig. 4, and comparing plots (i), (ii) and (iii) corresponding to P, PI and PID control, it is seen that the time for $|r|$ to reach zero for the first time is reduced from 25 s (P) to 13.1 s (PI) and 11.1 s (PID). The corresponding plots of $|r|$ vs time are given in Fig. 4(i), (ii) and (iii) respectively. Notice that the time scales for their plots are not the same. Notice that the above results correspond to column 3 results of Table 1 against rows (1, 0, 0) (P-Control, existing), (1, 3, 0) (PI-Control, proposed) and (1, 3, 3) (PID-Control, proposed) respectively.

Similarly plots (iv), (v) and (vi) of Fig. 4 correspond to column 4 of Table 1 and the plot $|r|$ under P (existing), PI (proposed) and PID (proposed) control. The corresponding time for $|r|$ to reach zero are 21.8 s, 3.4 s and 3.2 s respectively. The plots of (vii), (viii), (ix) of Fig. 4 correspond to column 6 of Table 1 in a similar way against the rows (1, 0, 0), (1, 3, 0) and (1, 3, 3) respectively. In all the cases it is seen that PI/PID (proposed) provide a much faster response for $|r| \rightarrow 0$ compared to P (existing) alone.

Figure 5 provides the plot Γ vs time for different initial points and under P, PI, PID control. Comparing plots (i), (ii) and (iii) of Fig. 3, for P (existing), PI (proposed) and PID (proposed) control corresponding to column 3 of Table 1, it is seen that the time to reach manifold Γ = −1 is again much improved with PI/PID (proposed) compared to P

Table 2. Time to reach $|r| = 0$ and $\Gamma = -1$ for different combination of μ_1, μ_2, μ_3 for different initial values of r_p and x_p.

μ_1, μ_2, μ_3	$(r_{px}, r_{py}), (x_{px}, x_{py})$ (4, −6), (0.735, 0.678) (1)		$(r_{px}, r_{py}), (x_{px}, x_{py})$ (−4, 9), (0.327, 0.945) (2)		$(r_{px}, r_{py}), (x_{px}, x_{py})$ (0, 8), (0.530, 0.848) (3)		$(r_{px}, r_{py}), (x_{px}, x_{py})$ (−1, 1), (0.632, 0.775) (4)		$(r_{px}, r_{py}), (x_{px}, x_{py})$ (−2, 10), (0.335, 0.942) (5)		$(r_{px}, r_{py}), (x_{px}, x_{py})$ (10, −1), (0.9, 0.436) (6)	
	Time at r = 0	Time at Γ = −1	Time at r = 0	Time at Γ = −1	Time at r = 0	Time at Γ = −1	Time at r = 0	Time at Γ = −1	Time at r = 0	Time at Γ = −1	Time at r = 0	Time at Γ = −1
1, 0, 0	5.2	3.5	10.1	5.86	10.1	6.88	NA	0.88	13.8	8.4	8	4.28
1, 1, 0	5.07	1.1	8.8	2.2	9	3.6	2.59	1	9.46	3.2	7.41	2.02
1, 2, 0	5	0.83	8.46	1.62	7.67	2.62	2.44	2.25	8.84	2.6	7.2	1.57
1, 3, 0	5	0.8	8.2	1.3	7.28	2.5	2.36	2.33	8.5	2.57	7.072	1.34
1, 1, 1	5	0.9	8.86	2.2	7.64	2.52	2.65	0.88	9.73	3.4	7.38	2
1, 1, 2	4.94	0.7	8.96	2.28	7.33	2.2	2.7	0.91	10.4	3.3	7.36	2
1, 1, 3	4.9	0.5	9	2.28	7.1	1.96	2.8	1	9.5	2.6	7.3	2
1, 2, 1	4.99	0.7	8.5	1.7	7.46	2.1	2.49	2.47	8.9	3	7.18	1.62
1, 2, 2	4.9	0.54	8.6	1.78	7.15	1.8	2.54	0.8	8.99	3.1	7.18	1.67
1, 2, 3	4.88	0.49	8.86	1.96	6.96	1.59	2.6	0.86	9	3.3	7.15	1.59
1, 3, 1	4.9	0.5	8.38	1.43	7.5	1.99	2.44	2.39	8.59	3	7.07	1.38
1, 3, 2	4.9	0.49	8.49	1.54	7	1.6	2.44	7.56	8.6	3	7	1.36
1, 3, 3	4.86	0.4	8.7	1.8	6.8	1.4	2.52	0.8	8.6	3.3	7.09	1.3

(existing) alone. Also, notice that the time scales of the plots in Fig. 5 are not all the same. Similar comments apply to plots (iv), (v) and (vi) which correspond to column 4 of Tables 1 as well as plots (vii), (viii) and (ix) which correspond to column 6 of Table 1.

Another feature noticed in Fig. 4 is that while the system reaches $|r| = 0$, the system stays at or near $|r| = 0$ line in the case of PI/PID control (proposed) as compared to P alone (existing). Similar case applies to Fig. 5 in regard to PI/PID control response settling closer to $\Gamma = -1$ line compared to P alone (existing).

4.2   Evader Motion Controlled

In this section, u_e is assumed to correspond to (3.10).

Table 2, Figs. 6 and 7 correspond to the case of evader following a classical control law. The results shows PI/PID (proposed) control strategies output correspond to much shorter time to reach $|r| = 0$ and the manifold $\Gamma = -1$ compared to P alone which is the existing result. Also the response curve of PI/PID (proposed) stay close to $|r| = 0$ and $\Gamma = -1$ line compared to P alone (existing).

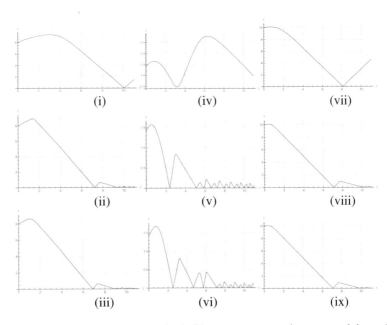

Fig. 6. Simulated plot of magnitude of 'r'. The row corresponds to control laws (P, with $\mu_1 = 1, \mu_2 = 0, \mu_3 = 0$), (PI, with $\mu_1 = 1, \mu_2 = 3, \mu_3 = 0$), (PID, $\mu_1 = 1, \mu_2 = 3, \mu_3 = 3$) respectively and columns corresponds to initial conditions for r_p and x_p {(0, 8), (0.530, 0.847)}, {(−1, 1), (0.631, 0.775)}, {(10, −1), (.9, 0.417)} respectively.

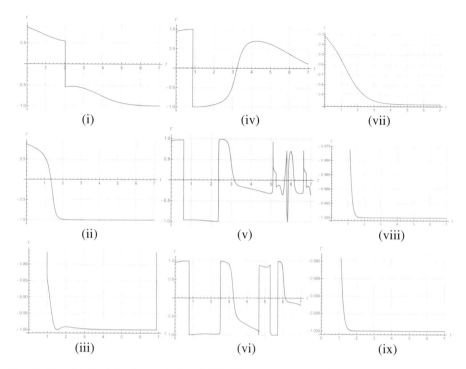

Fig. 7. Simulated plot of magnitude of 'Γ'. The row corresponds to control laws (P, with $\mu_1 = 1, \mu_2 = 0, \mu_3 = 0$), (PI, with $\mu_1 = 1, \mu_2 = 3, \mu_3 = 0$), (PID, $\mu_1 = 1, \mu_2 = 3, \mu_3 = 3$) respectively and columns corresponds to initial conditions for r_p and x_p {(0, 8), (0.530, 0.847)}, {(−1, 1), (0.631, 0.775)}, {(10, −1), (.9, 0.417)} respectively.

5 Conclusions

This paper has proposed PI and PID feedback control law for the pursuing strategies in a pursuit evasion game. It is seen that the time to reach the camouflage manifold as well as to capture the evader under the given assumption is much shortened with the introduction of proposed integral and derivative controls as compared to P alone (existing). The results are also applicable on both the cases where the evader is following a prescribed control motion.

While the results obtained seem to be encouraging, several questions arise for further investigation. How is the performance improvement of PI/PID controller affected by initial position and heading of the pursuer? These constitute the future work.

References

1. Isaacs, R.: Differential Games. Wiley, New York (1965)
2. Savec, P., Thakur, A., Shah, B.C., Gupta, S.K.: USV trajectory planning for time varying motion goals in an environment with obstacles. In: ASME 2012 International Engineering Technical Conferences (IDEC) and Computers and Information in Engineering conference (CIE), Chicago, USA, 12–15 August 2012
3. Wei, E., Justh, E.W., Krishnaprasad, P.S.: Pursuit and an evolutionary game. Proc. R. Soc. A **465**, 1539–1559 (2009). https://doi.org/10.1098/rspa.2008.0480
4. Li, W.: The confinement – escape problem of a defender against an evader escaping from a circular region. IEEE Trans. Cybern. **46**(4), 1028–1039 (2016)
5. Li, W.: A dynamics perspective pf pursuit-evasion: capturing and escaping when the pursuer runs faster than the agile evader. IEEE Trans. Autom. Control **62**(1), 451–457 (2017)
6. Wei, E., Justh, E.W., Krishnaprasad, P.S.: Steering laws for motion camouflage. Proc. R. Soc. A **462**, 3629–3643 (2006). https://doi.org/10.1098/rspa.2006.1742
7. Wei, E., Justh, E.W., Krishnaprasad, P.S.: Natural frames and interacting particles in three dimensions. In: Proceedings of the 44th IEEE Conference on Decision and Control, pp. 2841–2846, 12–15 December 2005
8. Pais, D., Leonard, N.E.: Pursuit and evasion: evolutionary dynamics and collective motion. In: AIAA Guidance, Navigation and Control Conference, Toronto, Ontario, Canada, pp. 1–14 (2010)

A Semantic Approach Towards Online Social Networks Multi-aspects Analysis

Asmae El Kassiri[✉] and Fatima-Zahra Belouadha

Mohammadia School of Engineers, Mohammed V University in Rabat,
Avenue Ibn Sina, B.P 765, Agdal, Rabat, Morocco
asmaeelkassiri@research.emi.ac.ma, belouadha@emi.ac.ma

Abstract. The semantic web uses the domains ontologies related to different topics on the web. Its potential is making the data on the web understandable by the machine and automatically treatable by algorithms without explicit human intervention. In this context, this paper proposes a semantic approach through a generic and intelligent framework to respond to different analytical needs applicable to Online Social Networks (OSN) data. This semantic approach consists in reusing and aligning with the standard ontologies, recommended by the W3C consortium, to formalize and synthetize OSN data, exploiting the ontologies inference potential, to calculate and represent useful indicators for OSN different analytical needs.

Keywords: OSN · Social media · SNA · Social mining · OWL · FOAF · SIOC
SKOS · Semantic web · Ontologies

1 Introduction

The semantic web formalize and standardize web data to ensure reusability, interoperability and modularity. The Online Social Networks (OSN) are at the heart of the second generation web and tend to be in the new generations, namely, the Semantic Web, the Objects Web and the Everything Web. The semantic web uses the domains ontologies related to different topics on the web. Its potential is making the data on the web understandable by the machine and automatically treatable by algorithms without explicit human intervention. This advantage is favored thanks to the standard ontologies exploitation that favors the discovery and the automatic selection from shared data on the Web thanks to their semantic expression force.

In this context, the proposed approach offer a generic and intelligent platform to respond to different OSN analysis. This approach semantic aspect consists in the reusing and alignment with W3C standard and good ontologies to formalize OSN data and a synthetization ontology exploiting ontologies inferring potential to compute and present utile indicators for the different OSN analysis.

The presented approach is based on two principal aspects. The first one is capturing the OSN data and represent it with an ontologies set composed from a standard, two recommended ontologies as good ontologies: Semantically-Interlinked Online Communities (SIOC) and Friend Of A Friend (FOAF); and its extensions. The second

© Springer International Publishing AG, part of Springer Nature 2018
A. Abraham et al. (Eds.): IBICA 2017, AISC 735, pp. 157–168, 2018.
https://doi.org/10.1007/978-3-319-76354-5_15

one is a synthetization ontology modelling multiple metric used in the different analysis. The objective is having on one hand, a Unified Semantic Model (USM) aligned with standard and recommended ontologies allowing presenting the most popular and analyzed social media data. On the other hand, the objective is having an ontology able to capture and present the deducible metrics and concepts from the OSN that are the analysis algorithms data used for different objectives (see Fig. 1).

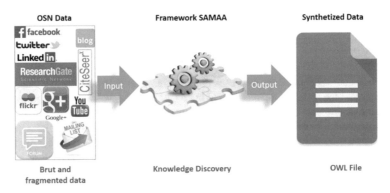

Fig. 1. The framework semantic approach for multi-aspects analysis of OSN (SAMAA of OSN)

This paper is organized in five sections. The second section presents the approach main founding directors, the third and fourth ones expose, respectively, the general demarche with ontological engineering methods, and the approach positioning compared to other approaches. The perspective of this work and a synthesis are the subject of the fifth section.

2   Principal Directors

In keeping with self-learning, ontological engineering and openness perspectives, the basic idea is based on four directors:

1. Knowledge discovery
2. Modelling by ontologies
3. Ontologies transformation
4. Alignment with standards

2.1   Knowledge Discovery

The first OSN analysis need is capturing the social media knowledge to execute the targeted analysis. This process have to respond to four challenges. The first one is that different social media owners stock this knowledge fragmentally. Actually, the users can generate data on multiple social networks that are members and so, each media have a knowledge fragment of a given user (74% of adult internets users use OSN and the mean accounts by user is 5.54 whose 2.82 active accounts (Lim et al. 2015). More than

56% of online adults use more than social media (Lister 2017), Instagram has reached 600 Million of users (Roberts 2016), 22% of the international population use Facebook (Lister 2017). LinkedIn reached 450 Million profiles (Lister 2017)).

The second challenge is the social networks data dynamism. On a social network, thousands of data are generated in each second fragment leading to the network state instability and big data volumes (per example, Snapchat has reached 10 Billion of video views by day (Roberts 2016). Facebook have reached 100 Million views videos hours by day (Lister 2017) knowing that its active users have reached 1.871 Milliard (Chaffey 2017)). Capturing knowledge on OSN changing over time, is consequently complicated. It depends on the giving period and have to be updated in real time.

The third challenge is the OSN knowledge nature that are explicit and implicit data. The explicit data are directly available on the social media while implicit knowledge are deducible from the explicit knowledge. The OSN knowledge discovery is a process that must be able to capture the explicit and implicit data.

As shown on the Fig. 2, the knowledge discovery aspect consist in using API allowing extracting explicit data from each social media, and then using an inference motor to infer the implicit knowledge from the explicit ones.

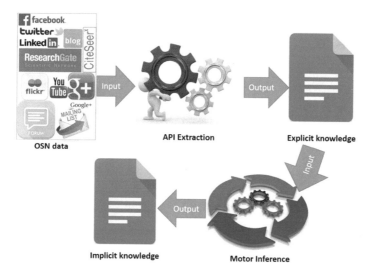

Fig. 2. The knowledge discovery process

2.2 Modelling by Ontologies

The captured or inferred knowledge from the OSN must be presented according a formalism allowing its exploitation by different analysis algorithms. In this context, the approach exploits the ontologies potential to formalize these knowledge. Actually, ontologies improve communication and offer a larger frame of reusability, share and interoperability frame in addition to a more flexibility to system (Uschold and Grüninger 1996).

1. The reusability. The ontology offers a modular modelling and independent frame from software implementation. The ontology module used in the application domain for a determined need, leads to knowledge reusable in other domains for other needs. Ontologies allow creating reusable knowledge bases.
2. The share and interoperability. One of ontologies objectives is the knowledge presentation standardization. This aspect guaranty the reusability independently from platforms and languages, it favorites the standardized knowledge interoperability and share. The interoperability concept denotes the possibility that many entities are able to interoperate, in other words, to communicate. It's a standardization and semantic presentation results ensured thanks to ontologies. The ontology offers a structuration, storage and share semantic model allowing to different systems exploiting the common ontological modules to understand each other and exchange knowledge allowing its cooperation.
3. The flexibility. A system using ontologies is flexible. It can evaluate and fit new needs without braking due to ontologies. Per example, introducing new knowledge is possible by simple ontologies extension, knowing that ontologies are naturally modular and extensible. The ontologies fit the scaling requirements and can be distributed over multiple systems.

To respond to the OSN knowledge formalization objective, the approach considered two ontologies. The first one is the social media explicit knowledge presentation USM. The second one is the implicit knowledge formalization (see Fig. 3).

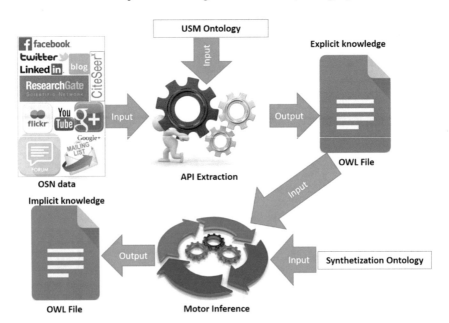

Fig. 3. Modelling OSN knowledge by ontologies

2.3 Ontologies Transformation

In the engineering domain directed by models, the transformation is an important mechanism allowing generating automatically transformation rules. These rules specifying mapping between sources models and targeted models elements. A transformation motor allowing transforming source models elements by mapping in a targeted model executes these rules. A sources element can have one or more targeted correspondents.

The approach defines two ontologies according two semantic models. The first one presents the OSN explicit knowledge, and the second one presents implicit knowledge deserving analysis objectives. This transformation is interesting because it allow generating the analysis ontology from the OSN ontology thanks to rules transformation (see Fig. 4).

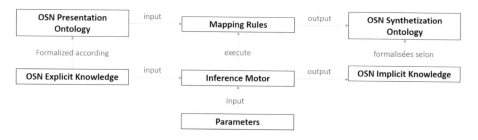

Fig. 4. Knowledge models transformation

These transformation rules are inference rules allowing peopling automatically the analysis ontology from social media description ontology instances. The obtained instances correspond to metrics needed for a multi-aspects analysis.

2.4 Alignment with Standards

The approach's aim is contributing to ensure different OSN analyses types by reusing knowledge modeled according ontologies. These ontologies, the OSN Unified Semantic Model (USM) ontology and the Synthetization ontology, must cover different knowledge manipulated on the social media and the indicators needed for different analysis types. Particularly, the USM ontology have to be aligned with the existent standards respecting norms and good practices to present the OSN data and to offer a generic frame covering the most popular social media.

Studying the social ontologies used in the literature, the authors have noted that multiple ontologies have been proposed to model OSN different aspects or data types. A unique standard ontology named Simple Knowledge Organization System (SKOS)[1] has been identified. Two other ontologies are considered standards by its large use and

---

[1] https://www.w3.org/TR/skos-reference/.

are recommended by W3C as good ontologies[2]. There are the FOAF Ontology and the SIOC ontology created by a W3C team. This works type is generally standardized[3].

FOAF was a project aiming allowing connecting persons and information on the web. It integrates three usage categories: the user, his friends groups and his associations' description independently from the web; the online accounts, addresses books and basic activities on the web description; and sharing these descriptions thanks to interconnected RDF documents.

The SIOC ontology gives the principal concepts and properties to describe online communities (wikis, weblogs, Forums, etc.) data on the semantic web.

Concerning the SKOS ontology, it's a model founded on the RDF language to present thesaurus, classifications or author vocabularies controlled or documentaries languages. Its' used on social web modeling to present semantic relationship between vocabularies used in shared contents.

The proposed OSN USM ontology and the synthetization ontology are aligned with the FOAF, SIOC and SKOS ontologies. Therefore, its creation and interrogation process respects the W3C standards, such, OWL2[4], RDF[5] and SPARQL[6].

The RDF (*Resource Description Framework*) framework presents information on the Web based on triplets <subject-predicate-object>, where elements (subjects and objects) can be URI (Uniform Resource Identifier), virgin nodes or data literal types (numeric, string, date, etc.).

The OWL2 is an ontology language defining classes, properties, individuals and data values and are stored as semantic web documents. OWL2 ontologies and its data (knowledge formalized by these ontologies) can be defined in RDF and are principally exchanged as RDF documents (Hawke et al. 2012).

The SPARQL language is used to express requests over diverse data sources.

3   Methodology

An ontology is an operational system component responding to precise functional needs. Its development respects the same Software Genie principles from which its life cycle is inspired. The Fig. 5 shows that this cycle is composed of an initial evaluation step, a construction or elaboration step, a diffusion step and a using step (Fürst 2002).

The state of the art associated to ontologies identifies multiple proposed methodologies (Grüninger and Fox 1995; Uschold and Grüninger 1996; Bachimont 2000; Gómez-Pérez et al. 2004; Kotis and Vouros 2006; Lima et al. 2010; Iqbal et al. 2013; Stadlhofer et al. 2013; Nicola and Missikoff 2016; Sawsaa 2017). A principal works synthetization allow noting that even if the proposed methodologies consider different names, the principal steps are similar by its objectives and definitions.

---

[2] https://www.w3.org/wiki/Good_Ontologies.
[3] https://www.w3.org/Submission/.
[4] https://www.w3.org/TR/owl2-quick-reference/.
[5] https://www.w3.org/TR/rdf11-concepts/.
[6] https://www.w3.org/TR/sparql11-query/.

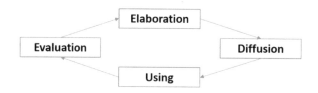

Fig. 5. Ontology life cycle (Fürst 2002)

Per example, Grüninger and Fox (1995) propose four steps: (i) defining the ontology objective, (ii) determining the terminology, (iii) coding the ontology by defining relationships and constraints between terms, and (iv) evaluating the ontology to prove its capacity to respond to the initial objectives.

Uschold and Grüninger (1996), propose two principal steps to create an ontology: (i) defining the objective, and (ii) constructing the ontology passing by identifying the concepts and its coding by integrating existent ontologies. They propose also evaluating it and documenting it over the creation cycle.

Similarly, Bachimont (2000) proposes three steps corresponding by its definition and objectives to the three last steps of Grüninger and Fox (1995). The same process have been adopted by Kotis and Vouros (2006) proposing the HCOME (Human-Centered Ontology Engineering Methodology) methodology. The three steps were named, respectively, specification, conceptualization and exploitation.

The adopted approach adopts the methodology that's the fundaments are aligned with the principal steps used in the literature.

3.1   Adopted Methodology for the Ontologies Creation

The adopted methodology to create the targeted ontologies is a process composed of three principal steps. A general view of these steps objectives is given below:

1. The semantic engagement. The objective is identifying the terminology by studying the different social media and OSN analysis works;
2. The ontological engagement. This second step elaborates the ontologies formal conceptualization by adapted semantic model to targeted needs. It also integrates and extends existent ontologies like FOAF and SIOC, and elaborates transformation rules based on inference;
3. The operational engagement. This last step describes the conceived sematic model in an ontologies description language and demonstrates its utility by projecting a set of scenarios, corresponding to the different OSN analysis needs, on presented data semantically in the proposed ontologies.

3.2   The Semantic Engagement

Under the targeted ontologies objectives and its application domain, the semantic engagement is a conceptualization step aiming principally identifying the semantic concepts of the explored domain and the targeted application domain terminology.

It corresponds to the first creation ontology step. In the literature, this step is founded on interviewing domain experts and analyzing the documents corpus of the domain.

To extract the candidate terms, relationships and verbalizations that correspond to norms and practices of the OSN analysis domain, the authors have specified two corpus categories to analyze. Relatively to the OSN description and analysis objectives, the two corpus categories are shown on Fig. 6:

1. The first corpus is the different most popular social media. The media study was made by practical tests and mapping between the different concepts that are used by different media to synthetize it to identify an aggregated terminology of OSN concepts and its relationships;
2. The second corpus corresponds to the research works that are the expertise researchers' result in the domain. The set of research works studied is inherent to different and principal types of OSN analysis. Studying it allow identifying the necessary knowledge for the targeted analysis.

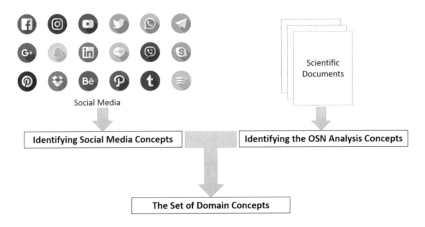

Fig. 6. The semantic engagement step illustration

It should be noted that, in this case, the specified corpus nature doesn't favorite its automatic analysis. On one hand, the OSN terminology perimeter is familiar, known and simple to determinate without an automatic processing. On the other hand, the connected research works analysis can't be made by a simple automatic extraction of these documents terms, it needs an analytic study of the algorithms and proposed approaches to discover the utile and reusable concepts used by the different OSN analysis.

3.3   The Ontological Engagement

Presenting a domain by ontologies is formalizing its data according these semantic models. In this case, it's a semantic normalization of the domain-connected concepts identified in the previous step. This step classifies identified terms and concepts and establishes a semantic tree for each targeted ontology illustrating its different concepts with its semantic relationships. The approach has two levels; the choice is justified thanks to the targeted objectives:

1. The first level proposes a USM ontology to present social media data; it models the online users behavior on the different most popular social media. The approach prevent using one ontology per media because users maintain multiple accounts online on different social media, and a user can be more active on a media than on another; so, aggregation their data from each media can be difficile and inexact. While aggregating complementary data from multiple social media has proven its efficiency to refine analysis results in many research works (Gilbert and Karahalios 2009; Zhang et al. 2013; Kong et al. 2013; Kosinskia et al. 2013).

2. The second level proposes an OSN synthetization ontology. It's an application ontology dedicated to different aspects of OSN analysis. It must modeling concepts and key metrics of the OSN analysis process, offering an interrogation semantic frame allowing on one hand, reusing these data, and on the other hand, automatizing and optimizing the analysis treatments.

The synthetization ontology offers a generic frame dedicated to OSN analysis by regrouping and synthetizing different usable knowledge by different algorithms ensuring the principle analysis (e.g. community detection, expert detection, privacy, similarity analysis, influence propagation, trusting, links prediction and recommendation).

The aggregated knowledge in this ontology are deducible from the explicit knowledge formalized by the USM ontology. The deducing process is guided by inference rules.

To align with the standards, the approach proposes elaborating the USM ontology based existent ontologies by ontological reengineering process. The Fig. 7 shows that the USM ontology is a global semantic model, principally constituted by fusing the SIOC, FOAF and SKOS ontologies, and four other new modules ensuring its extension to cover concepts and relationships in social media not supported by the proposed ontologies in the literature.

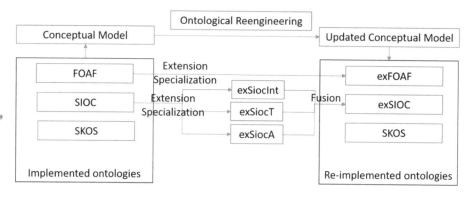

Fig. 7. The USM ontology elaboration by ontological reengineering

The OSN synthetization ontology is an original global semantic model dedicated to different analysis objectives. The authors propose modelling the identified implicit concepts from the studied corpus to elaborate the conceptual model that must be implemented to lead to the final synthetization ontology.

3.4 The Operational Engagement

Making an ontology operational is describing it first in a formal and operational language; and then enrich it to allow machines manipulating knowledge expressed by this ontology. Actually, enriching an ontology is developing mechanisms allowing the machine operating and interrogating its knowledge (Fürst 2002).

In this context, the presented approach proposes a set of measures to implement the targeted ontologies and making it operational:

1. The targeted ontologies implementation with a standard language. The authors have choose the OWL2 Language (Web Ontology Language) (Hawke et al. 2012) and the Protégé environment (Musen 2015);
2. The automatic social media knowledge extraction. As mentioned above, the extraction of social media data will be made thanks to API to people the USM ontology;
3. The transformation rules implementation. These rules allow transforming, by deducing, the USM knowledge into the synthetization ontology knowledge;
4. The inference rules implementation. These rules allow inferring new knowledge from the synthetization ontology into the ontology itself.
5. The ontologies test and validation. The objective is evaluating some analysis processes exploiting the synthetization ontology.

4 Conclusion

This paper introduces an approach founded on ontologies for an aggregated and multi-aspects analysis of users behaviors on the social web. It exposes its founder principles and describe the adopted methodology to elaborate the proposed ontologies. It presents also the exploitation demarche of these ontologies for OSN analysis objectives. Finally, it situates the proposed approach compared to the associated approaches.

To conclude, the proposed approach is founded on the ontological reengineering using particularly the fusion, the specialization and extension of standard and recommended ontologies in the domain. It exploits also the ontologies potential allowing its knowledge reusing. It's an original approach since it allows aggregating data from multitude of the most popular OSN. In addition, six principle aspects characterize it: the W3C standards and recommendations alignment, the oriented synthetization ontology reasoning, the genericity, the granularity, the extensibility, the quality guarantee.

To resume, the approach belongs to an OSN analysis demarche with three dimensions guided by the transformation (based on inference) of presented data by ontologies (see Fig. 8):

1. Aggregating users data and behaviors from multiple social media by a USM ontology;
2. Synthetizing the USM ontology data into metrics to aliment the synthetization ontology offering a generic frame to respond to the different analysis;
3. Multi-aspects analysis of the studied OSN by data mining technics to compute the Synthetization ontology analysis.

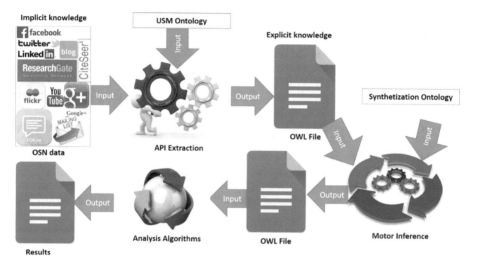

Fig. 8. A synthetic view of the semantic approach towards online social networks multi-aspects analysis

In future works, the authors will detail each step in a dedicated paper.

References

Bachimont, B.: Engagement sémantique et engagement ontologique: conception et réalisation d'ontologies en ingénierie des connaissances. In: Charlet, J., Zacklad, M., Kassel, G., Bourigault, D. (eds.) Ingénierie des connaissances: évolutions récentes et nouveaux défis (p. chapitre 19). Eyrolles, Paris (2000)

Chaffey, D.: Global social media research summary 2017, 27 February 2017. Smart Insights. http://www.smartinsights.com/social-media-marketing/social-media-strategy/new-global-social-media-research/

Fürst, F.: L'ingénierie ontologique. Institut de Recherche en Informatique de Nantes. Université de Nantes (2002)

Gilbert, E., Karahalios, K.: Predicting tie strength with social media. In: Proceedings of the SIGCHI Conference on Human Factors in Computing Systems, CHI 2009, pp. 211–220 (2009)

Gómez-Pérez, A., Fernández-López, M., Corch, O.: Methodologies and methods for building ontologies. In: Gómez-Pérez, A., Fernández-López, M., Corch, O. (eds.) Ontological Engineering, pp. 107–197. Springer, London (2004)

Grüninger, M., Fox, M.S.: Methodology for the design and evaluation of ontologies. In: Workshop on Basic Ontological Issues in Knowledge Sharing, IJCAI 1995, pp. 6.1–6.10, 13 April 1995

Hawke, S., Horridge, M., Parsia, B., Schneider, M.: OWL 2 web ontology language: W3C recommendation, 11 December 2012. W3C. https://www.w3.org/TR/owl2-overview/

Iqbal, R., Murad, M.A., Mustapha, A., Sharef, N.M.: An analysis of ontology engineering methodologies: a literature review. Res. J. Appl. Sci. Eng. Technol. **6**(16), 2993–3000 (2013)

Kong, X., Zhang, J., Yu, P.S.: Inferring anchor links across multiple heterogeneous social networks. In: Proceedings of the 22nd ACM International Conference on Conference on Information and Knowledge Management, pp. 179–188 (2013)

Kosinskia, M., Stillwella, D., Graepel, T.: Private traits and attributes are predictable from digital records of human behavior. Proc. Natl. Acad. Sci. **110**(15), 5802–5805 (2013)

Kotis, K., Vouros, G.A.: Human-centered ontology engineering: the HCOME methodology. Knowl. Inf. Syst. **10**(1), 109–131 (2006)

Lim, B.H., Lu, D., Chen, T., Kan, M.-Y.: #mytweet via Instagram: exploring user behaviour across multiple social networks. In: Proceedings of the 2015 IEEE/ACM International Conference on Advances in Social Networks Analysis and Mining, ASONAM 2015, pp. 113–120, 25–28 August 2015

Lima, J.F., Amaral, C.M., Molinaro, L.F.: Ontology: an analysis of the literature. In: Proceedings of the International Conference, CENTERIS 2010 - Part II, pp. 426–435, 20–22 October 2010

Lister, M.: 40 essential social media marketing statistics for 2017, 20 January 2017. Word Stream. http://www.wordstream.com/blog/ws/2017/01/05/social-media-marketing-statistics

Musen, M.A.: The protégé project: a look back and a look forward. AI Matters **1**(4), 4–12 (2015)

Nicola, A.D., Missikoff, M.: A lightweight methodology for rapid ontology engineering. Mag. Commun. ACM **59**(3), 79–86 (2016)

Roberts, P.: Social media statistics for 2017 (2016). Our Social Times. http://oursocialtimes.com/7-social-media-statistics-for-2017/

Sawsaa, A.F.: Methodology of creating ontology of information science (OIS). In: Lu, J., Xu, Q. (eds.) Ontologies and Big Data Considerations for Effective Intelligence, pp. 435–442. IGI Global, Hershey (2017)

Stadlhofer, B., Salhofer, P., Durlacher, A.: An overview of ontology engineering methodologies in the context of public administration. In: The Proceedings of the Seventh International Conference on Advances in Semantic Processing, SEMAPRO 2013, pp. 36–42, 09 October–29 March 2013

Uschold, M., Grüninger, M.: Ontologies: principies, methods an applications. Knowl. Eng. Rev. **11**(2), 93–155 (1996)

Zhang, J., Kong, X., Yu, P.: Predicting social links for new users across aligned heterogeneous social networks. In: Proceeding of 2013 IEEE 13th International Conference on Data Mining (ICDM), pp. 1289–1294, 7–10 December 2013

Towards a New Generation of Wheelchairs Sensitive to Emotional Behavior of Disabled People

Mohamed Moncef Ben Khelifa[1]([✉]), Hachem A. Lamti[1], and Adel M. Alimi[2]

[1] Bio-modélisation et Ingénierie des Handicaps (HANDIBIO) Laboratory,
South University, Toulon-Var, France
khelifa@univ-tln.fr
[2] Research Group on Intelligent Machines (REGIM) Laboratory,
National Engineering School of Sfax, Sfax, Tunisia

Abstract. The purpose of this article is to present a new alternative to handle wheelchair command and control especially for palsied patients. This project proposes a new framework based on visual and cerebral activities which are mapped into command/control and security blocks. While the former deals with the migration from a joystick-based navigation to a brain/gaze-based one, the latter enhances security by accounting for human factors. Those are assessed through emotions. Four emotions were induced and measured (relaxation, nervousness, excitement and stress) in three navigation scenarios where the introduction of the detection block was assessed. Based on those findings, an emotion block is built.

1 Introduction

The concept of "shared paradigm" consists on sharing the control between the wheelchair and its driver. Naturally, this approach requires personalization: Each pathology has its specific requirements (for example Locked-in patients can rely only on intellectual and visual faculties while tetraplegic ones can provide some limited muscular activities). Consequently, several approaches are proposed to ensure the navigation safety.

Vander Poorten et al. [8] proposed a communication channel between the subject and the wheelchair controller. The concept can help the wheelchair user to avoid mode confusions. Ren et al. [6], suggest a map matching algorithm between GPS positions or other sensors on a sidewalk network. This process of map matching in turn assists in making decisions under uncertainty. Peinado et al. [7] present a collaborative wheelchair control. The system estimates how much help the person needs at each situation and provides the correct amount of help based on his skills. Besides, the system combines between the human and the machine control by weighting them according to their local efficiency. Consequently, the better the person drives, the more control he/she is awarded with.

© Springer International Publishing AG, part of Springer Nature 2018
A. Abraham et al. (Eds.): IBICA 2017, AISC 735, pp. 169–177, 2018.
https://doi.org/10.1007/978-3-319-76354-5_16

During navigation, as the type of error committed cannot be predicted, this latter could lead to fatal accidents. These systems lack an anticipative behavior that could modify the wheelchair parameters: example decreasing velocity in order to prevent a misbehavior could occur. During our interviews with doctors, experts, occupational therapist and psychologists, they stated that many human factors have a direct effect on the navigation performance. Among them, mental workload [2,3] and emotions [4] are the most influent. However, these two parameters are not easy to cope with, as they need much more investigations. In the next section, we will focus on the influence of emotions on ElectroEncephaloGraphy (EEG) patterns.

2   Emotion Integration

The detection of emotions is very important as it could be integrated for two purposes: first, it could be a basis for EEG command patterns to be detected (example: when performing a motor imagery task, the EEG pattern manifestation depends on the user mental state). Second, in order to enhance wheelchair navigation, its velocity should be enslaved to the user's emotions (example, it should decrease if the user is frustrated). But it's still not evident that emotions can bring the expected results and especially that EEG reliability in this context is still not proved yet. Besides, measuring emotions tends to be very challenging. In this paper, different algorithms used for extraction, selection and classification are compared (Fig. 1).

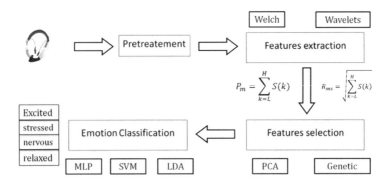

Fig. 1. Methodology work flow to detect emotion

In order to collect data to supply emotional database, 40 healthy subjects took part in the experiment. Aged from 22 to 55 years old, they were asked to complete and sign a consent form with personal information. Afterwards, they chose five audio/video excerpts sensitive to impact their emotions. Next, the chosen sequences were clipped to 63 s and projected one by one. At the end of each session, the subject gives his rating according to the SAM scale [5].

2.1   Extraction Methods

Welch Method. The welch approach aims at estimating spectral density at different frequencies. Its based on the concept of the periodogram spectrum estimates [9]. It is defined as the result of converting signal from time to frequency domain. Consider $x_i(n), i = 1, 2, \ldots, K$, K uncorrelated measurements of a randomized process x(n), over an interval of $0 \leq n < L$. Suppose that successive sequences are decaled by $D(\leq L)$ samples and that each of them is of length L, the i-th sequence is given by:

$$x_i(n) = x(n + iD), n = 0, 1, \ldots, L - 1 \tag{1}$$

Though, the overlapping quantity between $x_i(n)$ and $x_{(i+1)}(n)$ is L-D points, and if K sequences cover N data of the signal, then

$$N = L + D(K - 1) \tag{2}$$

This means, if we consider that sequences are overlapped by 50% ($D = \frac{L}{2}$), then we can form $K = 2NL - 1$ sections of length L. The Welch method is expressed by:

$$\hat{S}_w(w) = \frac{1}{KLU} \sum_{i=0}^{K-1} |\sum_{n=0}^{L-1} w(n)x(n + iD)exp(-jwn)|^2 \tag{3}$$

where $U = \frac{1}{N} \sum_{i=0}^{N-1} |w(n)|^2$, N is the length of the window $w(n)$. This method reduces the noise in power spectra. However, the resolution R depends on window type and length:

$$R = \frac{1}{LT_s} \tag{4}$$

where T_s the sampling period. The lower L is, the smoother Welch periodogram becomes.

The raw signal was filtered between 1 and 64 Hz (to cover all band wave lengths). Those are classified depending on the band interval limits. In the current study, we focus on: δ (up to 4 Hz), θ (4 Hz–8 Hz), α (8 Hz–13 Hz) β (13 Hz–30 Hz) and γ (30 Hz–64 Hz). The power spectral density (PSD) was computed on successive intervals of 1 s per trial per user. The Welch periodogram was computed using 512 points FFT and various Hamming window lengths: 128, 64, 32 and 16 points with a 50% overlapping. Finally, for each band, two parameters were computed: the mean Power (P_m) and the Root Mean Square (R_{ms}):

$$P_m = \sum_{k=i}^{j} S(k) \tag{5}$$

$$R_{ms} = \sqrt{\sum_{k=i}^{j} S(k)} \tag{6}$$

Where $S(k)$ are the sampled values of the periodogram and i and j are the indexes of the higher and lower sampled frequencies for each band.

Discrete Wavelets Transform. Discrete Wavelets Transform (DWT) is defined by the function:

$$\psi_{a,b}(t) = 2^{\frac{a}{2}}\psi(2^{\frac{a}{2}}(t-b)) \tag{7}$$

Where a are the scales and b the shifts. To approximate any function, $\psi(t)$ is dilated with the coefficient 2^k while the resulting function on interval is shifted proportionally to 2^{-k}. To obtain a compressed version of a wavelet function, a high-frequency component must be applied while a low frequency one is needed for dilated version. The correlation of the original signal with wavelet functions of different sizes, results in signal details obtained for many scales. These correlations are managed in hierarchical framework, the so called multi resolution decomposition algorithm can proceed by separating the signal into details at different scales from coarser representation named approximation.

2.2   Selection Features Methods

As it was mentioned before, extracted features need to be selected to avoid high dimensionality curse. For this purpose, two selection techniques are introduced.

Principal Component Analysis (PCA). Principal Component Analysis (PCA) transforms correlated variables into sub spaces called principal components. Some of the common applications include data compression, blind source separation and de-noising signals. PCA uses a vector space transformed to reduce the dimensionality of large data sets. Using mathematical projection, the original data set, with its high dimension, can be interpreted in fewer variables (principal components). This is important as it can reduce the processing time during classification and let the user interpret outliers, patterns and trends in the data. The aim of this section is to explain how to apply PCA on the feature vector in each second. The first step is to rescale data to obtain a new vector Z:

$$z^i_j = \frac{x^i_j - \bar{x}^j}{s^j} \tag{8}$$

Where $\bar{x}^j = \frac{1}{n}\sum_{i=1}^{n} x^j_i$ is the mean of the jth variable, s^j is the corresponding standard deviation. The correlation matrix R contains correlation coefficients between each pair of variables. Its symmetric and composed by ones in its diagonal. Its defined as follows

$$R = D_{\frac{1}{s}}VD_{\frac{1}{s}} \tag{9}$$

Where V is the variance matrix of X and $D_{\frac{1}{s}} = (\frac{1}{s1}, \frac{1}{s2}, \ldots, \frac{1}{sp})$. The next step is to calculate the eigen values of the correlation matrix. In fact, the eigen values contain the projection inertia of the original space on the sub space formed by the eigen vectors associated to each of them. Eigen vectors constitute the loadings of the Principal Component; thats the strength of the relation between the variable and principal component. The following figure gives an example about Eigen values, the associated principal component and the projection inertia (in percent) for each of them.

The selection of principal components axis was made empirically using Kaiser Criterion. The latter consists on retaining only the Eigen values greater than the mean value.

Genetic Algorithm (GA). Genetic Algorithm (GA) is an adaptive search technique [1]. The power of such a method is its ability to fine tune an initially unknown search space into more convenient subspaces. The main issue for a GA problem is the selection of a suited representation and evaluation function. This is very well suited to solve a feature selection problem. In this case, each feature is considered as a binary gene and each individual a binary string representing the subset of the given feature set. For a feature vector X of length s, the feature inclusion or elimination is processed as follow: if $X_i = 0$, then the feature is eliminated otherwise, 1 indicates its inclusion. As our purpose is to estimate the number of optimal features to keep, the proposed fitness function on the correlation matrix C is expressed as follows:

$$C = \begin{pmatrix} 1 & \cdots & c_{1,n} \\ \vdots & c_{h,l} & \vdots \\ c_{n,1} & \cdots & 1 \end{pmatrix} \tag{10}$$

Where $C_{h,l}$ represents the correlation coefficient between feature h and feature l. This value varies between -1 and 1. If $C_{h,l}$ tends to 1 or -1, then the features are highly correlated otherwise, if it tends to 0 then there is no correlation. Starting from the initial population which is constituted by the correlation of the pairs of features, the proposed fitness function F could be defined:

$$F = min_{h,l}|c_{h,l}| \tag{11}$$

For each iteration, the least correlation crossing chromosomes are kept for the next generations.

2.3   Classification

The output of different classification techniques are four classes each of which corresponds to one emotion (stressed, excited, nervous, and relaxed). For this purpose three classification techniques are deployed: the Linear Discriminate Analysis (LDA), the Multi Layer Perceptron (MLP) and Support Vector Machine (SVM).

Linear Discriminant Analysis (*LDA*). The *LDA* combines linearly variables:

$$y_{rn} = u_0 + u_1 X_{1rn} + u_2 X_{2rn} + \ldots + u_p X_{prn} \tag{12}$$

where: y_{rn} is the discriminant function for the case n on the group r as well as for X_{irn} which is the discriminate variable X_i for the case n on the group r, and u_i are the required coefficients. This implies that the number of discriminant functions is determined by the number of considered groups.

Multi Layer Perceptron (*MLP*). *MLP* is divided into three layers: an input layer with length, the selected features of the input vector, a hidden layer with 20 neurons and an output layer with 4 neurons. Sigmoid function is adopted as transfer and test set validation technique for cross-test; the database is separated into 3 sets: 70% for training, 15% for testing and 15% for validation (to avoid over fitting).

Support Vector Machine (*SVM*). SVM maps input vectors into higher dimensional space to ease classification. Then it finds a linear separation with the maximal margin in the new space. It requires the solution of the following problem:

$$min_{w,b,\epsilon} \frac{1}{2} w^T w + C \sum_{i=1}^{l} \epsilon_i \; subject \; to \; y_i(w^T \Phi(x_i) + b) \geq 1 - \epsilon_i \epsilon_i \geq 0 \tag{13}$$

where: C is the penalty parameter of the error ϵ_i with Gaussian radial basis function as kernel. This could be expressed as follows:

$$K(x_i, x_j) = e^{-\gamma |x_i - x_j|^2} \tag{14}$$

C and γ are fixed using a cross-validation technique.

2.4   Results

The results summarized in Figs. 2, 3, 4 and 5 present the different crossings between extraction (welch and wavelets), selection (genetic algorithm and PCA) over classification techniques (LDA, MLP and SVM). The comparison includes also the performance between R_{ms} and P_m. The results show that the highest classification recorded is assigned to the combination $MLPRms$ with wavelets as extraction technique and genetic algorithm as selection process with a rate of 93%. This being said, wavelets is suggested to be better than welch method thanks to its duality in time and frequency domains. Genetic algorithm is better than PCA. The latter suffers from subjectivity especially when choosing the adequate number of eigen values to be hold.

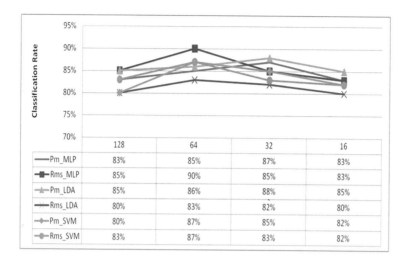

	128	64	32	16
—— Pm_MLP	83%	85%	87%	83%
—■— Rms_MLP	85%	90%	85%	83%
—▲— Pm_LDA	85%	86%	88%	85%
—✳— Rms_LDA	80%	83%	82%	80%
—◆— Pm_SVM	80%	87%	85%	82%
—●— Rms_SVM	83%	87%	83%	82%

Fig. 2. PCA/Welch periodogram classification with different window lengths

The 7% of misclassified data could be explained as follows: it is assumed that the rate given by the subject reflects his emotion during the whole visualization process which is not taken for granted. In fact, emotion could trigger from state to state at a certain period (which could be limited to a limited amount of time as well as it could start at the beginning or at the end of the session).

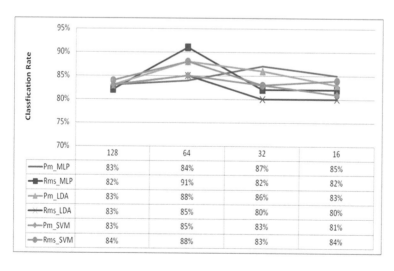

	128	64	32	16
—— Pm_MLP	83%	84%	87%	85%
—■— Rms_MLP	82%	91%	82%	82%
—▲— Pm_LDA	83%	88%	86%	83%
—✳— Rms_LDA	83%	85%	80%	80%
—◆— Pm_SVM	83%	85%	83%	81%
—●— Rms_SVM	84%	88%	83%	84%

Fig. 3. GA/Welch periodogram classification with different window lengths

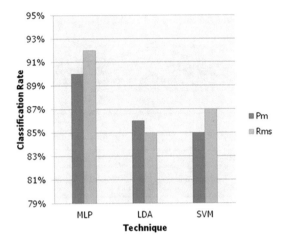

Fig. 4. PCA/Wavelets classification performance

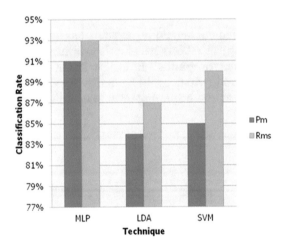

Fig. 5. GA/Wavelets classification performance

3   Conclusion and Perspectives

In this project, an investigation about the influence of emotions on cerebral activity was assessed. For this purpose, different techniques were deployed combining extraction, selection and classification phases. The overall performance showed that the combination between wavelets, genetic algorithm and MLP provides the highest classification rate. Those approaches were applied only on a simulated environment but not in real world. This could make the biggest shortage as challenges imposed by real wheelchair navigation differ to challenges faced in a simulated one. For example, in a simulated environment, all objects coordinates are already known as they are communicated by the software but this is not the case in real world. Another problem is that vibrations, wheelchair real velocity and synchronization issue biases much more the results. What's more, emotion study wasn't precise enough to account many other emotion cases; this is done intentionally because the chosen emotions were representative for each quadrant in order to not complicate the study. But in reality for each quadrant, many emotions could be situated in the same region which can create some confusion during classification.

References

1. Jong, K.: Learning with genetic algorithms: an overview. Mach. Learn. **3**(2–3), 121–138 (1988)
2. Lamti, H.A., Ben Khelifa, M.M., Alimi, A.M., Gorce, P.: Effect of fatigue on ssvep during virtual wheelchair navigation. J. Theor. Appl. Inf. Technol. **65**, 1–10 (2014)
3. Lamti, H.A., Ben Khelifa, M.M., Alimi, A.M., Gorce, P.: Emotion detection for wheelchair navigation enhancement. Robotica **34**(6), 1209–1226 (2016)
4. Lamti, H.A., Gorce, P., Ben Khelifa, M.M., Alimi, A.M.: When mental fatigue can be distinguished by event related potential (p300) during virtual wheelchair navigation. Comput. Methods Biomech. Biomed. Eng. **19**(16), 1749–1759 (2016)
5. Molina, G., Tsoneva, T., Nijholt, A.: Emotional brain-computer interfaces. In: 3rd International Conference on Affective Computing and Intelligent Interaction and Workshops, ACII 2009, pp. 1–9, September 2009
6. Ren, M., Karimi, H.: A fuzzy logic map matching for wheelchair navigation. GPS Solut. **16**(3), 273–282 (2012)
7. Urdiales, C., Perez, E., Peinado, G., Fdez-Carmona, M., Peula, J., Annicchiarico, R., Sandoval, F., Caltagirone, C.: On the construction of a skill-based wheelchair navigation profile. IEEE Trans. Neural Syst. Rehabil. Eng. **21**(6), 917–927 (2013)
8. Vander Poorten, E.B., Demeester, E., Hüntemann, A., Reekmans, E., Philips, J., De Schutter, J.: Backwards maneuvering powered wheelchairs with haptic guidance. In: Proceedings of the 2012 International Conference on Haptics: Perception, Devices, Mobility, and Communication - Part I, EuroHaptics 2012, pp. 419–431. Springer, Berlin, Heidelberg (2012)
9. Yuen, C.: On the smoothed periodogram method for spectrum estimation. Sig. Process. **1**(1), 83–86 (1979)

A Comprehensive Technical Review on Security Techniques and Low Power Target Architectures for Wireless Sensor Networks

Abdulfattah M. Obeid[2(✉)], Manel Elleuchi[1], Mohamed Wassim Jmal[1],
Manel Boujelben[1], Mohamed Abid[1], and Mohammed S. BenSaleh[2]

[1] CES Research Unit, National School of Engineers of Sfax, Digital Research Center (CRNS),
Technopark Sfax, Sfax, Tunisia
manelelleuchi@gmail.com, mohamed.wassim-jmal@ceslab.org,
boujelben_manel@yahoo.fr, mohamed.abid@enis.rnu.tn
[2] National Center for Electronics, Communications and Photonics, King Abdulaziz City
for Science and Technology, Riyadh, Kingdom of Saudi Arabia
{obeid,mbensaleh}@kacst.edu.sa

Abstract. Advancements in wireless sensor networks (WSNs) technologies have enabled their introduction in various application fields. A large number of these applications use sensitive data that require securing algorithms. In this paper, we present a comprehensive survey on the most commonly used security techniques in wireless sensor networks. In this survey, we also present the different implementations on numerous platforms used to realize these security algorithms with special attention to power consumption. Based upon our findings, we propose the main characteristics and parts of a new solution to realize a low power wireless sensor node with high level of security.

Keywords: Architecture · Communication system security
Cryptographic protocols · Energy consumption

1 Introduction

Nowadays, the environment around us is becoming increasingly ubiquitous. More and more elements sensing the environment and communicating useful data are being introduced every day. The system of these interconnected elements is called wireless sensor network (WSN), where each element is referred to as a node or a mote. Rapid advancements in electronics and wireless communication technologies have enabled the realization of more complex WSN nodes. WSNs are significantly improved as compared to traditional wired sensor networks and have been considered among the most important technologies of this century [1]. WSN technology has unlimited potential applications including military, medical, environmental, industrial, agricultural, etc. The majority of the aforementioned applications process sensitive data that need suitable security mechanisms to ensure confidentiality and integrity of data.

Several studies have been presented concerning the implementation of various security techniques to protect WSNs. Among the implemented security techniques,

© Springer International Publishing AG, part of Springer Nature 2018
A. Abraham et al. (Eds.): IBICA 2017, AISC 735, pp. 178–199, 2018.
https://doi.org/10.1007/978-3-319-76354-5_17

cryptographic techniques and hash functions play a key role in the majority of security protocols and frameworks, such as advanced encryption standard (AES) [2, 3], and elliptic curve cryptography (ECC) [4, 5]. Realization of these techniques has generally been achieved in software running on the microcontrollers of the nodes, such as TelosB [6, 7] and Mica family [8–10]. However, these solutions suffer from limited battery life and restricted memory capacity.

Considering the abovementioned limitations, there is a need to have a platform that can efficiently achieve these applications. The major challenge in WSNs is, to maximize the battery life of the nodes due to the difficulty of replacing scores of these nodes batteries. Thus, the trend now is toward the design of specialized WSNs platforms that can meet these power and applications-pace constraints.

Many research works in the open literature have recommended the implementation of platforms incorporating accelerators [11, 12]. These platforms have been implemented using application specific integrated circuits (ASICs) [13–17], field programmable gate arrays (FPGAs) [16, 18–21], system-on-a-chip (SoC) and system-on-a-programmable chip (SoPC) [22, 23].

In this paper, we review the popular security techniques used to prevent attacks that can be a threat to the WSNs. In addition, we survey their implementations on several platforms covering the architectural spectrum of these platforms from the microcontroller (μC) to the SoC/SoPC. The objective of this paper is also to propose the main characteristics of a feasible solution to realize a low power wireless sensor node.

The rest of the paper is organized as follows: Sect. 2 presents a comprehensive review of security techniques proposed in the literature. Section 3 discusses the implementations in low power consumption of security techniques studied so far. A detailed analysis of different implementations, algorithms and a suggestion of new solution are made in Sect. 4. Finally, conclusion and future work are presented in the last section.

2 Security in WSNs

2.1 Description

There are specific attacks to WSNs that address more specifically the limited energy of sensors. In fact, WSNs are applied in many critical applications such as health monitoring (collecting vital signs of a patient), building protection, pollution detection, battlefield management and military application (supervision of a war zone, recording condition or the position of the troops). These applications need strong security mechanisms. To guarantee the security of WSNs, several requirements such as availability, authenticity, confidentiality and integrity, should be achieved. These requirements are presented in the next paragraph.

2.2 Security Requirements of WSNs

The security techniques and algorithms used in WSNs should protect the exchanged data between sensor nodes from attacks while respecting the security requirements.

In general, security requirements change with the type of application [24]. The important security requirements for WSNs are presented here.

Availability: The accessibility of data is very important in a secure network. Availability ensures the survivability versus attacks of Denial-of-Service (DoS). DoS attacks can be initiated in whatever layer of wireless sensors networks and may exhaust battery charge of a sensor device with the excessive computation and communication operation.

- Authentication: The malicious or spoofed packets should be detected in WSNs with checking the sender. It is essential for a receiver to have a mechanism to verify that the received packets have indeed come from the actual sender node. Several authentication schemes designed for WSNs have been proposed by researchers. An authentication scheme should ensure that the communicating node is the one that it claims to be. A Message Authentication Code (MAC) is used as a solution in the case of communication between two nodes and the data authentication can be computed from the shared secret key.
- Confidentiality: In WSNs, it is very important to keep the sensitive data collected and transmitted secretly especially in case of critical applications. Data confidentiality guarantees that any unauthorized user or entity can never detect sensitive data and ensures that channels between nodes are secure. Sensor identities and routing information should be confidential to protect a network against traffic analysis.
- Integrity and freshness: To guarantee the integrity of data, it is very important in WSNs to prevent extermination change or unauthorized introduction of data. Data integrity is based on the quality of solidity, accuracy, completeness, and accordance for the promoter of the data [12]. The network functionality can be perturbed by changing messages by malicious node. In WSNs, integrity ensures that transferring message is never corrupted. Replay attack can be eliminated with data freshness by ensuring that the transmitted data is the recent one. In order to minimize the damage of different type of attacks those threat the integrity, authentication, confidentiality, etc., several cryptographic and other techniques are proposed and used in WSNs. A detailed study of these different techniques is presented in the next section.

2.3   Security Mechanisms for WSNs

Many attacks menace WSNs. Hence several research teams have tried to find ideal solutions to secure WSNs taking into account their specificities. The security solutions applied in the conventional ad hoc networks cannot be used in WSNs owing to the limited batteries. In addition, the sensors have a low computing capability that prevents using complex algorithms. In fact, the practical cryptographic solutions used currently, as symmetric and asymmetric algorithms, are too heavy to be calculated by the processors of current sensors.

Nevertheless, cryptographic algorithms are very efficient and used in the most works. In fact, several security protocol and frameworks designed for WSNs are based on symmetric cryptographic techniques such as SAODV [25], LHAP [26], SPIN [27], LEAP [23] and TinySec [28]. Asymmetric cryptographic techniques are also used in protocols and frameworks such as ARAN [28], SEAD [29], SPINS [30], TinyPK [31],

TinyECC [32], TinyPBC [33], Tiny Pairing [34], and SecFleck [35]. In addition, hybrid cryptography techniques are employed in various protocols and frameworks such as MASA [36] and SCUR [37]. Furthermore, hash functions are used in the majority of security protocol as [38, 40] to guarantee the integrity and authenticity of data exchanged in WSNs. Therefore, cryptographic techniques always have a central role in security schemes proposed for WSNs.

To meet the basic security requirements in WSNs, the cryptographic schemes are used while taking into account low cost and low power consumption. Cryptographic schemes provide security in WSNs through symmetric key techniques, asymmetric key techniques, hybrid cryptographic techniques and hash functions. The selection of light-weight cryptographic algorithm is essential in WSNs due to the limitations in the computational and memory capabilities, battery power and different constraints of sensor nodes. We present in the next subsections various cryptographic systems used today to secure WSNs. We start with the symmetric cryptographic techniques, then we discuss the asymmetric cryptographic techniques, hybrid cryptographic techniques, and finally hash functions.

Symmetric cryptographic techniques

In symmetric cryptographic techniques, encryption and decryption are executed with the shared key between two sensors. This key must be kept secret. This can be difficult due to the environment where sensor nodes may be placed. Block ciphers and stream ciphers are used as symmetric cipher types. When a block cipher is used, the clear text will be divided into blocks before being encrypted. A stream cipher can be considered as a block cipher with a block length of 1-bit. Most of the security schemes for WSNs have focused on testing and evaluating cryptographic algorithms in WSNs and suggesting energy efficient ciphers. Asymmetric algorithms take a lot of time in computing operations compared to symmetric key algorithms. So that, the encryption process used in symmetric algorithms is less complicated. We present in this section the symmetric encryption algorithms that are well known and widely used as symmetric ciphers in WSNs. We study then, DES [17], AES [16], RC5 [41] and Skipjack [42].

Asymmetric cryptography techniques

In asymmetric cryptography, a private key and a public key are used. Public key are used to encrypt and verify the signature of data and private key are used to decrypt and sign. The private key needs to be kept confidential while the public key can be published freely. There are various public key algorithms include Rabin's Scheme, Ntru-Encrypt, RSA, ECC, and Identity-Based Cryptography (IBC). RSA, ECC and IBC are among the popular asymmetric cryptography schemes used in most studies related to WSNs.

Hybrid cryptographic techniques

To take advantages of the two approaches; symmetric and asymmetric cryptographic techniques, many researchers apply the combination of the two approaches. Among the popular schemes based on hybrid cryptographic techniques for WSNs, we find ARIADNE [11], DSAA [47] and SCUR [37].

Hash function

Ahash function is a particular function that calculates a footprint from an inputted data to identify the initial data. A good hash function provides a significant change in the footprint in spite of a small change in the original text have been introduced. Usually in WSNs, Message-Digest5 (MD5) and Secure Hash Function 1(SHA-1) are the most popular hash functions [48] but there are also SHA2 and SHA3 [49]. Hash functions are central to compute signatures and MACs which allow users to verify authenticity and integrity of data.

Other functions

In addition to cryptographic techniques, there are also many others techniques to ensure security in WSNs. Most of these techniques are used to eliminate the risk of physical attacks. The other technique used is trust managements on sensor nodes in WSNs. These techniques include the mechanism proposed by [30] against search-based physical attacks. SWATT is a mechanism to control the memory of a sensor node against an abrupt change [50]. Furthermore, there is another approach of applications of trust-based frameworks to enforce level of security in WSNs [52]. Trust-based schemes help to protect against attacks untreated by cryptographic security. The trust-based framework can help to address many issues like evaluating the quality and capacity of sensor nodes [52]. Moreover, to achieve security in WSNs, there are also the intrusion detection systems (IDS) used in many applications to prevent some types of attack. There are two important classes of IDS, the signature-based IDS, and the anomaly-based IDS [53]. Security of routing is also one of the interesting topics of security in WSNs [54]. Therefore, among the basic techniques used to guarantee the WSNs security, we focus in this paper on cryptographic techniques that are applied in the majority of security protocols and frameworks proposed in the literature. These techniques also confirm their efficiency in preventing several types of attacks and achieve security requirements in WSNs. Figure 1 illustrates the different security techniques studied in this paper.

In order to assess the effectiveness of these proposed algorithms, several works implemented these algorithms in different platforms. These implementations will facilitate selecting directions to optimize the cost of different algorithms in low power WSNs. The choice of the type of implementation depends on the complexity of algorithm, memory and energy consumption results. We introduce in the next section the most commonly used implementation of security algorithms in WSNs.

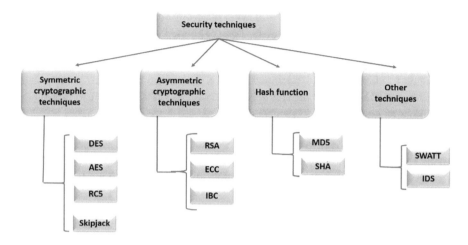

Fig. 1. Security techniques for WSNs

3   Implementation of Security Algorithms and Techniques

Several approaches propose numerous implementations for cryptographic functions using different platforms such as microcontrollers (µCs) [8, 14, 48, 56], Digital signal processors (DSPs) [59], ASICs [9, 22, 32, 62, 63] and SoCs [72]. A study of the implementation of security techniques based on µC is detailed in the following section.

3.1   Implementations of Security Techniques in WSNs Based on µC

During the past few years, the trend is to use wireless sensor nodes based on µC. The majority of security techniques are implemented in these platforms due to their, relatively low cost and acceptable energy consumption.

In [14], Ganesan et al. utilized different embedded processors in order to evaluate the performance of the cryptographic functions with Atmel AVR Atmega 103 and 32-bit Intel XScale. A 440 MHz 64-bit SPARC CPU, operated in 32-bit mode is also used to evaluate the performance of a workstation. The execution times of five cryptographic schemes RC4, RC5, IDEA, SHA-1, and MD5 on the various µC architectures (Atmega 103, Atmega 128, M16C/10, SA-1110, PXA250, and UltraSparc2) are presented. The objective was to find the most efficient cryptographic algorithm. They conclude that RC4 is better than RC5 on encryption in low-end processors. Nonetheless, the energy consumption is not discussed in this work.

In [55], the authors implemented ECC in MICA2 mote. Also, in this implementation, the energy cost has not been taken into account. In the last few years, the energy consumption was considered in different works. They studied the energy cost of using security techniques in a WSNs environment. For example in [44, 45], Mica2 motes (with a CC1000 radio and Ember sensors with an EM2420 radio) are used to measure the consumed energy of hash functions and symmetric-key algorithms. Different subsets of symmetric-key algorithms have likewise been examined.

Jinwala et al. had studied in [46] the performance of AES and corrected block tiny encryption algorithm (XXTEA) when performing it on Mica2. By comparing these algorithms to Skipjack, they concluded that XXTEA is the best security combination for WSNs. AVRORA simulator was used in the experiments. The energy consumption, the throughput and the CPU cycles was provided.

In 2006, Passing et al., made the experiments of some cryptographic algorithms such as MD5, SHA-1 and AES to examine the runtime behavior of these algorithms [48]. Typical sensors were employed. They demonstrated that MD5 outperforms SHA-1 in different sizes hashing tables. The implementations were accomplished to compare the time performance of cryptographic algorithms. The authors found that the amount of data and the time required are linearly dependent.

In 2007, Choi and Song [56] implemented various cryptographic algorithms such as AES, Blowfish, DES, IDEA, MD5, RC4, RC5, SEED, SHA-1 and SHA-256 based on MICAZ platform running TinyOS, to study their feasibility in WSNs. For each cryptographic algorithm, memory, computation time and power were experimentally analyzed. As a result, they conclude that RC4 and MD5 are the most suitable algorithms for MicaZ-type motes.

In [57], the authors had recourse to an 8-bit µC platform (Atmel ATmega128 processor) to quantify the energy cost of authentication and key exchange based on RSA and ECC. The Berkeley/Crossbow motes platform and specifically the Mica2dots are used for the experience setup. They demonstrated that RSA and ECC are viable for WSNs.

In [7], Piotrowski and Peter, implemented RSA and ECC operations in MICA2DOT, MICA2, MicaZ and TelosB to estimate the power consumption of the signature generation and verification operations, as well as the concerned transmissions. The result in [7] shows that public key cryptography is feasible for WSNs.

Szczechowiak et al., discussed the results of ECC and Pairing-Based Cryptography (PBC) on two sensor nodes MICA2 (8bit/7.3828-MHz ATmega128L) and Tmote Sky (16-bit/8.192-MHz MSP-430) [58].

In [59], the authors studied RC5, AES and Skipjack cryptographic techniques. They compared the memory consumption of these algorithms in both MicaZ and TelosB platforms (Table 1). The memory consumption is an essential factor to choose the best cryptographic techniques.

Table 1. Memory consumption (RC5, AES and Skipjack) [16].

Encryption algorithm	MICAZ		TELOSB	
	RAM (KB)	ROM (KB)	RAM (KB)	ROM (KB)
RC5 [16]	0.2	2.5	0.2	6
AES [22]	2	10	1.8	9
Skipjack [41]	0.6	10	0.04	7.5

Table 1 summarizes that RC5 consume very little memory in both MicaZ and TelosB platform. The memory efficiency of Skipjack is better than AES but worse than RC5.

Should be noted that memory must be in efficient way. Indeed, the memory consumption has an effect on energy. According to [42], Skipjack and RC5 are much faster than AES. Nevertheless, Skipjack is not as flexible as AES or RC5 [42] because all of its parameters are constants.

In [60], the authors have studied energy consumption and memory requirements in MicaZ and TelosB motes of four cryptographic techniques AES, RC5, Skipjack, and XXTEA. The influence of the different algorithm parameters, especially the key size, on the energy consumption was analyzed. Further, they confirmed that a number of block cipher modes of operations are suitable for WSNs applications.

To measure the energy consumption of the algorithms in MicaZ and TelosB sensor, the authors used Tektronix MSO 4034 oscilloscope. They concluded that RC5 has less energy than AES in encryption phase on MicaZ. However, in decryption phase AES consumes energy 10 times more than RC5 on TelosB mote. The authors assert that the most energy-efficient algorithms among the four block ciphers are Skipjack and XXTEA. These two algorithms could be considered the most secure. Table 2 presents energy consumption of AES, Skipjack and XXTEA and Hardware AES encryption for MicaZ and TelosB sensor motes [60].

Table 2. Energy consumption and execution time of AES and Skipjack on Micaz and Telosb sensors motes [60]

		Platform	Key size (bits)	Encryption		Decryption	
				Time (ms)	Energy (μJ)	Time (ms)	Energy (μJ)
Software	Skipjack [41]	MicaZ		0.22	5.52	0.22	5.52
		TelosB		0.35	2.63	0.35	2.63
	XXTEA [60]	MicaZ		1.94	49.55	1.86	47.50
		TelosB		2.34	17.48	2.39	17.85
	AES [22]	TelosB	128	3.77	28.16	43.20	322.70
			192	4.60	34.36	51.00	380.97
			256	5.58	41.68	66.00	493.02
		MicaZ	128	1.53	39.08	3.52	89.9
			192	1.82	46.48	4.25	108.55
			256	2.11	53.89	4.98	127.19
Hardware	AES [22]	MicaZ		0.023	1.83	–	–
		TelosB		0.225	14.30	–	–

Table 2 summarizes the effect of the key size on the energy consumption for AES algorithm. Indeed, the energy consumption increases when the key size increases. They also noted that skipjack consume less energy than AES in both MICAZ and TELOSB sensors nodes.

Pairing is the most computationally intensive function in IBC. Pairing is also used to provide efficient solutions to several long-standing problems in cryptography, such as three-way key exchanges and non-interactive zero-knowledge proof systems. Therefore, many related works focus on implementing the pairing function.

In [33], TinyPBC was proposed. ATmega128L had been used for the implementation. They demonstrated that TinyPBC takes only 5.45 s to compute pairings in comparison to the NanoECC (10.96 s). Memory cost to evaluate the pairing on ATmega128L using TinyPBC was also presented [33].

Aranha et al. in [76] used MICAz sensor node platform (8-bit/7.3828-MHz ATmega128L) to implement Pairings. They demonstrated the efficiency of their proposed implementation NanoPBC (2.54 s) and compared to the implementation in [33].

In this section, the implementation of cryptographic functions based on μC were presented and discussed. The main objectives of these studies were to choose the more energy efficient security algorithm. Nevertheless, minimizing energy consumption in WSNs is very challenging. In fact, several requirements should be considered such as real-time and low-cost. All the system aspects should be carefully and simultaneously analyzed. The use of low-energy accelerators could be suggested to tackle the challenges of minimizing energy consumption. Such solutions had been achieved in several works through different platforms, which include DSP, ASIC, FPGA, SoC and SoPC. Several researches exploited these platforms for implementation of security techniques for WSNs that are reviewed and presented in the next section. We start with works based on DSP implementations.

3.2   DSP Based Implementations of Security Techniques in WSNs

Digital signal processors (DSPs) are the basis for manipulating several signal-processing algorithms. With the lower prices of DSP chips and their density, they could be used as a base for the implementation of security techniques in WSNs with Energy Efficiency. In literature, DSP is not widely used for security techniques implementation. In [59], the TMS320VC5416 Fixed-Point Digital Signal Processors (DSP) was employed to perform ECC and it takes 63.4 ms. Simple signal processing could be achieved directly via μC which include basic DSP operations in their instruction set without resorting to special DSP. Despite of their evolution, DSPs may not meet the requirements of many complex security algorithms that need more performance and low power consumption. One alternative solution is to adopt the ASIC to implement security algorithms. This study is detailed in the next section.

3.3   ASIC Based Implementations of Security Techniques in WSNs

Recently, ASIC implementation is proposed as a solution. An ASIC is tailored to a specific application. Using ASIC helps to minimize power consumption and improves execution time due to the optimization of the number of transistors and clock cycles. However, the cost of the device and the time to market increase. Several security technique implementations based on ASIC had been studied. Among these techniques, ECC, hash functions and IBC are detailed here.

ECC: Implementation Based on ASIC

Hilal et al. presented a study of some implementation of ECC in WSNs in [13]. Table 3 summarizes the occupied chip area, the consumed power and the time performance.

From this table we conclude that the less power consumption (less than 30 µW) was obtained using 0.13 µm CMOS Technology at 500 kHz frequency for one point multiplication. For the same frequency, better time and energy performances are achieved with smaller chip area (6718 gates). A low-power and compact ASIC core for ECC is reported in the majority of references.

Table 3. Ecc implementation based on ASIC.

Ref.	CMOS Tech. (µm)	Chip area (Gates)	Power consumption	Timing performance (ms)
[62]	0.13	18720	Under 400 µW @500 kHz	410.45
[64]	0.35	23000	500 µW @ 68.5 MHz	6.67
[18]	0.13	6718	less than 30 µW @ 500 kHz	115
[65]	0.18	11957	305 µW @ 8 MHz	17
[66]	0.35	10 k to 18 k	NA @ 13.56 MHz	38.8

Hash Function: Implementation Based on ASIC

ASIC has been used as a platform to implement hash function. We summarize in Table 4 below different implementations of several hash functions in High-speed ASIC and low-area ASIC [61].

This table shows the throughput with various clock frequencies. The energy consumption is directly propositional to the clock frequency. In fact, minimizing the energy consumption depends on minimizing throughput and clock frequency. Skein-256, which implemented hash function using UMC 0.18 µm technology, could be chosen since it reduces the energy consumption [32].

IBC: Implementation Based on ASIC

In [79], the implementation of IBC is presented. The design was implemented using VHDL and incorporated in an ASIC [79]. Synthesis and physical synthesis is performed using Synopsys Design Compiler and Physical Compiler tools respectively. The authors presented results for the Tate pairing accelerator (Latency = 0.7 ms, Area = 0.574 mm^2, Energy = 29. 600 NJ).

ASIC platform could be adopted to achieve acceptable solution of implementing cryptographic functions especially in large-scale production. Nevertheless, the rapid evolution of security applications with new standards and services make this platform not effective. For this reason, other alternatives such as FPGA platforms could help to improve performances when implementing security techniques. This is explained in details in the following sections.

Table 4. Implementation of hash functions using ASIC (high-speed ASIC)

Ref.	CMOS Tech.	Hash function	Size (kGates)	Throughput (Mbit/s)	Clock frequency (MHz)
[32]	UMC 0.18 μm	BLAKE256	45.64	2836	170.64
		Grøstl256	58.40	6290	270.27
		Keccak256	56.32	21229	487.80
		Skein256	58.61	1882	73.52
		Grøstl256	14.62	145.9	55.87
		Skein256	12.89	19.8	80
		Grøstl256	135	16254	667
		JH256	80	9134	760
		Keccak256	50	43011	949
		Skein256	50	3558	264
[32]	STM 90 nm	BLAKE256	53	3196	96.15
		Skein256	369	3126	12.21
[15]	UMC 0.13 μm	BLAKE256	43.52	3318	200
		Grøstl256	110	9606	188
		JH256	62.42	4334	391
		Keccak256	47.43	15457	377
		Skein256	40.9	1941	159

3.4   FPGA Implementation of Security Techniques

The time to market and energy consumption are frequently at the top of concerns when implementing security techniques. As the design of ASICs could take more than a year, FPGAs are used to minimize the overall development time. In fact, as ASICs, FPGAs are used to make very specific functions in hardware. The flexibility is one of the main advantages of FPGAs. In addition, these kinds of circuits are characterized by their rapid configuration and the ability of reconfiguration for specific feature. Thus, FPGAs are specialized chips that execute complex operations and functions while maintaining the high level of performance. Therefore, FPGAs have been used in several security techniques implementations in WSNs that will be detailed in next subsections.

Elliptic Curve Cryptography: Implementation Based on FPGA

FPGAs are employed to accelerate arithmetic operations during modular multiplications and reductions in the Galois Field (GF) [67] that is the case of ECC [4, 5, 68]. In [68], the authors present an implementation of ECC in WSNs. They describe the possibility to adapt ECC parameters for increasing or reducing the security level according to the application scenario or the energy [68].

The authors implemented the scalar point multiplication considered as the fundamental operation of the ECC. This function could be used in encryption/decryption (ECIES), in digital signature (ECDSA) and in key establishment (ECDH) protocols.

To implement ECC, a mixed solution has been used including a Xilinx XC3S200 Spartan-3 FPGA and an 8052 compliant μC with an additional XC2V2000 Virtex-2 FPGA attached to the custom platform [68].

The authors demonstrated that a combination of a μC and FPGA is much faster and energy efficient than software solution. In fact, they explain that the implementation based on FPGA is three times better than the implementations based on low-power μC [68].

Hash Function: Implementation Based on FPGA

Numerous implementations of hash functions have been proposed for sensor nodes based on FPGAs. Several authors have proposed FPGA implementations of SHA256 and SHA3 hash functions. In [69], the authors proposed a secure protocol AP to secure the communication in WSNs based on the HMAC algorithm. The goal of their work was the implementation of SAH-256 hash function using FPGA. The number of operation and application of hash function was minimized in the HMAC equation in order to reduce its computational complexity.

The architecture of the hash function implementation in FPGA is based on three main hardware components i.e., Control unit, Memory unit and SHA-256 processing unit. The proposed SHA-256 core is described using VHDL code, with structural architecture logic. It is synthesized, placed and routed on Xilinx Virtex-5 XC5VLX50T FPGA using Xilinx ISE 11.4 tool. They used a block RAM memory to implement the SHA-256 ROM memory instead of using the slices available in the FPGA. The synthesis results show that the implementation utilizes only 6% of the total FPGA slices.

In [49], Beuchat et al. describe the compact FPGA implementation of the SHA-3 hash function. They described the compact coprocessor based on 8-bit data path. They detailed the steps of implementation: Sub Bytes, Mix Columns, and BIG.MixColumns and AddRoundKey steps. VHDL was used to describe the architectures and Virtex-5 FPGA was employed for prototyping the coprocessors. SHA-3 was compared with other candidates implemented on Virtex-5 FPGAs. The main properties of implementations of those SHA-3 candidates are summarized in [50]. They describe results of high-speed and low-area implementations for FPGAs. For each work, the implementation details, the technology used, the throughput and the clock frequency are presented. Table 5 summarizes several implementations of hash function based on FPGA.

From Table 5, we conclude the feasibility of implementing hash functions in FPGAs. The energy consumption depends on throughput and clock frequency. We note that Virtex-5 could be a platform that reduces energy consumption with the BLAKE-256 hash function candidate.

AES: Implementation Based on FPGA

In [71], Mohda et al. implemented a crypto-processor in hardware platform. The main elements of this secure platform were an external memory and a transceiver. The communication between WSNs components on board and the traffic in the wireless medium were also integrated. Modelsim digital simulator and functional verification tools were used to code the algorithms using VHDL. The described platform did not implicate an operating system when providing hardware-based key generation, encryption, and stocking. The crypto-processor designed was based on AES algorithm. Synthesis results for the Spartan-6 FPGA were utilized to compute the energy and time for the crypto-processor architecture. According to the authors, the proposed design is

Table 5. Implementation of hash functions in FPGA.

Ref.	FPGA technology	Hash function	Size (slices)	Throughput (Mbit/s)	Clock (MHz)
[9]	Stratix III	BLAKE256	5435	1562	46.97
[70]	Virtex-5	BLAKE256	1660	1911	115
		Grøstl256	4057	5171	101
		Skein256	854	1482	115
		BLAKE512	3064	3080	99.7
		Grøstl256	1597	7885	323.4
		Grøstl512	3138	10314	292.1
		JH256	1018	4578	380.8
		JH512	1104	4742	394.5
		Keccak256	1272	12817	282.7
		Keccak512	1257	6845	285.2
		Skein512	1621	3178	118.0
[22]	Virtex-5	BLAKE256	1118	835	118.06
		BLAKE512	1718	1137	90.91
		Grøstl256	2391	3242	101.32
		Grøstl512	4845	3619	123.4
		JH	1291	1641	250.13
		Keccak224	1117	5915	189
		BLAKE256	1660	1911	115
[10]	Virtex-5	Grøstl256	2616	7885	154
		JH256	2661	2231	201
		Keccak256	1433	8397	205
		Skein256	854	1402	115

efficient in terms of computation time and energy consumption. The computation results indicate that the design consumed only 53% of the device resources.

Pairing: Implementation Based on FPGA

Pairing-based cryptography (PBC) had been applied in FPGA platform. Due to the complexity of the pairing computation, many researchers proposed pairing coprocessor. In [77], the pairing coprocessor isimplemented on a Xilinx Virtex-6 XC6VLX240T-1 FPGA, which embeds 2518 DSP slices. Theauthors compared their work to several implementations of pairings at around 128-bit security. An example of comparison between the work in [77] and another one [80] is exhibited in Table 6.

Table 6. Performance comparison of the implementations of pairings at around 128-bit security

Design	Security (bit)	Platform	Algorithm	Area	Freq. (MHz)	Delay (ms)
[77]	126	Virtex-6	RNS (parallel)	5237 slices 64 DSPs	210	0.338
	128					0.358
[80]	126	Virtex-6	RNS (parallel)	7032 slices 32 DSPs	250	0.573
[80]	126	Stratix III	RNS (parallel)	4233 ALMs 72 DSPs	165	1.07

3.5 SoC/SoPC Based Implementation of Security Techniques

Microelectronics has recently evolved considerably thanks to higher integration levels on the same silicon chip. SoC benefited with other integrated circuits of this evolution. In fact, a SoC consist of one or multiple processors with other hardware subsystems. Various complex security techniques are implemented in the literature based on SoC such as [72, 73].

DES: Implementation Based on SoC

In [72], DES Algorithm had been used to secure data transmission and reception. Authors implemented this technique on Altera NIOS II embedded soft-core processor. An RTL view of this processor was presented.

RC5: Implementation Based on SoC

In [74], the authors designed and implemented a security coprocessor based on RC5 algorithm with 128-bit key and initialization vector for encryption and decryption. The RC5-FKM coprocessor was integrated on a SoC. A finger print-based key management algorithm was implemented in the SoC to build secret keys for cryptographic coprocessor. RC5-FKM cryptographic coprocessor was compared to Atmega128 processor and other AES coprocessors through real experiences to prove the efficiency of their proposal. Xilinx Spartan-3E FPGA was used as test platform. The comparison results are presented in Table 7.

Table 7. Comparison of encryption efficiency.

	Key expansion time (µs)	Encryption time (µs)	Decryption time (µs)	Total time (µs)
Atmega128 [23]	17037	1450	1528.5	20015.5
AES coprocessor [23]	22.1	16.5	16.5	55.1
EasiSOC [23]	30.7	4.2	4.3	39.2

RSA: implementation Based on SoC

In [75], RSA was implemented with different methods. The comparison results indicated that time complexity and power consumption of RSA-1024 could be ameliorated (reduced up to 30 times) when using hardware technology. In fact, time demands and power consumption could be reduced respectively up to 88× and up to 70× compared to software implementation (case of 2048-bit). It should be noted that, authors did not mention any details about the used platform or the applied architecture. Only implementation results were presented.

As described above, Table 8 confirmed that hardware implementation is more efficient than software one. In [30], Yusnani et al. developed a sensor node platform using ARM11. Security parameter was implemented in a single chip using ARM11 trust zone feature. RSA was saved in the On-chip SoC memory thus prevented from hack and lab attacks.

Table 8. Time complexity and power consumption for RSA algorithm.

Key length	Implementation	Time (s)	Power (mW)
1024	RSA software	22.03	726.99
	RSA hardware	0.75	27.15
2048	RSA software	166.85	5506.05
	RSA hardware	1.89	79.09

Pairing: Implementation Based on SoC

Sharif et al., in [78] proposed hardware-software co-design for Pairing-Based Crypto-systems. Xilinx Zynq-7000 SoC was used. It integrates a dual-core ARM Cortex-A9 processing system with a 28 nm Xilinx programmable logic (equivalent to Artix-7 FPGA). They made use of RELIC library for software implementation. The Orup-Suzuki Montgomery multiplier based on DSP units of modern FPGAs had been applied to the hardware part. They used DMA to exchange data between the ARM processing system and the hardware accelerator. They measured execution time with a hardware counter implemented in reconfigurable logic. The Zynq Evaluation and Development Board, ZedBoard, has been used to perform experimental test. Combining multi-core and coprocessor technology helps to reduce energy consumption for security in the embedded area. The recourse to hardware implementations allows improving and accelerating security algorithms and adapting the dynamical behavior of these applications.

The different implementations of security algorithms studied in this paper are presented in Fig. 2 below:

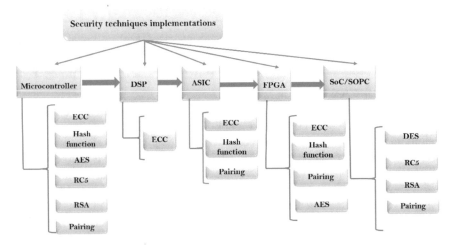

Fig. 2. Implementations of security algorithms for WSNs

4 Analysis

Different implementations of security techniques for WSNs using μC, DSP, FPGA, ASIC and SoC/SoPC platforms have been studied in this paper. These implementations represent different tradeoffs aimed at optimizing time to market, real-time operation and energy consumption. Moreover, optimization on the algorithmic level has been addressed as well. The choice for any particular platform and algorithmic techniques must be well studied. In most cases, the choice does not depend on a single requirement, but rather it is the combination of several factors. Choosing a mixed architecture platform is often an optimal solution to maximize the battery life of the WSNs node.

Typically, μC based platforms (Mica family, iMote, TelosB, TMote, Wasp mote) are preferred choices for security applications. However, μC based platforms suffer from limitations in memory and computational capacity. Therefore, they would not be well suited for cryptographic applications especially asymmetric cryptographic algorithms. To remedy the aforementioned problem, [59] proposed a mixed μC/DSP based platform. Nonetheless, the complexity of modern cryptographic algorithms requires more computational capacity and low power consideration. ASIC based platforms represent an attractive solution when low power consumption, high performance and lower cost for mass volume production are needed. On the other hand, although ASICs are very efficient, they suffer from rigidity and poor adaptation and upgrade capabilities. Reconfigurable technologies present a solution bridging flexibility, cost, power consumption and performance gaps. In this context, [49, 50, 68, 69] have chosen to utilize FPGA supported platforms. However, taking the above concept one-step further it may be of greater value to consider a SoPC solution to overcome the shortcomings of the solutions discussed previously. In addition to the platforms, the choice of security technique is an essential issue. In fact, in this study, several works have done many comparisons between security techniques. Several elements affect the energy consumption such as the key size and the change of platforms. Security algorithms are tested in different aspects in order to choose the most suitable security algorithms for sensor node.

In conclusion, security in WSNs is a grouping of efficient security algorithms that achieve security requirements and efficient platforms which ensure low power consumption. Therefore, software/hardware implementations are one of the best solutions and exceed the software applications that are less secure and less efficient in terms of time and energy.

In Table 9, we compare a number of proposed WSNs solutions showing platforms, energy and algorithms to ensure security in WSNs according to our study made in previous sections. The platform presents the different types of implementation done for such algorithm in software or hardware/software. The energy column indicates that there are proposals for energy consumption for each algorithm. The "+" means that there are works which have computed the energy and evaluated the degree of security for each algorithm.

Table 9. Analysis.

Algorithms	Platforms		Energy	
	Software	Hardware	Software	Hardware
AES	Mica family, TelosB	FPGA, ASIC	+	+
RC5	Mica family, TelosB	SOC	+	+
Skipjack	Mica family, TelosB	–	+	+
RSA	Mica family, TelosB	SOC	+	+
ECC	Mica family, TelosB	ASIC, FPGA, DSP	+	+
Hash Function	Mica family, TelosB	ASIC, FPGA	+	+
DES	Mica family, TelosB	SOC	–	–
MD5	Mica family, TelosB	–	–	–
XXTEA	Mica family, TelosB	–	+	+
IBC	Mica family, TmoteSky	ASIC, FPGA, SOC	+	+

5   Conclusion

WSNs have emerged in everyday life. A secure and efficient energy system to monitor sensitive data has become an important issue. The majority of related works presented in this paper aim to reduce energy consumption of sensor networks and ensure their security. Therefore, they take into account the low consumption in order to have the possibility to implement a complex algorithm and increase the lifetime of the sensors nodes. The choice of security solution depends on the achievement of security requirements. The intended solution should consume low energy while maintaining a high level of security. In this paper, a survey of several studies based on µC, DSP, ASIC and FPGA had been reviewed. Our study has shown that software/ hardware solutions could not only improve time efficiency and security but also decrease power and energy consumption of sensor nodes in case of using strong cryptography.

In fact, there are many solution based on µC, but the commercial nodes run at low frequency and they can be overloaded due to the considerable amount of data used in applications, thus depleting the battery. Therefore, there are many proposed implementations in ASIC. Despite the amelioration in energy consumption with ASIC implementation, it is still a very costly solution. The implementation based on FPGA is also used as a solution to reduce energy consumption and many results in related works confirm the amelioration. However, this solution offers a high-speed encryption and generates a key with a higher price and a higher energy consumption. Yet, for the important of several applications such as commercial, medical and military applications, software/hardware implementations is still the most used in literature. Consequently, the vital components of a designed system on a single chip is proposed in this paper to allow the execution of the entire complex security algorithms. This solution guarantees security in WSNs with minimum energy consumption. As future work, we will develop the different parts of the proposed solution.

References

1. Culler, D., Estrin, D., Srivastava, M.: Overview of sensor networks. Computer **37**, 41–49 (2004)
2. Drimer, S., Guneysu, T., Paar, C.: DSPs, BRAMs and a pinch of logic: new recipes for AES on FPGAs. In: Proceedings of the 2008 16th International Symposium on Field-Programmable Custom Computing Machines, Stanford, CA, USA, 14–15 April 2008, pp. 99–108 (2008)
3. Hamalainen, P., Hannikainen, M., Hamalainen, T.D.: Review of hardware architectures for advanced encryption standard implementations considering wireless sensor networks. In: Proceedings of the 7th International Conference on Embedded Computer Systems: Architectures, Modeling, and Simulation, Samos, Greece, 16–19 July 2007, pp. 443–453 (2007)
4. Guneysu, T., Paar, C.: Ultra high performance ECC over NIST primes on commercial FPGAs. In: Proceeding of the 10th International Workshop on Cryptographic Hardware and Embedded Systems, Washington, DC, USA, 10–13 August 2008, pp. 62–78 (2008)
5. Peter, S., Stecklina, O., Portilla, J., de la Torre, E., Langendoerfer, P., Riesgo, T.: Reconfiguring crypto hardware accelerators on wireless sensor nodes. In: Proceedings of the 6th Annual IEEE Communications Society Conference on Sensor, Mesh and Ad Hoc Communications and Networks Workshops, Roma, Italy, 22–26 June 2009, pp. 1–3 (2009)
6. Healy, M., Newe, T., Lewis, E.: Analysis of hardware encryption versus software encryption on wireless sensor network motes. In: Smart Sensors and Sensing Technology, vol. 20, pp. 3–14. Springer, Heidelberg (2008)
7. Piotrowski, K., Peter, S.: How public key cryptography influences wireless sensor node lifetime. In: Proceedings of the 4th ACM Workshop on Security of Ad Hoc and Sensor Networks, New York, NY, USA, pp. 169–176 (2006)
8. Sharma, G., Bala, S., Verma, A.K.: Security frameworks for wireless sensor networks-review. In: 2nd International Conference on Communication, Computing and Security, Rourkela, pp. 978–987 (2012)
9. Namin, A.H., Hasan, M.A.: Hardware implementation of the compression function for selected SHA-3 candidate. In: 2010 IEEE International Symposium on Parallel & Distributed Processing, Workshops and Ph.D. Forum (IPDPSW), Atlanta, GA, 19–23 April 2010, pp. 1–4. IEEE (2010)
10. Matsuo, S., Knezevic, M., Schaumont, P., Verbauwhede, I., Satoh, A., Sakiyama, K., Ota, K.: How can we conduct fair and consistent hardware evaluation for SHA-3 candidate. In: Second SHA-3 Candidate Conference, University of California, Santa Barbara, 23–24 August 2010
11. Hu, Y., Perrig, A., Johnson, D.: Ariadne: a secure on-demand routing for ad hoc networks. In: Proceedings of the 8th Annual International Conference on Mobile Computing and Networking, Atlanta, 23–28 September 2002, pp. 12–23 (2002)
12. Daniluka, K., Szynkiewicza, E.N.: A survey of energy efficient security architectures and protocols for wireless sensor networks. J. Telecommun. Inf. Technol. **3**, 64–72 (2012)
13. Houssain, H., Badra, M., Al-Somani, T.F.: Comparative study of elliptic curve cryptography hardware implementations in wireless sensor networks. Int. J. RFID Secur. Cryptogr. (IJRFIDSC), **2**(1) (2013)
14. Ganesan, P., Venugopalan, R., Peddabachagari, P., Dean, A., Mueller, F., Sichitiu, M.: Analyzing and modeling encryption overhead for sensor network nodes. In: 2nd ACM Wireless Sensor Networks and Applications, pp. 151–159. ACM Press, New York (2003)

15. Guo, X., Huang, S., Nazhandali, L., Schaumont, P.: Fair and comprehensive performance evaluation of 14 second round SHA-3 ASIC implementations. In: Second SHA-3 Candidate Conference. University of California, Santa Barbara, 23–24 August 2010
16. Kaps, J.P., Sunar, B.: Energy comparison of AES and SHA-1 for ubiquitous computing. In: Proceedings of the 2006 International Conference on Emerging Directions in Embedded and Ubiquitous Computing, Korea, 1–4 August 2006, pp. 372–381 (2006)
17. Fournel, N., Minier, M., Ubeda, S.: Survey and benchmark of stream ciphers for wireless sensor networks. In: Sauveron, D., Markantenorkis, K., Bilas, A., Quisquater, J.-J. (eds.) Information Security Theory and Practices. Smart Cards, Mobile and Ubiquitous Computing Systems, pp. 202–214. Springer, Heidelberg (2007)
18. Batina, L., Mentens, N., Sakiyama, K., Preneel, B., Verbauwhede, I.: Low-cost elliptic curve cryptography for wireless sensor networks. In: Proceeding of the Third European Conference on Security and Privacy in Ad-Hoc and Sensor Networks, Chicago, 17–21 September 2006, pp. 6–17 (2006)
19. Koo, W.K., Lee, H., Kim, Y.H., Lee, D.: Implementation and analysis of new lightweight cryptographic algorithm suitable for wireless sensor networks. In: International Conference on Information Security and Assurance, Busan, 24–26 April, pp. 73–76. IEEE (2008)
20. Rivest, R.L.: The RC5 encryption algorithm. In: Fast Software Encryption, Bart Preneel, vol. 1008, pp. 86–96. Springer, Heidelberg (1995)
21. Uluagac, A.S., Lee, C.P., Beyah, R.A., Copeland, J.A.: Designing secure protocols for wireless sensor networks. In: Li, Y., Huynh, D.T., Das, S.K., Du, D.Z. (eds.) Wireless Algorithms, Systems, and Applications, vol. 5258, pp. 503–514. Springer, Heidelberg (2008)
22. Baldwin, B., Hanley, N., Hamilton, M., Lu, L., Byrne, A., O'Neill, M., Marnane, W.P.: FPGA implementations of the round two SHA-3 candidates. In: Second SHA-3 Candidate Conference, University of California, Santa Barbara, 23–24 August 2010
23. Zhu, S., Setia, S., Jajodia, S.: LEAP: efficient security mechanisms for large-scale distributed sensor networks. ACM Trans. Sens. Netw. **2**, 500–528 (2003)
24. Daidone, R., Dini, G., Tiloca, M.: STaR: a reconfigurable and transparent middleware for WSNS security. In: Proceedings of 4th International Workshop on Networks of Cooperating Objects for Smart Cities, 8 April 2013, Philadelphia, USA, pp. 73–88 (2013)
25. Zapata, M.: Secure ad hoc on-demand distance vector (SAODV). Mob. Comput. Commun. Rev. **6**, 106–107 (2002)
26. Zhu, S., Xu, S., Setia, S., Jajodia, S.: LHAP: a lightweight hop-by-hop authentication protocol for ad-hoc networks. In: Proceeding of the 23rd International Conference on Distributed Computing Systems Workshops, 19–22 May 2003, pp. 749–755. IEEE (2003)
27. Perrig, A., Szewczyk, R., Wen, V., Culler, D., Tygar, J.: SPINS: security protocols for sensor networks. Wirel. Netw. **8**, 521–534 (2002)
28. Karlof, C., Sastry, N., Wagner, D.: TinySec: a link layer security architecture for wireless sensor networks ACM. In: Proceeding of the 2nd International Conference on Embedded Networked Sensor System, 03–05 November 2004, Baltimore, MD, USA, pp. 162–175 (2004)
29. Hu, Y., Johnson, D., Perrig, A.: SEAD: secure efficient distance vector routing in mobile wireless ad-hoc networks. In: Proceedings of the 4th IEEE Workshop on Mobile Computing Systems and Applications, 20–21 June 2002, pp. 3–13. IEEE (2002)
30. Yussoff, Y.M., Hashim, H.: Trusted wireless sensor node platform. In: Proceedings of the World Congress on Engineering, 30 June–2 July 2010, London, U.K., pp. 774–779 (2010)

31. Watro, R., Kong, D., Cuti, S., Gardiner, C., Lynn, C., Kruus, P.: TinyPK: securing sensor networks with public key technology. In: Proceedings of the 2nd ACM Workshop on Security of ad hoc and Sensors Networks, Washington, DC, USA, 25–29 October 2004, pp. 59–64. ACM (2004)
32. Liu, A., Ning, P.: TinyECC: a configurable library for elliptic curve cryptography in wireless sensor networks. In: Proceedings of the International Conference on Information Processing in Sensor Networks, St. Louis, MO, 22–24 April 2008, pp. 245–256. IEEE (2008)
33. Oliveira, L., Scott, M., Lopez, J., Dahab, R.: TinyPBC: pairings for authenticated identity-based non-interactive key distribution in sensor networks. In: Proceedings of the 5th International Conference on Networked Sensing Systems, Kanazawa, 17–19 June 2008, pp. 173–180. IEEE (2008)
34. Xiong, X., Wong, D., Deng, X.: TinyPairing: a fast and lightweight pairing-based cryptographic library for wireless sensor networks. IACSIT Int. J. Eng. Technol. **5**, 320–324 (2013)
35. Hu, W., Corke, P., Shih, W., Overs, L.: secFleck: a public key technology platform for wireless sensor networks. In: Proceedings of the 6th European Conference on Wireless Sensor Networks, Ireland, 11–13 February 2009, pp. 296–311 (2009)
36. Alkaid, H., Alfaraj, M.: MASA: end-to-end data security in sensor networks using a mix of asymmetric and symmetric approaches. In: Proceedings of the 2nd IEEE International Conference on New Technologies, Mobility and Security, 5–7 November 2008, Tangier, pp. 1–5. IEEE (2008)
37. Tahir, R., Jived, M., Ahmad, A., Iqbal, R.: SCUR: secure communications in wireless sensor networks using rabbit. In: Proceedings of the World Congress on Engineering, London, U.K., 2–4 July 2008, pp. 489–493 (2008)
38. Zhang, Y., Liu, W., Lou, W., Fang, Y.: Securing mobile ad hoc networks with certificate less public keys. IEEE Trans. Dependable Secure Comput. **3**, 386–399 (2006)
39. Dener, M.: Security analysis in wireless sensor networks. Int. J. Distrib. Sens. Netw. 9, Article ID 303501 (2014)
40. Clausen, T., Adjih, C., Jacquet, P., Laouiti, A., Muhlethaler, A., Raffo, D.: Securing the OLSR protocol. In: Proceeding of IFIP Med-Hoc-Net, June 2003
41. Kukkurainen, J., Soini, M., Sydanheimo, L.: RC5-based security in wireless sensor networks: utilization and performance. WSEAS Trans. Comput. **9**, 1191–1200 (2010)
42. Mansour, I.: Contribution à la sécurité des communications des réseaux de capteurs sans fil. Doctoral thesis, Université Blaise Pascal - Clermont-Ferrand II, 05 July 2013
43. Knudsena, L., Wagner, D.: On the structure of Skipjack. Discret. Appl. Math. **111**, 103–116 (2001)
44. Shamir, A.: Identity-based cryptosystems and signature schemes. In: Advances in Cryptology, vol. 196, pp. 47–53. Springer, Heidelberg (1985)
45. Boneh, D., Franklin, M.: Identity-based encryption from the weil pairing. SIAM **32**, 586–615 (2003)
46. Tharani, M., Senthilkumar, N.: Integrating wireless sensor networks into internet of things for security. In: Proceedings of the International Conference on Global Innovations in Computing Technology, 6–7 March 2014, India, pp. 467–473 (2014)
47. Jailin, S., Kayalvizhi, R., Vaidehi, V.: Performance analysis of hybrid cryptography for secured data aggregation in wireless sensor networks. In: Proceedings of International Conference on Recent Trends in Information Technology, 3–5 June 2011, Chennai, Tamil Nadu, pp. 307–312. IEEE (2011)

48. Passing, M., Dressler, F.: Experimental performance evaluation of cryptographic algorithm on sensors nodes. In: Proceedings of the IEEE International Conference on Mobile Adhoc and Sensor Systems, pp. 882–887. IEEE (2006)
49. Beuchat, J.L., Okamoto, E., Yamazaki, T.: A compact FPGA implementation of the SHA-3 candidate ECHO. In: IACR Eprint archive (2010)
50. Rivest, R.: The MD5 Message-Digest Algorithm. RFC 1321, April 1992
51. Eastlake, D., Jones, P.: US Secure Hash Algorithm 1 (SHA1). RFC 3174, September 2001
52. Sen, J.: A survey on wireless sensor network security. Int. J. Commun. Netw. Inf. Secur. (IJCNIS) **1**, 55–78 (2009)
53. Alrajeh, N.A., Khan, S., Shams, B.: Intrusion detection systems in wireless sensor networks: a review. Int. J. Distrib. Sens. Netw. (2013)
54. Sharmila, S., Umamaheshwari, G., Ruckshana, M.: Hardware implementation of secure AODV for wireless sensor networks. ICTACT J. Commun. Technol. **1**, 218–229 (2010)
55. Malan, D., Welsh, M., Smith, M.: A public-key infrastructure for key distribution in TinyOS based on elliptic curve cryptography. In: Proceedings of the 1st IEEE International Conference on Sensor and Ad Hoc Communications and Networks, Santa Clara, California (2004)
56. Choi, K., Song, J.: Investigation of feasible cryptographic algorithms for wireless sensor network. In: Proceedings of the ICACT, Phoenix Park, 20–22 February, pp. 1381–1382. IEEE (2006)
57. Wander, A., Gura, N., Eberle, H., Gupta, V., Shantz, S.: Energy analysis of public-key cryptography for wireless sensor networks. In: Proceedings of the 3rd IEEE International Conference on Pervasive Computing and Communication, 8–12 March 2005, pp. 324–328. IEEE (2005)
58. Szczechowiak, P., Oliviera, L., Scott, M., Collier, M., Dahab, R.: NanoECC: testing the limits of elliptic curve cryptography in sensor networks. In: Proceedings of the 5th European Conference on Wireless Sensor Networks, Bologna, Italy, 30 January–1 February 2008, pp. 305–320. Springer, Heidelberg (2008)
59. Malik, M.Y.: Efficient implementation of Elliptic Curve Cryptography using low-power Digital Signal Processor. In: The 12th International Conference on Advanced Communication Technology (ICACT), Phoenix Park, 7–10 February 2010, pp. 1464–1468. IEEE (2010)
60. Lee, J., Kapitanova, K., Sonb, S.H.: The price of security in wireless sensor networks. Int. J. Comput. Telecommun. Netw. **54**, 2967–2978 (2010)
61. http://ehash.iaik.tugraz.at/wiki/SHA-3_Hardware_Implementations
62. Ozturk, E., Sunar, B., Savas, E.: Low-power elliptic curve cryptography using scaled modular arithmetic. In: Proceedings of 6th International Workshop on Cryptographic Hardware in Embedded Systems (CHES), pp. 92–106 (2004)
63. Gaubatz, G., Kaps, J.P., Öztürk, E., Sunar, B.: State of the art in ultra-low power public key cryptography for wireless sensor networks. In: Proceedings of the Third IEEE International Conference on Pervasive Computing and Communications Workshops, pp. 146–150 (2005)
64. Wolkerstorfer, J.: Scaling ECC hardware to a minimum. In: ECRYPT Workshop - Cryptographic Advances in Secure Hardware - CRASH 2005, 6–7 September 2005
65. Bertoni, G., Breveglieri, L., Venturi, M.: Power aware design of an elliptic curve coprocessor for 8 bit platforms. In: Proceedings of the Fourth Annual IEEE International Conference on Pervasive Computing and Communications Workshops, Pisa, 13–17 March, pp. 337–341. IEEE (2006)
66. Kumar, S., Paar, C.: Are standards compliant elliptic curve cryptosystems feasible on RFID? In: Proceedings of Workshop on RFID Security, Graz, Austria, July 2006

67. de la Piedra, A., Braeken, A., Touhafi, A.: Sensor systems based on FPGAs and their applications: a survey. Sensors **12**, 12235–12264 (2012)
68. Portilla, J., Otero, A., de la Torre, E., Riesgo, T., Stecklina, O., Peter, S., Langendörfer, P.: Adaptable security in wireless sensor networks by using reconfigurable ECC hardware coprocessors. Int. J. Distrib. Sens. Netw. **6**, 1 (2010)
69. Errandani, A., Abdaoui, A., Doumar, A., Châtelet, E.: Reconfigurable hardware implementation of a simple authentication protocol for a wireless sensor networks platform. Int. Interdisc. J. **16**, 1739 (2013)
70. Kobayashi, K., Ikegami, J., Matsuo, S., Sakiyama, K., Ohta, K.: Evaluation of hardware performance for the SHA-3 candidates using SASEBO-GII. IACR Eprint report 2010/010
71. Mohda, A., Marzib, H., Aslama, N., Phillipsa, W., Robertsona, W.: A secure platform of wireless sensor networks. In: Proceedings of the 2nd International Conference on Ambient Systems, Networks and Technologies (ANT-2011), Canada, pp. 115–122 (2011)
72. Ramachandran, R., Prakash, T.J.J.: Design and implementation of SOC in NIOS-II soft core processor for secured wireless communication. Int. J. Comput. Appl. **53**, 1–5 (2012)
73. Raymond, G.K.: Data Encryption Standard (DES). Federal Information Processing Standards Publication 46-3, 25 October 1999
74. Wang, Y., Lu, S., Cui, L.: Design and implementation of a SoC-based security coprocessor and program protection mechanism for WSNS. In: Proceedings of the International Conference on Wireless Sensor Network, Beijing, 15–17 November 2010, pp. 148–153. IEEE (2010)
75. Alkalbani, A.S., Mantoro, T., Tap, A.O.M.: Comparison between RSA hardware and software implementation for WSNS security schemes. In: International Conference on Information and Communication Technology for the Muslim World (ICT4M), Jakarta, 13–14 December 2010, pp. E84–E89. IEEE (2010)
76. Aranha, D., Lopez, J., Oliveira, L., Dahab, R.: NanoPBC: implementing cryptographic pairings on an 8-bit platform. In: Frutillar, C. (ed.) Conference on Hyperelliptic Curves, Discrete Logarithms, Encryption, etc., CHiLE 2009 (2009)
77. Yao, G.X., Fan, J., Cheung, R.C.C., Verbauwhede, I.: Faster pairing coprocessor architecture. In: Abdalla, M., Lange, T. (eds.) Pairing-Based Cryptography, Pairing 2012, Germany, 16–18 May 2012, vol. 7708, pp. 160–176. Springer, Heidelberg (2012)
78. Sharif, M.U., Rogawski, M., George, K.G.: Hardware-software codesign of pairing-based cryptosystems for optimal performance vs. flexibility trade-off. Cryptographic architectures embedded in reconfigurable devices, France, June 2014
79. McCusker, K., O'Connor, N.E.: Low-energy symmetric key distribution in wireless sensor networks. IEEE Trans. Dependable Secure Comput. **8**, 363–376 (2010)
80. Cheung, R.C.C., Duquesne, S., Fan, J., Guillermin, N., Verbauwhede, I., Yao, G.X.: FPGA implementation of pairings using residue number system and lazy reduction. In: Cryptographic Hardware and Embedded Systems – CHES 2011, Bart Preneel, Tsuyoshi Takagi, Nara, Japan, 28 September–1 October 2011, vol. 6917 (2011)

A Closed Form Expression for the Bit Error Probability for Majority Logic Decoding of CSOC Codes over $\Gamma\Gamma$ Channels

Souad Labghough[1]($^{\boxtimes}$), Fouad Ayoub[2], and Mostafa Belkasmi[1]

[1] ICES Team, ENSIAS, Mohammed V University in Rabat, Rabat, Morocco
slabghough@gmail.com, m.belkasmi@um5s.net.ma
[2] LaREAMA Lab, CRMEF, Kenitra, Morocco
ayoubfouadn@gmail.com

Abstract. In this paper, we derive a closed form expression for the bit error probability for Majority Logic Decoding (MLGD) of convolutional self orthogonal codes (CSOC) over free space optical Gamma-Gamma $\Gamma\Gamma$ channels. We derive firstly a pairwise error probability (PEP) expression which will be used in conjunction with the probability generating function of CSOC codes to evaluate the bit error probability of MLGD decoding. Simulations of our communication system are carried out and confirm the analytical results obtained.

Keywords: Convolutional Self Orthogonal Codes (CSOC)
Free Space Optics (FSO) · Atmospheric turbulence · MLGD decoding
Gamma-Gamma $\Gamma\Gamma$ distribution

1 Introduction

Free-space optics (FSO) is a line-of-sight technology that uses invisible beams of light to provide optical bandwidth connections. It's able of sending and receiving data, voice, and video communications simultaneously through the air. This optical communication technology does not require costly fiber optic cables or securing spectrum licenses as in radio frequency (RF) communication. FSO technology requires only light [1]. Its unique properties make it appealing for a number of applications, including metropolitan area network extensions, enterprise/local area network connectivity, fiber backup, back-haul for wireless cellular networks, redundant link and disaster recovery [1].

Despite these advantages, there are some issues to be taken into account when deploying FSO-based optical wireless system such as: Fog, absorption, scintillation or atmospheric turbulence, which can alter light characteristics or completely obstruct the passage of light through a combination of absorption, diffusion and reflection, these can also provoke the power density (attenuation) of the FSO beam and directly affect the availability of the system, as they can also cause fluctuations in the amplitude of the signal, and therefore the system will experience a performance degradation in terms of the bit error rate (BER) [1, 2].

Error control coding techniques are used over FSO links to improve the bit error rate performance [3]. The objectif of these techniques is to add some redundancy to the original message to protect it against noise, in such a way that it is possible for the receiver using a decoding algorithm to detect the error and correct it. Many applications can use error-correcting codes, such as: Deep-space telecommunications, satellite broadcasting (DVB), data storage, error-correcting memory, satellite communication, wireless Networks LAN/WAN, Mobile telephony standards (3G, LTE ...), etc., [4].

In this work, we study the performance of Majority Logic Decoding (MLGD) for convolutional self orthogonal codes that were proposed by Massey [5] in 1963, and that are characterized by their encoding mechanism that keeps the coded symbols in memory. Furthermore, this coding technique has the advantage of simplicity of implementation, construction of a large number of codes and an ability to function at a very high speed. Similarly, MLGD decoding is a sub-optimum but simple decoding scheme that allows a high-speed implementation [4].

The paper is organized as follows. Section 2, introduces the channel model considered in this paper. In Sect. 3, we give a review on convolutional self orthogonal codes and their Majority Logic Decoding. In Sect. 4, the pairwise error probability (PEP) expression is derived to obtain then the closed form expression for the bit error probability of MLGD decoding. therefore, the Analytical results obtained are confirmed through Monte-Carlo simulation in Sect. 5. Finally, Conclusion is provided in Sect. 6.

2   System Model

Considering an FSO communication system using IM/DD (Intensity Modulation/Direct Detection) as we can see in Fig. 1, the laser beam is propagated along a direct path through Gamma-Gamma turbulence channel damaged by additive white Gaussian noise (AWGN). The received electrical signal of the FSO system is given by:

$$r = \eta Z_s + n \tag{1}$$

where:

- Z_s is the received signal light intensity, which can be write as $Z_s = Zx$ with x is the emitted light intensity.
- Z the channel atmospheric turbulence.
- n is the AWGN noise.
- η denotes the optical-to-electrical conversion coefficient.

Gamma-Gamma $\Gamma\Gamma$ distribution is considered as the most accepted statistical model for describing the effects of atmospheric turbulence, because of its

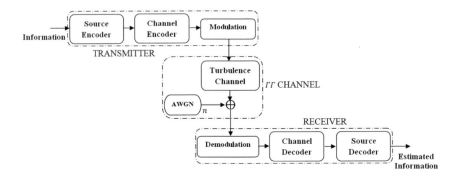

Fig. 1. Simplified diagram of a communication system

excellent agreement between theoretical and measurement data in a wide range of turbulence conditions [6].

The irradiance turbulence of Gamma-Gamma distribution is factored into the product of two independent random processes representing the large-scale and small-scale irradiance fluctuations, denoted by Z_x and Z_y respectively, both follows Gamma distribution [7].

The received irradiance $Z = Z_x Z_y$ follows a Gamma-Gamma distribution [6] with the probability density function (PDF) given by:

$$f_Z(Z) = \frac{2(\alpha\beta)^{\frac{\alpha+\beta}{2}}}{\Gamma(\alpha)\Gamma(\beta)} Z^{\frac{\alpha+\beta}{2}-1} K_{\alpha-\beta}(2\sqrt{\alpha\beta Z}) \tag{2}$$

where:

- Z is the signal intensity,
- $\Gamma()$ is the Gamma function,
- $K_\nu()$ is the modified Bessel function of second kind and of order ν,
- The PDF parameters $\alpha \geq 0$ and $\beta \geq 0$ are the effective numbers of small scale and large scale irradiance of the scattering environment respectively [6], expressed as:

$$\alpha = [exp(\frac{0.49\chi^2}{(1+1.11\chi^{12/5})^{7/6}}) - 1]^{-1} \text{ and } \beta = [exp(\frac{0.51\chi^2}{(1+0.69\chi^{12/5})^{5/6}}) - 1]^{-1}$$

with:

- $\chi^2 = 1.23 C_n{}^2 k^{7/6} L^{11/6}$ is the Rytov variance.
- $k = \frac{2\pi}{\lambda}$ is the wave number with λ the wavelength,
- L is the link distance,
- C_n^2 the index of refraction structure parameter.

3 Convolutional Self Orthogonal Codes (CSOC) and MLGD Decoding

3.1 Convolutional Self Orthogonal Codes (CSOC)

Considering a convolutional self orthogonal code $(n, n-1, m)$ with rate $R = \frac{n-1}{n}$ and with generator polynomials:

$$g_{(i)}^{(n)}(D) = g_{i,0}^{(n)} + g_{i,1}^{(n)}D + ... + g_{i,K_i}^{(n)}D^{K_i}, \quad i = 1, 2, ..., n-1 \tag{3}$$

where n here represents the number of encoder outputs, $(n-1)$ the number of entries, m the code memory that we can write as $m = max_{(1 \preceq i \preceq n-1)}K_i$ and the integers $g_{i,\alpha_1}^{(n)}, g_{i,\alpha_2}^{(n)}, ..., g_{i,\alpha_{J_i}}^{(n)}$ indicate the non zero elements of the generator polynomials.

CSOC codes are systematic self orthogonal codes, where each parity bit is obtained by a combination of $n-1$ systematic bits, this combination is specified by the generator polynomials above. These codes are characterized by a set of J parity check sums orthogonal on the first symbol of the received systematic information [5,8], these orthogonal parity check sums are obtained by syndrome equations alone, and not sums of syndrome equations.

Furthermore, denoting by Δ_i the positive difference set associated with a set of positive integers $\{\alpha_1, \alpha_2, ..., \alpha_{J_i}\}$. An $(n, n-1, m)$ CSOC code is self orthogonal if and only if the positive difference sets $\Delta_1, \Delta_2, ..., \Delta_{n-1}$ don't contain any differences in common, and if all the differences in Δ_i are distinct.

As an example, a CSOC code $(n, n-1, m)$ with $n = 2$, $m = 6$, $\alpha_1 = 0$, $\alpha_2 = 1$, $\alpha_3 = 4$ and $\alpha_4 = 6$ is illustrated in Fig. 2, where u_i and p_i represent respectively the systematic symbol and the parity symbol of the output of this encoder.

At the receiver side, we can express the J parity check sums orthogonal on e_i^u by the following equation:

$$A_{j,i} = e_{i+\alpha_j}^p \oplus e_i^u \oplus \sum_{k=1}^{J} \oplus e_{i+\alpha_j - \alpha_k}^u, \quad j = 1, 2, ..., J, \; i = 0, 1, 2..., N-1 \tag{4}$$

where e_i^u and e_i^p represent respectively the systematic error symbol and parity error symbol and N is the block length of the transmitted information. We can

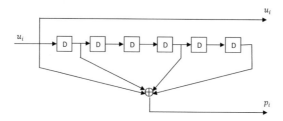

Fig. 2. CSOC code $(2, 1, 6)$

see that all these J parity check sums are orthogonal on e_i^u if in each check sum the error symbol is included, and the other symbols only appear once.

3.2   Majority Logic Decoding (MLGD)

Suppose u_i and p_i are transmitted with BPSK modulation over Gamma Gamma atmospheric turbulence channel.

At the reception, we considered a MLGD decoder without feedback, that accepts as inputs the demodulated received symbols denoted by y_i^u and y_i^p which represent the systematic symbol and the parity symbol respectively.

Algorithm 1. MLGD Decoding

---
for $i = 0, 2, ..., N - 1$ **do**
- Calculate $A_{j,i}$, $j = 1, ..., J$
  if $\sum_{j=1}^{J} A_{ji} > \frac{J}{2}$ **then**
  $\quad \hat{e}_i^u = 1$
  else
  $\quad \hat{e}_i^u = 0$
  end if
end for

As illustrated in Algorithm 1, the MLGD decoder forms at first the J parity equations A_{ji} orthogonal to e_i^u, and then its decision rule according to [5,8] is based on a majority logic such that its decided $e_i^u = 1$ if and only if more than $J/2$ orthogonal parity equations are equal to 1, this means that $e_i^u = 1$ if and only if:

$$\sum_{j=1}^{J} A_{ji} > \frac{J}{2}. \tag{5}$$

4   Bit Error Probability Expression for MLGD Decoding

4.1   Derivation of Pairwise Error Probability (PEP)

The first step in evaluating the bit error probability is to compute the pairwise error probability associated with the transmitted BPSK symbol sequences. Thus, considering an IM/DD link using BPSK modulation and assuming that the noise is modeled as additive white gaussian noise (AWGN), the PEP p which represents the probability of choosing the coded sequence: $\hat{C} = (\hat{c}_1, \hat{c}_2, ...)$ when indeed $C = (c_1, c_2, ...)$ was transmitted [9], is calculated by averaging the gaussian Q-function over the PDF of Z, so, we have to evaluate the integral:

$$p = \int_0^\infty Q(\sqrt{2R\gamma}) f_Z(Z) dZ \tag{6}$$

with: $\gamma = \bar{\gamma} Z^2$

- Z: Is Gamma Gamma distributed in a Gamma Gamma turbulence channel
- γ: The instantaneous SNR
- $\bar{\gamma}$: The average SNR
- R: The code rate

where $Q(.)$ is the gaussian Q function which is related to the complementary error function $erfc(.)$ by:

$$erfc(x) = 2Q(\sqrt{2}x) \tag{7}$$

so:

$$p = \int_0^\infty \frac{1}{2} erfc(\sqrt{R\bar{\gamma}Z^2}) \frac{2(\alpha\beta)^{\frac{\alpha+\beta}{2}}}{\Gamma(\alpha)\Gamma(\beta)} Z^{\frac{\alpha+\beta}{2}-1} K_{\alpha-\beta}(2\sqrt{\alpha\beta Z}) dZ \tag{8}$$

By expressing the $K_v(x)$ and $erfc(.)$ as Meijer G-function respectively according to [10] and ([11], Eq. 07.34.03.0619.01), we obtain Eqs. (9) and (10) below:

$$K_v(x) = \frac{1}{2} \; G_{0,2}^{2,0}\left[\frac{x^2}{4}\,\Big|\,{- \atop \frac{v}{2}, \frac{-v}{2}}\right] \tag{9}$$

$$erfc(\sqrt{x}) = \frac{1}{\sqrt{\pi}} \; G_{1,2}^{2,0}\left[x\,\Big|\,{1 \atop 0, \frac{1}{2}}\right] \tag{10}$$

by substituting Eqs. (9) and (10) in Eq. (8), the integral (8) above becomes:

$$p = \frac{\frac{2(\alpha\beta)^{\frac{\alpha+\beta}{2}}}{\Gamma(\alpha)\Gamma(\beta)}}{4\sqrt{\pi}} \int_0^\infty Z^{\frac{\alpha+\beta}{2}-1} G_{1,2}^{2,0}\left[R\bar{\gamma}Z^2\,\Big|\,{1 \atop 0, \frac{1}{2}}\right] G_{0,2}^{2,0}\left[\alpha\beta Z\,\Big|\,{- \atop \frac{\alpha-\beta}{2}, \frac{\beta-\alpha}{2}}\right] dZ \tag{11}$$

and using Eq. (12) ([11], Eq. 07.34.21.0012.01), this yields a final closed form of the PEP as Eq. (13):

$$\int_0^\infty \tau^{\alpha-1} G_{u,v}^{s,t}\left[\sigma\tau\,\Big|\,{c_1, c_2, ..., c_u \atop d_1, d_2, ..., d_v}\right] G_{p,q}^{m,n}\left[\omega\tau^r\,\Big|\,{a_1, a_2, ..., a_p \atop b_1, b_2, ..., b_q}\right] d\tau = \sigma^{-\alpha}$$

$$H_{p+v,q+u}^{m+t,n+s}\left[\frac{\omega}{\sigma^r}\,\Big|\,{(a_1, 1), ..., (a_n, 1), (1-\alpha-d_1, r), ..., (1-\alpha-d_v, r), (a_{n+1}, 1), ..., (a_p, 1) \atop (b_1, 1), ..., (b_m, 1), (1-\alpha-c_1, r), ..., (1-\alpha-c_u, r), (b_{m+1}, 1), ..., (b_q, 1)}\right], r \succ 0 \tag{12}$$

$$p = \frac{\frac{2(\alpha\beta)^{\frac{\alpha+\beta}{2}}}{\Gamma(\alpha)\Gamma(\beta)}}{4\sqrt{\pi}} (\alpha\beta)^{\frac{-(\alpha+\beta)}{2}} H_{3,2}^{2,2}\left[\frac{R\bar{\gamma}}{(\alpha\beta)^2}\,\Big|\,{-, \alpha, \beta, (1,1) \atop (0,1), (\frac{1}{2}, 1), -, -}\right] \tag{13}$$

where $H(.)$ is the Fox H-function [12], and $G(.)$ the Meijer G-function [11]. Thus, our exact PEP expression Eq. (13), is obtained without simplifications, neither approximations.

4.2   A Closed Form Expression for the Bit Error Probability

For the hard Majority Logic Decoding (MLGD), a probability generating function (PGF) for the random variable $\sum_{k=1}^{J} A_k$ [13], knowing that A_k is the estimate of the parity check sum A_{ki} associated to the symbol u_i, is:

$$g(x) = \prod_{k=1}^{J}(1 - P_k + P_k x) \tag{14}$$

which can be written after development as:

$$g(x) = \sum_{j=1}^{J} g_j x^j \tag{15}$$

with:

- g_j: The probability that the random variable $\sum_{k=1}^{J} A_k$ is equal to j (the exponent of x^j).
- P_k: The probability that an odd number of errors occur in a kth parity check sum A_k.

With this technique, the probability of the first decoding error can be written as:

$$P_{fe} = (1 - p) \sum_{j>T}^{J} g_j + p \sum_{j \geq \bar{T}}^{J} g_j \tag{16}$$

with T is the error correction capability, $\bar{T} = J - T$ and p represents the channel transition probability.

Considering that the error symbols of each check sum are independent [13], which due to the memoryless channels considered, then, the probability generating function may be rewritten as:

$$g(x) = [1 - P + Px]^J \tag{17}$$

where P the probability that an odd number of errors occur in a parity check sum, and expressed as:

$$P = \frac{(1 - (1 - 2p)^\sigma)}{2}, \sigma = J(n - 1) \tag{18}$$

Further, since there is no feedback, each error symbol that is decoded depends only on a given set of neighboring channel error symbols and there is no dependency upon previous decoding decisions. Thus, the overall bit error probability is equal to P_{fe} and is given by:

$$P_b = (1 - p) \sum_{i=T+1}^{J} \binom{J}{i} P^i (1 - P)^{J-i} + p \sum_{i=J-T}^{J} \binom{J}{i} P^i (1 - P)^{J-i} \tag{19}$$

In this equation, the thresholds are $T = J/2$ for J even and $T = (J + 1)/2$ for J odd, with p is our PEP expression calculated previously.

5   Numerical Results

In the following, we consider a convolutionally coded FSO communication system considering the three typical cases of atmospheric turbulence, weak, moderate, and relatively strong turbulence.

As well, analytical results obtained in the previous section are compared with computer simulation results, where we used 200 as the minimum number of residual bit errors and 2000 as the minimum number of transmitted blocks [7], as shown in Table 1.

The exact bit error probability for MLGD decoding of CSOC code (2,1,17) under weak, moderate and strong turbulence of $\Gamma\Gamma$ channel, is illustrated along with the corresponding BER simulation in Fig. 3. The average BER of uncoded FSO system with BPSK modulation and modeled by $\Gamma\Gamma$ channel (Uncoded BPSK) is published in [14], we notice here that we have a performance gain of 22 dB in weak turbulence between the coded and uncoded FSO system at BER of 10^{-5}, while, for moderate turbulence conditions, we obtained a performance gain of 29 dB at the same BER, and finally for strong turbulence coditions and considering the same CSOC code, a coding gain of 37 dB is approximately achieved at BER of 10^{-5}.

As we can see clearly in Fig. 4, the exact bit error probability for MLGD decoding of CSOC codes with $R = 2/3$ under weak turbulence conditions, is ploted with the corresponding BER simulation. This figure shows the effect of the number of orthogonal parity check sums J on the performance of MLGD decoding of CSOC codes, we observe that the performance improvement of BER is great from CSOC code (3,2,13) with $J = 4$ to CSOC code (3,2,130) with

Table 1. Simulation parameters

Parameter	Value
Codes	CSOC(2,1,17), $J = 6$–CSOC(3,2,13), $J = 4$– CSOC(3,2,40), $J = 6$–CSOC(3,2,130), $J = 10$
Code rates	$R = 1/2$, $R = 2/3$
Modulation	BPSK
N (Block length)	100
Channel	Gamma-Gamma $\Gamma\Gamma$
	- Strong turbulence: $\alpha = 2.064$, $\beta = 1.342$
	- Moderate turbulence: $\alpha = 2.296$, $\beta = 1.822$
	- Weak turbulence: $\alpha = 2.902$, $\beta = 2.51$
Simulation method	Monte Carlo
Minimum number of transmitted blocks	2000
Minimum number of residual bit errors	200

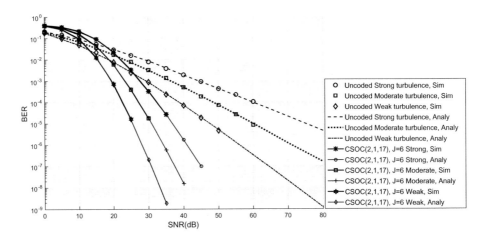

Fig. 3. Exact bit error probability for MLGD decoding of CSOC code (2,1,17) under $\Gamma\Gamma$ atmospheric turbulence channels with BPSK modulation

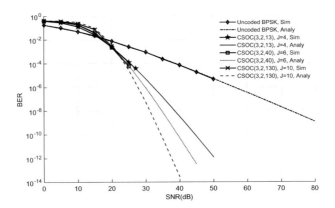

Fig. 4. Exact bit error probability for MLGD decoding of CSOC codes of $R = 2/3$ under weak $\Gamma\Gamma$ atmospheric turbulence channel with BPSK modulation

$J = 10$, and this is due to the minimum distance $d_{min} = J + 1$ and the error correction capability $T = J/2$ which increase both with J. Thus the bit error probability of MLGD decoding decreases if the number J increases.

For the three cases of turbulence we considered, the bit error probability calculated based on the derived PEP gives an excellent result to the true BER, and because of the long simulation time involved during execution, we can present simulation results only up to BER $= 10^{-5}$. Consequently, these simulation results demonstrate an excellent coincidence with the analytical ones.

6   Conclusion

In this paper, the performances of Majority Logic Decoding for convolutional self orthogonal codes, over free space optical Gamma Gamma $\Gamma\Gamma$ atmospheric turbulence channels, employing IM/DD with BPSK modulation are investigated. Moreover, we derived an exact PEP expression for coded FSO links with BPSK modulation and adopting a probability generating function technique associated with our exact PEP expression, we have obtained a closed form expression for the bit error probability for MLGD decoding. Extended simulations confirmed our analytical results through Monte-Carlo simulation. Consequently, comparing the results obtained with the uncoded BPSK system, it is shown that the FSO system using MLGD decoding of convolutional self orthogonal codes in the presence of atmospheric turbulence shows better performance than that of the uncoded signal. Our Future work consists of evaluating the bit error probability for threshold decoding of convolutional self orthogonal codes, and analyzing its performance.

References

1. Khalighi, M.A., Uysal, M.: Survey on free space optical communication: a communication theory perspective. IEEE Commun. Surv. Tutorials **16**, 2231–2258 (2014)
2. Kaushal, H., Kaddoum, G.: Free space optical communication: challenges and mitigation techniques, CoRR, vol. abs/1506.04836 (2015)
3. Djordjevic, I., Ryan, W., Vasic, B.: Coding for Optical Channels, p. 444. Springer, Heidelberg (2010). ISBN 10:1441955682
4. Wu, W.W.: New convolutional codes - part I. IEEE Trans. Commun. **23**, 942–956 (1975)
5. Massey, J.L.: Threshold decoding, Thesis, Massachusetts Institute of Technology, Cambridge, Massachusetts, 5 April 1963
6. Ghassemlooy, Z., Popoola, W., Rajbhandar, S.: Optical Wireless Communication: System and Channel Modelling with MATLAB. CRC Press, Boca Raton (2013). ISBN 13:978-1-4398-5235-4
7. Labghough, S., Ayoub, F., Belkasmi, M.: Majority logic coding schemes performance over FSO channels. In: International Conference on Advanced Communication Systems and Information Security (ACOSIS), pp. 1–6. IEEE, Marrakesh, 17–19 October 2016
8. Lin, S., Costello, D.J.: Error-Control Coding-Fundamentals and Applications, p. 603. Prentice-Hall, Englewood Cliffs (1983)
9. Goldsmith, A.: Wireless Communications. Cambridge University Press, New York (2004). ISBN 0521837162
10. Gradshteyn, I.S., Ryzhik, I.M.: Table of Integrals, Series, and Products, 7th edn. Academic Press, Orlando (2007)
11. Wolfram Research, Mathematica Edition, Version 7.0. Wolfram Research Inc., Champaign (2001–2008)
12. Kilbas, A.A., Saigo, M.: H-transforms, Theory and Applications, p. 408. CRC Press, Boca Raton (2004). ISBN 9780415299169

13. Clark Jr., G.C., Cain, J.B.: Error-Correction Coding for Digital Communications, p. 422. Springer, Heidelberg (1982). Plenum, ISBN 978-1-4899-2174-1
14. Ansari, I.S., Yilmaz, F., Alouini, M.-S.: Performance analysis of FSO links over unified Gamma-Gamma turbulence channels. In: Proceedings of the 81st IEEE VTC Spring, Glasgow, Scotland, pp. 1–8, May 2015. https://doi.org/10.1109/VTCSpring.2015.7145999

Straightforward MAAS to Ensure Interoperability in Heterogeneous Environment

Majda Elhozmari[(✉)] and Ahmed Ettalbi

IMS Team, ADMIR Laboratory, ENSIAS, Rabat IT Center,
Mohammed V University, Rabat, Morocco
elhozmari.majda@gmail.com, ettalbi1000@gmail.com,
a.ettalbi@um5s.net.ma

Abstract. Web services are the key of communication between different applications based on SOA architecture, nowadays, they still the principal key of communication in Cloud Computing and constrained environment. There are two famous Web service protocols; SOAP used in industry or education applications, and REST used to support Media data in web applications. The objective of this paper is to evolve and complete our proposed middleware SaaS solution, to allow communication between SOAP and REST Web services independent of user and environment (Cloud, constrained or on-premise application). The middleware is completed by appending three new components, REST/SOAP component that enable translation from SOAP-based and REST protocol, JSON/XML component used in case of communication between XML and JSON files, and mapping rules storage component.

Keywords: SOAP · REST · Cloud · MAAS · XML · JSON

1 Introduction

In our previous paper [1] we proposed an architecture of middleware SaaS (Software as a Service) to ensure interoperability between different Cloud providers and client in heterogeneous Cloud environment. The focus was on interaction between Cloud provider using REST Web service and client using SOAP. We defined the structure of the architecture and the five principal steps needed to interact automatically between Cloud provider using REST web service and client using SOAP fashion. We defined also how converter makes the mapping from WADL (Web Application Description Language) file to WSDL (Web Application Description Language) file, by presenting the method of conversion.

In this paper, we focus to make our preview architecture [1, 14] more global and flexible to client using JSON scheme, and we add new converter to ensure conversion of request and response messages from REST to SOAP and vice versa. Moreover, we split the architecture in many components to make the solution generalized to solve interoperability problems and be more flexible to maintenance; each component is responsible for managing a specific task.

© Springer International Publishing AG, part of Springer Nature 2018
A. Abraham et al. (Eds.): IBICA 2017, AISC 735, pp. 211–220, 2018.
https://doi.org/10.1007/978-3-319-76354-5_19

The rest of this paper is organized as follows: Sect. 2 provides background of the technology used. Section 3 presents our motivation and related works. In Sect. 4 we present our proposed solution. We conclude this paper by presenting our further works.

2   Background

2.1   REST and WADL

REST [2] is an architecture style particularly adapted to the World Wide Web but is not the only one used. REST is not a protocol such as HTTP (Hypertext Transfer Protocol). Roy Fielding defines these constraints, which can be used in other applications as HTTP protocols. This architectural style is not limited to performing application to a simple user. It is also used to realize SOA (Service Oriented Architecture) using Web services to enable communication between different machines. REST brings some benefits, such as using unique address for every process instance and client can have one generic listener interface for notifications [3].

Web Application Description Language [4] is a textual document, which is sometimes supplemented with some formal specifications such as XML scheme for XML-based data formats or JSON format. It describes the access to the internal data of REST applications. A number of Web-based enterprises such as (Google, Yahoo, Amazon ...) are developing HTTP-based applications that provide programmatic access to their internal data. Typically, these applications are described using WADL files. In few words, WADL provides a machine process description of HTTP-based Web applications.

2.2   SOAP and WSDL

SOAP [3] is a messaging protocol based on XML (eXtended Markup Language) using to exchange information in a decentralized environment. SOAP is also used to establish communication between Web services. It defines a set of message structure rules used in simple way transmissions, but it is particularly useful for performing RPC (Remote Procedure Call) in request and response dialogues. SOAP enables messaging protocols based on XML to exchange information between different applications. It brings some benefits like using a unique address for every operation, increased privacy and so complex operations can be hidden behind facade.

The Web Services Description Language is a language proposed by the World Wide Consortium (W3C) to describe Web Services. WSDL [5] is an XML document that describes different operations that a web service can perform. WSDL has some advantages such as the permission to separate the abstract description of the functionality offered by a Service from the description of details, like the message format or the communication protocol that could be SOAP, HTTP or MIME (Multipurpose Internet Mail Extensions).

2.3 Cloud Computing

NIST [6] defines Cloud Computing as "a model for allowing ubiquitous, convenient; on-demand network access to a shared pool of configurable computing that can be rapidly provisioned and released with minimal management effort or service provider interaction". According to NIST, Cloud Computing has five essential characteristics: on-demand self-service, broad network access, resource pooling, rapid elasticity, measured service. Cloud Computing has a lot of advantages, but also still some pitfalls such as:

- Security and privacy: Public Cloud providers form an attractive target for hackers because of the number of users [7]
- Portability: Many customers are attracted by the public Cloud because of its attractive cost, but recover data may even be impossible. Consumers can find themselves in vendor lock-in situation [7]
- Standardization: Absence of standardization, essentially to relate different Cloud providers. Even if it is possible to provide service for diverse Cloud interfaces through a middleware, there are no rules that can be supervised by Cloud providers [8].

3 Motivation and Related Work

In [1], we have studied the problem of interoperability between Cloud provider using REST web service, and client using SOAP web service, using a simple case study. We have proposed an architecture of middleware solution SaaS between different Cloud Providers and Client in heterogeneous Cloud environment. The focus was on interaction between Cloud Provider using REST Web service and Client using SOAP web service.

We have defined the structure of the architecture, different steps needed to interact automatically between Cloud Providers and Cloud client in heterogeneous web service environment. We have also defined how the converter makes translation from WADL file to WSDL file. We presented also a method of conversion, which consists to parse the WADL file, creates a simple WADL object and WSDL object using mapping rules. Finally, we have made a simple case study to validate our proposition.

The proposition includes request and response conversion from SOAP to REST standards and vice versa, also it withstands XML scheme, but in the case of JSON schema, the messages are not treated.

REST2SOAP [9] is a framework, which integrates SOAP service and RESTful service semi-automatically, using JAVA2WSDL to generate a WSDL file by wrapping RESTful service automatically to make RESTful cross a BPEL-Based composite service. The main weakness of this solution is that the integration is semi-automatic, and the objective of this framework is just wrapping the RESTful service into a SOAP service, not to make an equivalent SOAP file by mapping a RESTful file.

STORHM [10] is a protocol that enables an existing SOAP client to interact with a REST service. The translation is done using a configuration wizard which takes as an input, WSDL file, WSDL schema and optionally a WADL file. The user must interact with a form interface to choose parameters. In addition, the conversion from SOAP to

REST is semi-automatic. The translation is on one sense from SOAP service to RESTful service; in the inverse case, this approach is not useful.

DreamFactory [11] is also a REST API middleware platform wraps a SOAP service into a REST API to make it simple to use and create a REST endpoints, instantly turn any SOAP into a live automatically. This conversion can make a request with JSON, calls the legacy SOAP service and then the SOAP response is converted back to JSON for the client application. The translation from SOAP to REST is automatic but it requires developer interaction to make this conversion available and understands what the mapping methods are talking about and how to use it in the application.

Migration of SOAP to Restful service [12], in this approach, SOAP-based services are converted to RESTful services. The resources are identified from analyzing the WSDL file and mapping the contained operations to resources and HTTP methods. The main limitation of this approach is that the user can just manually validate and modify the resource.

MicroWSMO [13] is used to support dynamic replacement of SOAP and REST services inside a service composition, even in presence of syntactic mismatches between service interfaces. It is based on a running prototype implemented in order to apply this approach to lightweight processes execution. The main limitation of this solution is that the translation is semi-automatic.

Migration of SOAP-based services to RESTful services [12] "An approach generates automatically the configuration files needed to deploy the RESTful services, and wrappers for accessing SOAP based services in the REST architecture" [12]. This is to allow the interaction between SOAP-based services and RESTful services; this approach convert dynamically Messages between SOAP and REST web services. This approach permit the translation from SOAP to REST protocol. It is based on provider side, so; the provider has to have RESTful and WSDL based services to send then to the right client, WSDL to SOAP client and RESTFul to client using REST web service.

4   Proposed MAAS Architecture

Our proposed MaaS as shown in Fig. 1, is an extension to our previous work [1], the objective is to enable our architecture to support not only the translation of WSDL and WADL files, but also the translation of the requests and responses, between REST and SOAP messages. In addition, it involves the transmission between XML and JSON scheme. Moreover, we tried to make it straightforward, by outsourcing mapping rules from the converters, which is unfair and reasonable way to simplify update, and adding new rules, for other standards. To ensure Security and intelligence of our middleware, we combine Client interface and Provider interface from [1] in one Proxy interface orchestrator as shown in Fig. 1.

The architecture is composed of five main components, IOP (Interface Orchestrator Proxy), MRS (Mapping Rules Storage), 2WC (WSDL/WADL to WADL/WSDL converter), 2RC (Request and Response Converter) and XJC (XML/JSON Converter). We explain in detail these components in next subsections.

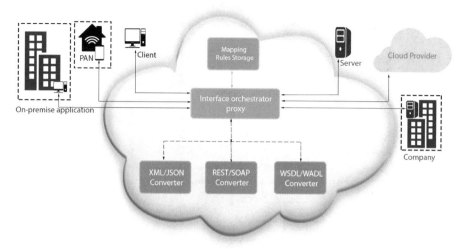

Fig. 1. Global architecture of our proposed MAAS

4.1 Interface Orchestrator Proxy

To make our architecture more intelligent, we merge Client interface and Provider interface into one secure Interface Orchestrator. The main objective of this component is to Orchestrate and manage data transformation and movement. In the case of REST and SOAP interaction, the Interface makes the difference between Invoke, Request, and Response messages, and makes decision to each component should send the message. As mentioned in Fig. 2, in a (or the) case of Invoke message, the Interface determine communication protocol of Provider. Using the collected information about transmitter and receiver, the interface makes sure of type of heterogeneity of environment. In a case of heterogeneous protocols, the interface sends the XML file to 2WC to convert WSDL to WADL file and vice versa, and imports concerning mapping rules from Mapping rules storage. In a case of REST and SOAP Request, it sends the message to 2RC with its appropriate mapping rules, if the scheme file is not the same, interface uses XML to JSON converter.

Figure 2 shows an example of interaction between Client using SOAP web service and REST server, in this case IOP treats the heterogeneity of services and sends the WADL file to translator to have the description in WSDL format. In the case of request and response message, the IOP detects schemas and protocol heterogeneity, so the proxy imports mapping rules of this case and sends the message and mapping rules to 2RC to make transmission as shown in Fig. 3 by step.

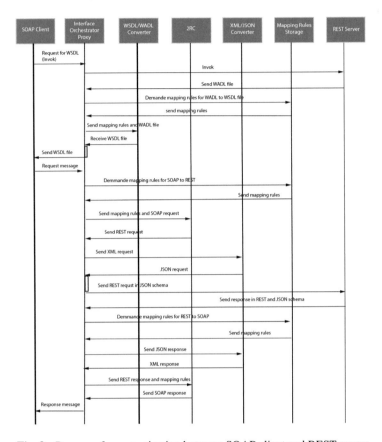

Fig. 2. Process of communication between SOAP client and REST server

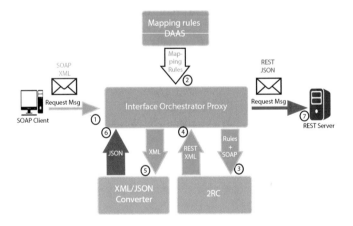

Fig. 3. Request message translation road

4.2 WSDL/WADL Converter

As mentioned in [1] WSDL/WADL Converter, is an internal component of our proposed solution, it ensures standardization and interoperability in heterogeneous environments. Using this component, the receiver using the REST interface can easily communicate with users using the SOAP interface; by converting the WADL file to a WSDL file and vice versa, respecting defined mapping rules sent by the interface. Figure 4 shows an example of a WADL file transformed to WSDL file using mapping rules imported from mapping rule storage.

```xml
<?xml version="1.0"?>
<application xmlns:xsi="http://www.w3.org/2001/XMLSchema-instance"
xsi:schemaLocation="http://research.sun.com/wadl/2006/10 wadl.xsd"
xmlns:tns="urn:yahoo:yn"
xmlns:xsd="http://www.w3.org/2001/XMLSchema"
xmlns:yn="urn:yahoo:yn"
xmlns:ya="urn:yahoo:api"
xmlns="http://research.sun.com/wadl/2006/10"> <resources base=
"http://api.search.yahoo.com/NewsSearchService/V1/">
<resource path="newsSearch">
<method name="GET" id="search">
<request>
<param name="appid" type="xsd:string" style="query" required="true"/>
<param name="query" type="xsd:string" style="query" required="true"/>
<param name="type" style="query" default="all">
</param>
<param name="results" style="query" type="xsd:int" default="10"/>
<param name="start" style="query" type="xsd:int" default="1"/>
<param name="sort" style="query" default="rank">
</param>
<param name="language" style="query" type="xsd:string"/>
</request>
<response>
<representation mediaType="application/xml" element="yn:ResultSet"/>
</response>
</method>
</resource>
</resources>
</application>
```

```xml
<definitions name=" YahooNewsSearchService"
xmlns="http://schemas.xmlsoap.org/wsdl/"
xmlns:soap="http://schemas.xmlsoap.org/wsdl/soap/"
xmlns:tns="http://www.examples.com/wsdl/YahooNewsSearchService.wsdl"
xmlns:xsd="http://www.w3.org/2001/XMLSchema">
<message name="GetNewsSearchRequest">
<part name="appid" type="xsd:string"/>
<part name="query" type="xsd:string"/>
<part name="type" type="xsd:string"/>
<part name="results" type="xsd:string"/> 33 Projet de Fin d'Année 2015-2016
<part name="start" type="xsd:string"/>
<part name="sort" type="xsd:string"/>
<part name="language" type="xsd:string"/>
</message>
<message name="GetNewsSearchResponse">
<part name="NewsSearch" type="xsd:string"/>
</message>
<portType name="YahooNewsSearchServicePortType">
<operation name="GetNewsSearch">
<input message="tns:GetNewsSearchRequest"/>
<output message="tns:GetNewsSearchResponse"/>
</operation>
</portType>
<binding name="YahooNewsSearchServiceBinding" type="tns:wsname_PortType"
<soap:binding style="rpc"
transport="http://schemas.xmlsoap.org/soap/http"/>
<operation name="GetNewsSearch">
<soap:operation soapAction="GetNewsSearch"/>
<input>
<soap:body
encodingStyle="http://schemas.xmlsoap.org/soap/encoding/"
namespace="urn:examples:YahooNewsSearchService"
use="encoded"/>
</input>
<output>
<soap:body
encodingStyle="http://schemas.xmlsoap.org/soap/encoding/"
<output>
<soap:body
encodingStyle="http://schemas.xmlsoap.org/soap/encoding/"
namespace="urn:examples:YahooNewsSearchService"
use="encoded"/>
</output>
</operation>
</binding> 34 Projet de Fin d'Année 2015-2016
```

Fig. 4. Example of conversion from WADL file to WSDL

4.3 XML/JSON Converter

XML/JSON Converter is an internal component used by IOP to make translation between JSON and XML files in the case of heterogeneous schema environment. Figure 3 illustrates in 5 and 6 the translation flow.

4.4   2RC Converter

This component is responsible of translation from SOAP to REST messages and vice versa. As mentioned in Fig. 3, 2RC translates the Request sent by interface from SOAP-base to corresponding REST service operation, and convert back REST response to SOAP-based operation. The 2RC defines the resources and http methods from Request message, and makes sure that the resources are valid by using original WADL or WSDL xml and mapping rules sent by IOP before translation, and generates automatically equivalent message. Concerning the identification of operation (GET, PUT, POST and DELET), and how to identify similar operations, semantic, Input and Output parameters, we are based on [14]. Also for being more effective, the 2RC hides the translation of each request to reuse it in the case of other request.

4.5   Mapping Rules Storage

Mapping rules Storage is Data as service component, in this data the administrator can add, remove or update mapping rules. This component is accessible only by IOP, it sends the mapping rules demanded by IOP. It contains mapping rules concerning WSDL to WADL conversion, and REST to SOAP request and response conversion. The objective of separation between mapping rules and converter is to make our system easy to maintain and more efficient, and ensure mapping rule security.

4.6   Benefits of the Architecture

Our proposed architecture shows the main features and benefits of the whole process. Some of these benefits are:

- **Standardization:** IT environment knows interoperability challenge especially between Cloud Providers because of the miss of standardization. Many of the interfaces offered are unique to a particular vendor, thus raising the risk of vendor lock-in. For that, our proposed MAAS solution enables communication between both of REST and SOAP interfaces by using many components that can make a translation from SOAP to REST interfaces and vice versa and make schemas conversion between JSON and XML. The lack of standardization will not be eliminated definitely from its environment, but our proposed MAAS could minimize this problem but it can be evolved easily using separated components.
- **Environment maintenance:** Every component is independent from other, which help to facilitate maintenance of mapping rules in mapping rule storage and converters.
- **Environment security:** We integrate a Proxy security in interface Orchestrator to make sure of authentication of users for every connection. The independence of component helps to increase the level of security between components, every information turned in MAAS is controlled by Interface Orchestrator Proxy.
- **Management of access user:** Management of users is controlled, and Interface orchestrator proxy governs every data access. Thus, the control of interaction between Cloud providers and consumers is not lost.

- **Availability of data:** In this case, data is available with both technology REST and SOAP thanks to Converter components, which can translate REST message to SOAP message adapted to Cloud consumer and provider, in addition to JSON and XML files.
- **Independent environment:** Contrariwise, other systems like StoRHm and REST2SOAP, our proposed solution brings an automatic conversion from REST to SOAP, without changing or adapting client and Cloud provider side.

5 Conclusion

In this paper, we have proposed a new version of our previous architecture. The objective is to complete the architecture, and make it global and flexible to every client using REST or SOAP web service to communicate with any Cloud Provider or other System using SOAP or REST web service. Also the proposed MAAS tolerate XML and JSON schema files, we also made our MAAS proposition more intelligent and secure using IOP, To ensure easy maintenance of mapping rules, we add a new mapping rule storage component to our previous architecture. The independent translation, generated by convertors, respecting and applying mapping rules defined in mapping rules storage.

Concerning our perspectives, we will make alive our system by implementing the new component, strengthening the security of data, and extending this architecture to support constrained environment.

References

1. Elhozmari, M., Ettalbi, A.: Towards a Cloud Service Standardization to ensure interoperability in heterogeneous Cloud based environment. Int. J. Comput. Sci. Netw. Secur. (IJCSNS) **16**(7), 60–70 (2016)
2. Fielding, R.T.: Architectural styles and the design of network-based software architectures. Ph.D. dissertation, University of California, Irvine (2000)
3. Zur Muehlen, M., Nickerson, J.V., Swenson, K.D.: Developing web services choreography standards—the case of REST vs. SOAP. Decis. Support Syst. **40**(1), 9–29 (2005)
4. Hadley, M.J.: Web application description language (WADL) (2006)
5. World Wide Web Consortium (W3C) Working Draft 3: Web Services Description Language (WSDL) Version 1.2 (2003). http://www.w3.org/TR/wsdl12/
6. Hogan, M., Liu, F., Sokol, A., Tong, J.: Nist Cloud computing standards roadmap, vol. 35. NIST Special Publication, Gaithersburg (2011)
7. Johan Loeckx, G.O.: Cloud computing concept vaporeux ou relle innovation? (2011). http://documentatie.smals.be
8. Sammes, A.: Computer Communications and Networks. Springer, Heidelberg (2014)
9. Peng, Y.-Y., Ma, S.-P., Lee, J.: REST2SOAP: a framework to integrate SOAP services and restful services. In: 2009 IEEE International Conference on Service-Oriented Computing and Applications (SOCA), pp. 1–4. IEEE (2009)
10. Kennedy, S., Stewart, R., Jacob, P., Molloy, O.: STORHM: a protocol adapter for mapping SOAP based web services to restful http format. Electron. Commer. Res. **11**(3), 245–269 (2011)
11. https://www.dreamfactory.com/resources

12. Upadhyaya, B., et al.: Migration of SOAP-based services to RESTful services. In: 2011 13th IEEE International Symposium on Web Systems Evolution (WSE), pp. 105–114. IEEE (2011)
13. De Giorgio, T., Ripa, G., Zuccalà, M.: An approach to enable replacement of SOAP services and REST services in lightweight processes. In: International Conference on Web Engineering. Springer, Berlin (2010)
14. Elhozmari, M., Ettalbi, A.: Using cloud SaaS to ensure interoperability and standardization in heterogeneous Cloud based environment. In: 2015 5th World Congress on Information and Communication Technologies (WICT), pp. 29–34. IEEE (2015)

A Capability Maturity Framework for IT Security Governance in Organizations

Yassine Maleh[1](✉), Abdelkbir Sahid[2], Abdellah Ezzati[1],
and Mustapha Belaissaoui[2]

[1] FST, Univ Hassan 1, B.P. 577, 26000 Settat, Morocco
{yassine.maleh, abdezzati}@uhp.ac.ma
[2] ENCG, Univ Hassan 1, B.P. 577, 26000 Settat, Morocco
{ab.sahid, belaissaoui}@uhp.ac.ma

Abstract. There is a dearth of academic research literature on the practices and commitments of information security governance in organizations. Despite the existence of referential and standards of the security governance, the research literature remains limited regarding the practices of organizations and, on the other hand, the lack of a strategy and practical model to follow in adopting an effective information security governance. This study aims to propose ISMGO a practical maturity framework for the information security governance and management in organizations. The findings will help organizations to assess their capability maturity state and to address the procedural, technical and human aspects of information security governance and management process.

Keywords: IT governance · Maturity · Capability framework
Security metrics · Organization · Use case

1 Introduction

The threat to technology-based information assets is greater today than in the past. The evolution of technology has also reflected in the tools and methods used by those attempting to gain unauthorised access to the data or disrupt business processes [1]. Attacks are inevitable, whatever the organization [2]. However, the degree of sophistication and persistence of these attacks depends on the attractiveness of this organization as a target [3], mainly regarding its role and assets. Today, the threats posed by some misguided individuals have been replaced by international organized criminal groups highly specialized or by foreign states that have the skills, personnel, and tools necessary to conduct secret and sophisticated cyber espionage attacks. These attacks are not only targeted at government entities. In recent years, several large companies have infiltrated, and their data have been "consulted" for several years without their knowledge. In fact, improving cyber security has emerged as one of the top IT priorities across all business lines [4].

To address these concerns, some practice repositories (ITIL, Cobit, CMMi, RiskIT) and international standards (ISO 27000 suite, ISO 15408) now include paragraphs on security governance. The first reports or articles in academic journals that evoke the governance of information security date back to the early 2000s. The proposed

© Springer International Publishing AG, part of Springer Nature 2018
A. Abraham et al. (Eds.): IBICA 2017, AISC 735, pp. 221–233, 2018.
https://doi.org/10.1007/978-3-319-76354-5_20

referential and best practices designed to guide organizations in their IT security governance strategy. However, does not define the practical framework to implement or to measure the organization engagement in term of IS security governance.

The paper is structured as follows. Section 2 presents the previous work on information security governance proposed in the literature. Section 3 describes the proposed capability maturity framework for information security management and governance ISMGO. Section 4 discuss the results of the implementation of the practical maturity framework for the information security governance and management ISMGO through a practical use case. Finally, Sect. 5 presents the conclusion of this work, and gives some limitations.

2 Related Works

In management sciences, several authors put forward the responsibilities and roles of management and in particular of the general management. Schou and Shoemaker [6] find that to provide a greater benefit to the organization, the information security governance can eventually coordinate with strategic approaches to economic intelligence, social responsibility or communication. In [7], Williams outlines the roles of management and the board of directors in the area of information security. Dhillon et al. [8] present the results of an empirical study to understand better the dimensions of IS security governance. For Kryuko et al. [9], the added value and the performance are two crucial elements of information security governance. Klaic et al. [10] discusses the need to define a level of governance in the organization and clarifies the link between that level and security programs. Michael et al. [11] propose in their article value to the Executive by first defining governance as it applied to the information security and the exploration of three specific governance issues. The first inspected how government can be used to the critical aspect of planning for both formal operations and contingency operations. The next issue describes the need for programs measurement and how it can develop an information security assessment and a continuous improvement. Finally, aspects of effective communication between and among the general security and information managers presented. William et al. [12] illustrate the malleability and heterogeneity of information security governance ISG across different organizations involving intra- and inter-organizational trust mechanisms. They identify the need to reframe ISG, adopting the new label information to protecting governance (IPG), to present a more multifaceted vision of the information protection integrating a vast range of technical and social aspects that constitute and are constituted by governance arrangements. The objective of Yaokumah et al. [13] is to assess the levels of implementation of information security governance (ISG) in the main sectors of the Ghanaian industry. The purpose is to compare the implementation of the ISG of the inter-industry sector and to identify areas that may require improvement. In their study, Horne et al. [14] argue for a paradigm shift from internal information protection across the organization with a strategic vision that considers the inter-organizational level. In the recent work Carcary et al. [15], a maturity framework presented help organizations assess their maturity and identify problems. It addresses the technical, procedural and human aspects of information security and provides guidelines for the implementation of information security management and related business processes.

3 Theoretical Framework

3.1 Framework Overview

We propose a global maturity framework to achieve an effective information security management and governance approach, as shown in Fig. 1. The path to security maturity requires a diversified range of layered endpoint protection, management and capabilities, all integrated and fully automated. The only practical and survivable defensive strategy are to move to a more mature security model that incorporate multiple layers of protective technology.

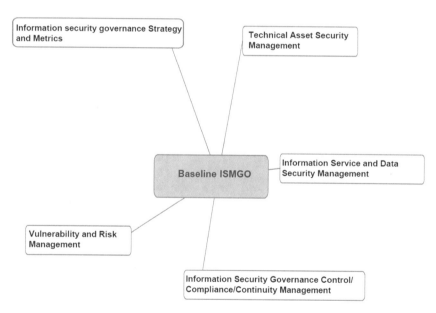

Fig. 1. The proposed maturity framework for information security management and governance in organizations ITSMGO

The ISMGO framework focuses on determining the capacity of an organization to direct oversee and monitor the actions and processes necessary to protect documented and digitised information and information systems and to ensure protection against access, unauthorized use, disclosure, disruption, alteration or destruction, and to guaranty confidentiality, integrity, availability, accessibility and usability of the data [16]. The framework extends the triad confidentiality, integrity and availability of commonly cited with accessibility and usability concepts. Concerning accessibility, a failure to support and understand how security can change work practices can impede how data and information are accessed, shared, and acted on in an increasingly dynamic, competitive environment. Similarly, usability is a one of a main key factor to engaging stakeholders in the business processes, independently of the availability of technology to support work practices, if the technology is difficult to interact and engage with, users might adopt other locally developed, less secure methods of access. The proposed Information Capability Maturity Framework is a comprehensive suite of proven management practices, assessment approaches and improvement strategies covering 5 governance capabilities, 21 objectives and 80 controls.

As Table 1 shows, these high-level function categories are decomposed into 21 security practice objectives (SPOs).

Table 1.

Governance functions	Security practice objective	Description
Information security governance strategy and metrics	Information security strategy and policies	Develop, communicate, and support the organization's information security objectives Establish and maintain security policies and controls, taking into account relevant security standards, regulatory and legislative security requirements, and the organization's security goals
	Strategic alignment of security	From risk analysis to the actual deployment of global policy, security must be aligned with the business priorities of the company while respecting regulatory and legal constraints
	Communication and training	Disseminate security approaches, policies, and other relevant information to develop security awareness and skills
	People roles and responsibilities	Document and define the responsibilities and roles for the security of employees, contractors and users, by the organization's information security strategy
	Security performance assessment	Report on the efficiency of information security policies and activities, and the level of compliance with them
	Assessment of security budget and investments	Provide Security related investment and budget criteria

(continued)

Table 1. (*continued*)

Governance functions	Security practice objective	Description
Technical asset security management	Security architecture	Build security measures into the design of IT solutions—for example, by defining coding protocols, depth of defence, the configuration of security features, and so on
	IT component security	Implement measures to protect all IT components, both physical and virtual, such as client computing devices, servers, networks, storage devices, printers, and smartphones
	Physical infrastructure security	Establish and maintain measures to safeguard the IT physical infrastructure from harm. Threats to be addressed include extremes of temperature, malicious intent, and utility supply disruptions
Information service/system/data security management	Incident management	Manage security-related incidents and near incidents. develop and train incident response teams to identify and limit exposure, manage communications, and coordinate with regulatory bodies as appropriate
	Resource effectiveness	Measure "value for money" from security investments; capture feedback from stakeholders on the effectiveness of security resource management
	Data identification and classifications	Define information security classes, and provide guidance on protection and access control appropriate to each level
	Access management	Manage user access rights to information throughout its life cycle, including granting, denying, and revoking access privileges
	System acquisition, development, and maintenance security policy	Ensure the management of security throughout the life cycle of information systems Reduce risks related to exploiting technical vulnerabilities and applications

(*continued*)

Table 1. (*continued*)

Governance functions	Security practice objective	Description
Vulnerability and risk management	Security threat profiling	Gather intelligence on IT security threats and vulnerabilities to better understand the IT security threat landscape within which the organization operates, including the actors, scenarios, and campaigns that might pose a threat
	Security risk assessment	Identify exposures to security-related risks, and quantify their likelihood and potential impact
	Security risk prioritization	Prioritize information security risks and risk-handling strategies based on residual risks and the organization's risk appetite
	Security monitoring	Manage the ongoing efficacy of information security risk-handling strategies and control options
Information security governance control/compliance/continuity management	Compliance control	Identify applicable law, statutory and contractual obligations that might impact the organization Establish security and compliance baseline and understand per-system risks
	Security testing and auditing	Adopt solution for information security audit Establish project audit practice. Derive test cases from known security requirements
	Business continuity planning	Continuity management Business continuity planning Provide stakeholders throughout the organization with security advice to assist in the analysis of incidents and to ensure that data is secure before, during, and after the execution of the business continuity plan

3.2   Framework Maturity Profile

We propose a mature and systematic approach to information security management and governance. Adopting a security maturity strategy requires a full range of protection, management and defensive features that must be integrated and capable of fully automated operation. Concerning each security practice objectives SPO outlined in Table 1, the framework defines a five-level of maturity that serves as the basis for understanding an organization's ISMGO capability and provides a foundation for capability improvement planning.

Level 0 - None: No process or documentation in place.

Level 1 - Initial: Maturity is characterised by the ad hoc definition of an information security strategy, policies, and standards. Physical environment and IT component security are only locally addressed. There is no explicit consideration of budget requirements for information security activities, and no systematic management of security risks. Access rights and the security of data throughout its life cycle are managed at best using informal procedures. Similarly, security incidents are managed on an ad hoc basis.

Level 2 - Basic: Maturity reflects the linking of a basic information security strategy to business and IT strategies and risk appetite in response to individual needs. It also involves the development and review of information security policies and standards, typically after major incidents. IT component and physical environment security guidelines are emerging. There is some consideration of security budget requirements within IT, and requirements for high-level security features are specified for major software and hardware purchases. A basic risk and vulnerability management process are established within IT according to the perceived risk. The access rights control and management depend on the solutions provided by the provider. Processes for managing the security of data throughout its life cycle are emerging. Major security incidents are tracked and recorded within IT.

Level 3 - Defined: Maturity reflects a detailed information security strategy that's regularly aligned to business and IT strategies and risk appetite across IT and some other business units.

Information security policies and standards are developed and revised based on a defined process and regular feedback. IT and some other business units have agreed-on IT component and physical environment security measures. IT budget processes acknowledge and provide for the most important information security budget requests in IT and some other business units. The security risk-management process is proactive and jointly shared with corporate collaboration. Access rights are granted based on a formal and audited authorization process. Detailed methods for managing data security throughout its life cycle are implemented. Security incidents are handled based on the urgency to restore services, as agreed on by IT and some other business units.

Level 4 - Managed: Maturity is characterized by regular, enterprise-wide improvement in the alignment of the information security strategy, policies, and standards with business and IT strategies and compliance requirements. IT component security measures on IT systems are implemented and tested enterprise-wide for threat detection and mitigation. Physical environment security is integrated with access controls and surveillance systems across the enterprise. Detailed security budget requirements are incorporated into enterprise-wide business planning and budgeting activities. A standardised security risk-management process is aligned with a firm risk-management process. Access rights are implemented and audited across the company. Data is adequately preserved throughout its life cycle, and data availability is effectively requirements. Recurring incidents are systematically addressed enterprise-wide through problem-management processes that are based on root cause analysis.

Level 5 - Optimized: Maturity reflects an information security strategy that is regularly aligned to business and IT strategies and risk appetite across the business ecosystem. Information security policies and standards are periodically reviewed and revised based on input from the business ecosystem. The management of IT component security is optimised across the security framework layers. Physical access and environmental controls are regularly improved. Security budget requirements are adjusted to provide adequate funding for current and future security purposes. The security risk-management process is agile and adaptable, and tools can be used to address the business ecosystem's requirements. The access rights control and management are dynamic and can effectively deal with the organizational restructuring of acquisitions and divestitures. Processes for managing data security throughout its life cycle are continuously improved. Automated incident prediction systems are in place, and security incidents are effectively managed.

4   Use Case: Applying the Proposed Framework for IT Security Management and Governance (ISMGO)

The framework proposed in Table 2 was applied to the IT department of a large leading of the port sector in Morocco. The organization manages more than 30 ports and sites with more than 1000 users. The information system department comprises a staff of 40 people of different profiles. The audit questionnaire consists of 100 questions divided into different objectives and control of the information security governance inspired by best practice guides ISO 27001 [17] and OWASP [18]. Each item is assigned a weighting coefficient on the effectiveness of the rule of the reference system to which the question relates regarding risk reduction. After the validation of the Questionnaire, the chosen answers were introduced in the software maturity framework that was used to allow the automation of the processing and to determine the maturity score. The treatment consists of calculating a weighted average of the scores obtained according to the chosen responses and the efficiency coefficient. The result is a numerical result (0 to 5 or expressed as a percentage) representing the level of security (maturity) of the audited IS.

4.1   Conducting Assessments

Scoring an organization using the evaluation spreadsheets is simple. After answering questions, assess the answer column to determine the score. Insurance programs may not always consist of activities that fall carefully over a limit between maturity levels.

An organization will receive credit for the different levels of work it has performed in practice. The score is fractional to two decimal places for each practice and one decimal for a response. Questions were also changed from Yes/No to four options related to maturity levels. Anyone who completed the assessment discussed whether to report a yes or no answer when it is honestly something in between.

Table 2. Target objectives of phase 1 (Months 0–6) to achieve the target maturity level

Governance functions	Target goals (Months 0–6)
Information security governance strategy and metrics	- Establish and maintain of assurance and protection program roadmap - Classify applications and information based on business-risk - Ensure data owners and appropriate security levels are defined
Technical asset security management	- Derive security requirements from business functionality - Ensure asset management system and process for hardware and software
Information service and data security management	- Identify, inventory and classify all assets needed for data management - Define and maintain appropriate security levels
Vulnerability and risk management	- Ensure that Standards are implemented on all machines, has current definitions and appropriate settings - Ensure users are periodically informed of unit virus prevention policies
Information security governance control/compliance/continuity management	- Ensure documented control processes are used to ensure data integrity and accurate reporting - Ensure periodic system self-assessments/risk assessments, and audits are performed - Ensure identification and monitoring of external and internal compliance factors

The toolbox worksheet contains contextual answers for each question in the assessment. The formulas in the toolbox will average the answers to calculate the score for each practice, a loop average for each business function and an overall rating. The toolkit also features dashboard graphics that help to represent the current score and can help show program improvements when the answers to the questions change. An example of an evaluation calculation can be found in Appendix A1 (Fig. 2).

Fig. 2. Assessment score

4.2   Assessing Capability Maturity

The framework's assessment tool provides a granular and focused view of an organization's current maturity state for each SPO, desired or target maturity state for each SPO, and importance attributed to each SPO. These maturity and significance scores are primarily determined by an online survey undertaken by the organization's key IT and business stakeholders. The survey typically takes each assessment participant 20 to 30 min to complete, and the data collected can be augmented by qualitative interview insights that focus on issues such as key information-security related business priorities, successes achieved, and initiatives taken or planned. The assessment provides valuable insight into the similarities and differences in how key stakeholders view both the importance and maturity of individual SPOs, as well as the overall vision for success. Figure 3 shows the results of an organization's ISMGO capability maturity assessment, outlining its current and target SPO maturity across all 22 SPOs. For each SPO, the maturity results are automatically generated by the proposed assessment tool, based on averaging the survey participants score across all questions about that SPO. Based on this average score achieved, the organization highlighted in Fig. 3 reflects a level 1.4 (initial) current maturity status for ISMGO overall, but it is less mature in some SPOs, such as security budgeting, resource effectiveness, security threat profiling, and security risk handling. Based on the average across all SPOs, its desired target ISMGO maturity state is maturity level 2.4 (Basic).

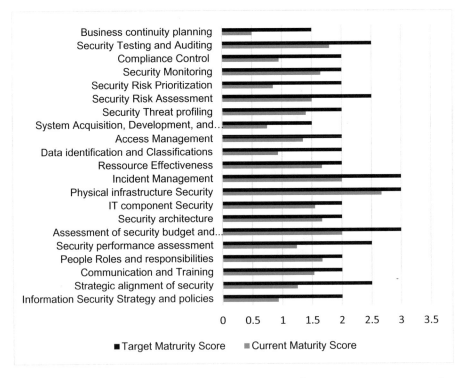

Fig. 3. The proposed information security management and governance assessment results

4.3   Developing Improvement Action Plans

The output from the framework's assessment supports understanding the actions necessary to drive improvement and enable the organization to transition from its current to target maturity state systematically. This is achieved by implementing a series of industry-validated practices that allow organizations to improve incrementally, and monitoring and tracking progress over time using a number of industry-validated metrics. For each of these SPOs, the figure outlines the currently reported maturity and the practices required to transition to the next maturity state. Note that additional practices are available to support transitioning to the desired maturity state.

This phase also included the implementation of a number of concepts for the IT team to improve their security tools. IT teams already had a number of tools in place for quality assessments. An additional survey of code review and security testing tools was conducted. During this phase of the project, the organization will implement the following security maturity practices & activities as shown in Table 2:

5   Conclusion and Limitations

This paper proposes a framework for measuring the maturity of information security was proposed with the aim of providing a practical tool for measuring and improving governance of information security in the organization. ISMGO has been implemented in a medium organization to drive and improve ISMGO maturity. The results are satisfactory and prove that the model will be able to provide great support to organizations in different sizes and various sectors of activity in their governance and management of information security. Nevertheless, it is suggested that the scientific community and organizations adopt this framework and test it in different case studies.

Appendix A

See Table A1.

Table A1. Maturity assessment interview (sample)

Information security governance strategy and metrics		Current State	
Information security strategy and policies		*Answer*	*Rating*
SP1	Is there an information security policy and program in place?	**Yes in ad hoc basis**	**0,93**
	Do the security rules specify a clear definition of tasks, specific roles affecting information security officers?	**Yes a small percentage are/do**	
	A plan ensures that the review is conducted in response to changes in the baseline of the initial assessment, such as major security incidents, new vulnerabilities, or changes to organizational or technical infrastructure?	**Yes there is a standard set**	
	Is there a formal contract containing, or referring to all security requirements to ensure compliance with the organization's security policies and standards?	**Yes a small percentage are/do**	
SP2	Management actively supports the organization's security policy through clear direction, demonstrated commitment, explicit function assignment, and recognition of information security responsibilities?	**Yes at least half of them are/do**	
	Are risk ratings used to adapt security and insurance required?	**No**	
	Does the organization know what's required based on risk ratings?	**Yes at least half of them are/do**	

(*continued*)

Table A1. (*continued*)

Information security governance strategy and metrics		Current State	
Information security strategy and policies		*Answer*	*Rating*
Strategic alignment of security		*Answer*	*Rating*
SA1	Does the organization measure the contribution of IT security to its performance?	**Yes a small percentage are/do**	**1,25**
	Does the organization defined and managed the role of information security in the face of business and technological change?	**Yes but on an ad hoc basis**	
SA2	Are there formal processes in place that emphasize strengthening the partnership relationships between IT Security and Business (e.g. cross-functional teams, training, risk sharing/recognition)?	**Yes there is a standard set**	
	What is the degree of IT control of security or business changes (implementation of new technology, business process, and merger/acquisition)?	**Yes, a small percentage are/do**	
SA3	What is the degree of perception of IT Security by the organization?	**Yes, a small percentage are/do**	
	Does the organization periodically use audits to collect and control compliance conformity?	**Yes, localized to business areas**	
Communication and training		*Answer*	*Rating*
CT1	Have IT staff been given high-level security awareness training?	**Yes we do it every few years**	**1**
	Are system security items included with employee orientation?	**Yes at least half of them are/do**	
CT2	Are those involved and engaged in the IT process, given specific guidance and training on security roles and responsibilities?	**Yes at least half of them are/do**	
	Are users aware and equipped to comply with IS principles, policies and procedures	**Yes a small percentage are/do**	
CT3	Is ongoing security education of users planned and managed?	**Yes teams write/run their own**	
	There is any regular communication process with unit personnel (unit security newsletter/web page)	**Yes a small percentage are/do**	
People Roles and responsibilities		*Answer*	*Rating*
PR1	Do the security rules specify a clear definition of tasks, specific roles affecting information security officers?	**Yes we do it every few years**	**1.67**
	Are the roles and responsibilities for the safety of employees, contractors and third-party users defined and documented by the organization's information security policy?	**Yes at least half of them are/do**	
PR2	Are users, IT Staff and providers gave roles and responsibilities for throughout the organization?	**Yes at least half of them are/do**	
	Are information security responsibilities allocated to ensure accountability and responsibility for the implementation of IS initiatives?	**Yes a small percentage are/do**	
PR3	Is security-related guidance centrally controlled and consistently distributed throughout the organization?	**Yes teams write/run their own**	
	Are responsibilities identified at the unit and at the division or enterprise level?	**Yes we did it once**	

References

1. Goodhue, D.L., Straub, D.: Security concerns of system users: a study of perceptions of the adequacy of security. Inf. Manag. **20**, 13–27 (1991)
2. IT Governance Institute: Information Security Governance: Guidance for Boards of Directors and Executive Management Guidance for Boards of Directors and Executive Management (2006)

3. Rockart, J.F., Crescenzi, A.D.: Engaging top management in information technology, vol. 25 (1984)
4. Safa, N.S., Von Solms, R., Furnell, S.: Information security policy compliance model in organizations. Comput. Secur. **56**, 1–13 (2016)
5. Duffield, M.: Global Governance and the New Wars: The Merging of Development and Security (2014)
6. Schou, C., Shoemaker, D.P.: Information Assurance for the Enterprise: A Roadmap to Information Security. McGraw-Hill Inc., New York City (2006)
7. Dhillon, G., Tejay, G., Hong, W.: Identifying governance dimensions to evaluate information systems security in organizations (2007)
8. Kyukov, D., Strauss, R.: Information security governance as key performance indicator for financial institutions. Sci. J. Riga Tech. Univ. **38**, 161–167 (2009)
9. Klaic, A.: Overview of the state and trends in the contemporary information security policy and information security management methodologies. In: International Convention on Information and Communication Technology, Electronics and Microelectronics, MIPRO (2010)
10. Mattord, H.J., Whitman, M.E.: Roadmap to Information Security: For IT and Infosec Managers. Delmar Learning, Clifton Park (2011)
11. Williams, S.P., Hardy, C.A., Holgate, J.A.: Information security governance practices in critical infrastructure organizations: a socio-technical and institutional logic perspective. Electron. Mark. **23**(4), 341–354 (2013)
12. Yaokumah, W.: Information security governance implementation within Ghanaian industry sectors: an empirical study. Inf. Manag. Comput. Secur. **22**(3), 235–250 (2014)
13. Horne, C.A., Ahmad, A., Maynard, S.B.: Information security strategy in organisations: review, discussion and future research directions (2015)
14. Carcary, M., Renaud, K., McLaughlin, S., O'Brien, C.: A framework for information security governance and management. IT Prof. **18**(2), 22–30 (2016)
15. Kenneally, M., Curley, J.: IT capability maturity framework, p. 20 (2012)
16. Johnson, B.G.: Measuring ISO 27001 ISMS processes, pp. 1–20 (2014)
17. Deleersnyder, S., et al.: Software Assurance Maturity Model (2009)

System Multi Agents for Automatic Negotiation of SLA in Cloud Computing

Zineb Bakraouy[✉], Amine Baina, and Mostafa Bellafkih

STRS Laboratory, National Institute of Posts and Telecommunications,
2, Avenue Allal El Fassi, Madinat Al Irfane, Rabat, Morocco
Zineb.bakraouy@gmail.com, {baina,bellafkih}@inpt.ac.ma

Abstract. Over the past few years, more and more researchers have been involved in research on both agent technology and cloud computing. Efforts have mainly been made to reduce the border between the two technologies. The combination of agents and cloud computing is based on the challenges faced by both communities and the need to develop more intelligent systems. The technology of agents whose goal is to deal with complex systems has revealed opportunities for improving the Cloud Computing Domain. This article present our contribution that is devoted to the integration of SMA in Cloud Computing. The role and functionality of each which constitute our system is detailed. We realized an SMA-based Cloud Computing system for the classification and the automatic negotiation of SLAs.

Keywords: Agent · SMA · SLA · Negotiation · QOS · Availability · Web services
Service broker

1 Introduction

Cloud computing has a promising technology that facilitates the execution of scientific and commercial applications. It provides flexible and scalable services, at the request of users, via a pay-as-you-go model. Typically, it can provide three types of services: SaaS (Software as a Service), PaaS (Platform as a Service), IaaS (Infrastructure as a Service) and three deployment models: public cloud, private cloud, hybrid cloud. At the same time, multi-agent systems (SMAs) represent a new concept in distributed applications. SMAs are based on multiple agents interacting with each other to solve problems using a decentralized approach where several agents contribute to the solution by cooperating with each other. The client/server model is the most used model for construction of applications. Therefore, the disadvantage of this model has is increasing the traffic on the network and needs a permanent connection. Moreover the large and strong demand of services in cloud provokes and causes many collisions in the network. For this, we propose a framework that uses system multi agents, in order to improve availability of services with automatic negotiation based SMA. The use of SMAs has a number of advantages:

© Springer International Publishing AG, part of Springer Nature 2018
A. Abraham et al. (Eds.): IBICA 2017, AISC 735, pp. 234–244, 2018.
https://doi.org/10.1007/978-3-319-76354-5_21

- Dynamic system: SMAs inherit the benefits of AI in terms of symbolic (knowledge) processing. However, contrary to the traditional approaches of Artificial Intelligence that simulate, to a certain extent, the capacities of human behavior, SMAs allow to model a set of interacting agents [1].
- Large number of agents: a large number of agents are at the heart of the problem in this type of modeling, contrary to the theory of games where rarely more than three actors are represented.
- Flexibility of the IT tool: to modify the behavior of agents, to add or remove possible actions, to extend the information available to all agents, unlike models traditionally used in economics.
- Distributed problem solving: it is possible to decompose a problem into sub-parts to solve each independently to arrive at a stable solution [2].

In this paper we provide an overview of the SMA approach adopted in the QoS Negotiation platform to expose negotiation functionalities to Web services. SLA-based negotiation is a crucial support to handle the widely-ranging requirements that characterize Web services. The paper is organized as follows: Sect. 2 overviews the Cloud Computing concept, in Sect. 3 we present the System Multi Agents used in our architecture, in Sect. 4 we discuss a collaboration of SMA and Cloud computing from an analysis of existing Literature, in Sect. 5 is dedicated to present our solution with results and discussion, eventually in Sect. 6 we draw our conclusions.

2   Cloud Computing

2.1   Definition

There are many definitions of the term Cloud Computing (CC) and there is little consensus on a single and universal definition. This multitude of definitions reflects the diversity and technological richness of Cloud Computing. In what follows, we cite some of the most relevant. According to [7], based on a close-up view of the Grid computing grids [3, 4], Cloud Computing [6] is mainly based on the paradigm of distributed computing [5] on a large scale to ensure an on-demand service accessible through the Internet. A second definition, proposed in [8, 9] and which is more abstract, defines cloud computing by using the computing resources (hardware and software) that are offered as a service through a network (typically the Internet). A third definition, developed by a working group of the European Commission [10], considers Cloud Computing as an elastic performance environment for resources involving multiple actors to offer a service with a certain level of quality of service. This definition has been extended in [11] taking into account the perspectives of the different players in the Cloud Computing ecosystem (supplier, developer, user). However, the definition proposed by the National Institute of Standards and Technology (NIST) in [12], defines Cloud Computing as a model that allows access via a network in a simple and on-demand way to a set of shared and configurable computing resources. These IT resources can be allocated and released quickly with minimal management effort or interaction with service providers. In addition, NIST states that Cloud Computing is composed of five essential features, three service models and four deployment models. These elements are listed below.

2.2 Service Models of Cloud Computing

Behind the term Cloud Computing mainstream hides an economic and technological trend that concerns all levels involved in services between a provider and a customer. It is also an application (the supplier runs the application for the client and returns the results), as well as the hardware (the supplier offers its rental equipment as well as a software infrastructure to exploit it). The current taxonomy declined in more precise ways cloud computing into three main categories [13]: Software as a Service (SaaS): Provides to the consumer the capability to use the supplier's applications running on a cloud infrastructure. So the applications are accessible from various client devices through either a thin client interface, like a program interface, or a web browser. The consumer does not control or manage the underlying layers of cloud infrastructure including such us network, servers, operating systems, storage. Platform as a Service (PaaS). Provides to the client the capability to deploy into the cloud infrastructure consumer-created using libraries, programming languages, tools, and services supported by the provider. The consumer does not control or manage the underlying layers of cloud infrastructure including operating systems, network, servers, or storage, but has possibly configuration settings for the application-hosting environment and control over the deployed applications. Infrastructure as a Service (IaaS): Provides to the client the capability to provision processing, networks, storage, and other computing resources where the consumer is able to run and deploy arbitrary software, which can include operating systems and applications. The customer has control over operating systems, storage, and deployed applications; but he has possibly limited control of select networking components.

2.3 Deployment Models of Cloud Computing

Generally, there are three main types of deployment models for cloud computing: private cloud, public cloud, hybrid cloud, and Community Cloud [14]. The Private Cloud: The infrastructure of a private cloud is only used by a single client. It can be managed by this customer or by a service provider and can be located at the premises of the client company or at the service provider, if applicable. Using a private cloud ensures, for example, that the allocated hardware resources will never be shared by two different clients. The Public Cloud: The infrastructure of a public cloud is publicly accessible or for a large industrial group. Its owner is a company that sells computers as a service. The Hybrid Cloud: The infrastructure of a hybrid cloud is a composition of two or three of the cloud types mentioned above. The different clouds that comprise it remain independent entities in their own right, but are linked by standards or proprietary technologies that allow the portability of applications deployed on different clouds. A typical hybrid cloud usage is the load balancing across multiple clouds during peak utilization rates. The Community Cloud, is used by many organizations with common needs. It can host a very specialized business application that is common to several entities, who decide to federate their efforts by building a cloud to host and manage it.

3   SMA: System Multi Agents

3.1   Agent

An agent is a computer program that is located in an environment and that has autonomous behaviors (action) enabling it to achieve, in this environment, the objectives that were set during its conception [15]. A software agent is an autonomous entity capable of communicating, having a partial knowledge of its surroundings and private behavior, as well as its own capacity for execution. An agent acts on behalf of a third party (another agent, a user) that he represents without being necessarily connected to it, he reacts and interacts with other agents. The agent is capable of [16]: To act in an environment and to communicate directly with other agents, Is driven by a set of trends, and possibly to reproduce, also to own own resources and to perceive its environment, and to possess skills and to offer services, Behaviors tending to meet its objectives, moreover to perceive its environment in a limited way, having only one partial representation of this environment. Agent Features, An agent must be [17]:

- Autonomous: the agent is able to act without the influence or intervention of a human or agent and controls his own actions as well as his internal state;
- Proactive: the agent must exhibit opportunistic behavior;
- Social: the agent must be able to interact with other agents, especially when the situation requires it;
- Cooperation: able to coordinate with other agents to achieve the common objective;
- Mobility: the agent can be mobile, able to move to another environment;
- Rationality: the agent is able to act according to his internal objectives and his knowledge;
- Learning: the agent is able to evolve and learn; as a function of this learning, he is able to change his behavior.

3.2   System Multi Agent: SMA

The agent is the main component of multi-agent systems. According to Ferber [17] a multi-agent system is a system composed of the following elements:

- An environment E, that is to say a space generally having a metric.
- A set of objects O. These objects are located, that is to say that for any object it is possible, at a given moment, to associate a position in E. These objects are passive; That is, they can be perceived, created, destroyed and modified by agents.
- A set A of agents, which are particular objects, which represent the active entities of the system.
- A set of relations R that unite objects (and therefore agents) between them.
- A set of operations Op allowing the agents of A to perceive, produce, consume, transform and manipulate O objects.

4   Multi-agent System for Cloud Services Management

Several studies have established a mathematical model for modeling resource provisioning requirements in cloud computing, and these studies are based on the theory of queues mostly. The researchers proposed numerous multi-agent approaches in a Cloud context to overcome the limitations of the latter such as resource allocation and security threats. For example, in article [18], the authors proposed a combination of Agent technology with the Cloud to arrive at a method for calculating the resource management model. It reinforces the theory that the use of agents makes it possible to effectively achieve the management of resources in the Cloud. In Mansura's article [19], the integration of SMA in a Cloud context can allow high performance for complex systems and intelligent applications, proving that a reliable and scalable infrastructure can be Application on a large scale. SMAs have also been used in the cloud to provide an access management service. Indeed, in paper [20] emphasize that traditional systems are not sufficiently effective to support the functionality of access control in the cloud mainly due to the high scalability of the environment of cloud. They used a multi-agent system to define the accessibility and functionality of their model in order to improve the access control system. Other uses of multi-agent systems to provide Cloud-based security services have been proposed as for article [21] where the authors used SMA architecture to ensure privacy and availability for a collaborative storage service Hosted in the cloud, and also in a disaster management service presented in the paper. [22] The authors used a workflow model to help and maintain rescue and reorganization of disaster activities. Other work has addressed the concept of formal description of SMA.

Their studies aim to evaluate the integrated functionalities and to present the formal specifications of multi-agent systems. Typically the article [23] proposes a new collaborative formalism in SMA between agents combined with the Ferber model, characterized by the possibility of self-evaluation in the application of collaborative work. Their results show that the proposed formalism more effectively manages agent communication for better production.

5   SMAANQOS Framework (System Multi Agent for Automatic Negotiation of Quality of Services)

5.1   Motivation

The rise of embedded information systems in hosted mode has shown a marked decline since the beginning of the year 2010. This trend is explained by the arrival of an outsourced offering, notably in the market for business solutions on demand, driven by both the diversity and richness of the Cloud Computing operators 'offerings and the customers' financial, technological and operational flexibility. In an economic approach, the Cloud is emerging as a promising commercial offer that can satisfy both applicants and digital service providers. Thus, SMEs are required to contract some of the services offered by Cloud Computing operators in order to obtain operational guarantees, in addition to the usual services between customers and suppliers. In a customer/operator relational approach, the delicate concretization of a cloud service depends strongly on the service model targeted, ranging from the provision of software to that of infrastructures via the platform (Saas, PaaS, IaaS) and Model of deployment privileged (Public, Private, Hybrid, Community). So this conclusions cited above we decide to build a new framework of negotiation of QOS based on SMA in order to improve availability of services.

5.2 SMAANQOS Framework

SMAANQOS framework is based on JADE (Java Agent DEvelopment Framework) that is a multi-agent platform developed in Java by CSELT (Gruppo Telecom Research Group, Italy), which aims to build multi-agent systems and implement applications that comply with the FIPA standard (Platform Foundation for Intelligent Physical Agents, 1997). JADE consists of two basic components: a FIPA-compatible agent platform and a software package for the development of Java agents. Jade is a middleware that facilitates the development of multi-agent systems (SMA) (Fig. 1).

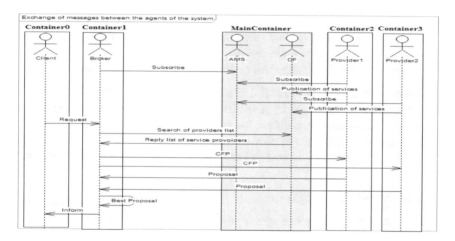

Fig. 1. SMAANQOS diagram of interactions

JADE contains: A runtime environment: the environment where agents can live. This runtime environment must be enabled in order to launch the agents, also a class library that developers use to write their agents and a suite of graphical tools: which facilitate the management and supervision of the agent platform. Each instance of the JADE is called a container, and can contain multiple agents. A set of containers is a platform. Each platform must contain a special container called the main container and all other containers register with the container as soon as they are launched. In our platform, Multi agents are compatible with FIPA, it includes three agents that are automatically created and activated when the platform is activated (Fig. 2):

- Agent Management System (AMS): is the agent who exercises supervisory control over the access and use of the platform; it is responsible for authenticating resident agents and checking records.
- Agent Communication Channel (ACC): is the agent that provides the route for the basic interactions between agents in the platform.
- Directory Facilitator (DF): is the agent that provides a yellow pages service to the multi-agent platform.
- Remote Management Agent (RMA): The RMA allows controlling the lifecycle of the platform and all its components.

Fig. 2. FIPA reference model of an agent [23].

Our framework is implemented using the JAVA language, under the environment Eclipse development. For the multi-agent system we used the JADE platform.

The choice of JAVA language was motivated by the following reasons:

- The agents developed under the JADE platform are entirely written in Java.
- Java ensures total independence of applications from the environment execution: any Java-enabled machine is capable of to run a program without any adaptation (neither recompilation nor parameterization of environmental variables).
- JAVA has a huge library of ready-to-use objects, which fully implement the implementation procedure.

Our framework runs as below:

- Step 1: The agents broker and agents providers subscribes by AMS.
- Step 2: Then the providers agents register their services.
- Step 3: Every second the broker contacts the DF in order to have information about the publication of the providers agents
- Step 4: When a client send a request for specific service, the broker contact the DF to have a list of providers of the service requested
- Step 5: After getting list of providers, the broker send CFP (call for proposal) to each one.
- Step 6: The providers responds to a CFP by sending their proposal.
- Step 7: The broker make a choice of the best proposal by calculate the best ratio availability, price.
- Step 8: At the end the broker inform client about the provider which has a best ratio.
- Step 9: if the client send another request for the same service while DF had not another entry for the same service the broker responds client with the same message if not the broker send another time CFP for each provider and calculate the best proposal.

N.B: If the broker will be not reachable or killed, the client switches automatically to another broker from the list.

Inside JADE container the message exchanged in JADE architecture components can be traced by a special JADE agent called SNIFFER Agent. (See Fig. 3) presents SNIFFER GUI.

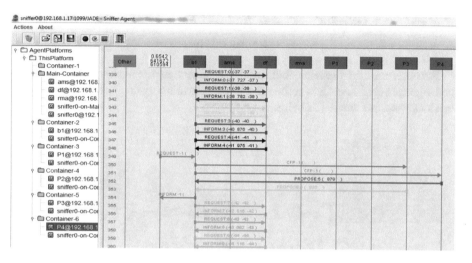

Fig. 3. Exchange of messages between the agents of the system.

An agent who wishes to publish a service must provide a description that includes: Its AID identifier, list of languages and ontologies that other agents must use to communicate with them, List of published services (Fig. 4).

Fig. 4. Container of provider agent.

The Class diagram of Broker agent is modeled as below (Fig. 5):

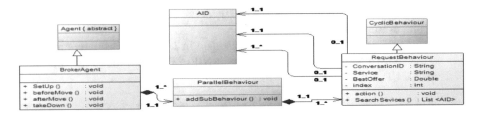

Fig. 5. Class diagram broker agent.

For each published service, indicate: Type of service, the name of the service, and the languages and ontologies to be used to operate this service. The Class diagram of Provider agent is modeled as below (Fig. 6):

Fig. 6. Class diagram provider agent.

6   Conclusion and Future Work

The Cloud Computing has appeared as a new computing concept which has a purpose to provide reliable, customized, dynamic computing environments towards better IT infrastructure availability and quality of service without much financial burden. Therefore, the large and strong demand of services in cloud provokes and causes many collisions in the network. In order to relieve network and improve availability of services by automatic negotiation we were able developed a new framework based system multi agent technology called SMAANQOS used in a cloud computing as it simulates above. This article can be considered as a proof of concept of the SLA negotiation using SMA in a micro cloud computing. As the SMA developed is flexible and scalable, it can be extended to appropriate cases for the cloud computing environment and impose new algorithms for the automatic negotiation of Quality of Service. As a Future work, we will compare the performance of our proposal Framework to other existing solutions for negotiating access to cloud services.

References

1. Chaib-Draa, B., Jarras, I., Moulin, B.: Systèmes multi-agents: principes généraux et applications, pp. 1030–1044. Edition Hermès (2001)

2. Le Bars, M.: Un Simulateur Multi-Agent pour l'Aide à la Décision d'un Collectif: Application à la Gestion d'une ressource Limitée Agroenvironnementale. Diss. Université Paris Dauphine-Paris IX (2003)
3. Rajaei, H., Wappelhorst, J.: Clouds & grids: a network and simulation perspective. In: Proceedings of the 14th Communications and Networking Symposium, CNS 2011, pp. 143–150. Society for Computer Simulation International (2011)
4. Foster, I., Zhao, Y., Raicu, I., Lu, S.: Cloud computing and grid computing 360-degree compared. In: Grid Computing Environments Workshop, vol. abs/0901.0, pp. 1–10. IEEE (2008)
5. Höfer, C.N., Karagiannis, G.: Cloud computing services: taxonomy and comparison. J. Internet Serv. Appl. 2(2), 81–94 (2011)
6. Buyya, R., Yeo, C.S., Venugopal, S., Broberg, J., Brandic, I.: Cloud computing and emerging IT platforms: vision, hype, and reality for delivering computing as the 5th utility. Future Gener. Comput. Syst. 25(6), 599–616 (2009)
7. Kushwaha, D.S., Maurya, A.: Cloud computing-a tool for future. Int. J. Math. Comput. Res. 1(1), 09–14 (2013)
8. Ahmed, M., Sina, A., Chowdhury, R., Ahmed, M., Rafee, M.H.: An advanced survey on cloud computing and state-of-the-art research issues. IJCSI Int. J. Comput. Sci. Issues 9(1), 201–207 (2012)
9. Schubert, L., Jeffery, K., Neidecker-Lutz, B.: The future of cloud computing - opportunities for european cloud computing beyond 2010. Technical report, European Commission (2010)
10. Schubert, L., Jeffery, K.: Advances in clouds: research in future cloud computing. Technical report, European Commission (2012)
11. Mell, M., Grance,T.: The NIST definition of cloud computing. Technical report 6, National Institute of Standards and Technology (2011)
12. Armbrust, M., Fox, A., Griffith, R., Joseph, A.D., Katz, R., Konwinski, A., Lee, G., Patterson, D., Rabkin, A., Stoica, I., Zaharia, M.: A view of cloud computing. Commun. ACM 53(4), 50–58 (2010)
13. The NIST Definition of Cloud Computing, September 2011. http://dx.doi.org/10.6028/NIST.SP.800-145. Accessed 20 Oct 2017
14. Wooldridge, M.: An Introduction to MAS (2009)
15. Ferber, J.: Les systèmes multi-agents: vers une intelligence collective. Inter Editions (1995)
16. Briot, J.P., Yves, D.: Principes et architecture des SMA, Hermès (2001)
17. Wu, W., Zhang, X., Zheng, Y., Liang, H.: Agent-based layered cloud resource management model. In: 2013 6th International Conference on Information Management, Innovation Management and Industrial Engineering (ICIII), vol. 2, pp. 70–74 (2013)
18. Talia, D.: Clouds meet agents: toward intelligent cloud services. IEEE Internet Comput. 16(2), 78–81 (2012)
19. Habiba, M., Islam, M.R., Ali, A.S.: Access control management for cloud. In: 12th IEEE International Conference on Trust, Security and Privacy in Computing and Communications (TrustCom), pp. 485–492 (2013)
20. Amir, A.M., Rodziah, A., Rusli, A., Masrah, A.M.: Security framework of cloud data storage based on multi agent system architecture-a pilot study. In: International Conference on Information Retrieval and Knowledge Management (CAMP), pp. 54–59 (2012)
21. Mansura, H., Shamim, A.: MAS workflow model and scheduling algorithm for disaster management system. In: 2012 International Conference on Cloud Computing Technologies, Applications and Management (ICCCTAM), pp. 164–173 (2012). 1-2-3-5-13-14-15

22. Khezami, N., Otmane, S., Mallem, M.: A new formal model of collaboration by multi-agent systems. In: International Conference Integration of Knowledge Intensive Multi-Agent Systems (KIMAS 2005), pp. 32–37 (2005)
23. Foundation for Intelligent Physical Agents. Specifications (1997). http://www.fipa.org. Accessed 20 Oct 2017

A Comparison Between Modeling a Normal and an Epileptic State Using the FHN and the Epileptor Model

R. Jarray[1], N. Jmail[2], A. Hadriche[1,3(✉)], and T. Frikha[4]

[1] Gabes University, Gabes, Tunisia
abir.hadriche.tn@ieee.org
[2] Miracl Laboratory, Sfax University, Sfax, Tunisia
[3] REGIM Laboratory, ENIS, Sfax University, Sfax, Tunisia
[4] CES Laboratory, ENIS, Sfax University, Sfax, Tunisia

Abstract. In spite of important technological developments in the medical field and particularly in neuroscience one, epilepsy remained a serious pathology that could affect the human brain. In this work, we modeled a healthy and an epileptic cerebral activity in rest state. We used, the virtual brain TVB toolbox to simulate the two states based on FHN and epileptor model. We compared phase plane spaces, electrophysiological time series (electroencephalogram EEG, magnetoencephalogram MEG and intracerabral EEG), specter of eigenvalues transition matrix and topographic maps for healthy and epileptic rest state. There is a unique metastable state for healthy cerebral dynamics convergence which disappears in epileptic cerebral dynamics. Epileptic rest state time series depicts several transitory activities that vanish in the normal state. Normal rest state topographic maps illustrate a limited dipolar activity; which is more extended in epileptic model. These prominent differences would have an important impact on real cerebral activities analysis.

Keywords: Epilepsy · Healthy · Modeling · TVB · Phase plane
EEG · MEG · IEEG · Topography

1 Introduction

Fisher and his colleagues define the epileptic seizure as "a transient occurrence of signs and/or symptoms due to excessive or abnormally synchronous cerebral neuronal activity" [1]. One of the most important techniques to understand cerebral human brain function, and to improve clinical diagnosis for several neurologic diseases specially, in our case epilepsy; are the non-invasive acquisition (non traumatic and good time precision). The variation of the electrophysiological signal through electroencephalography (EEG), magnetoencephalography MEG and intra cerebral EEG signal (IEEG) is required to restrict the epileptogenic zone (EZ) for seizure free.

© Springer International Publishing AG, part of Springer Nature 2018
A. Abraham et al. (Eds.): IBICA 2017, AISC 735, pp. 245–254, 2018.
https://doi.org/10.1007/978-3-319-76354-5_22

To understand the brain epileptic dysfunction, presented by abnormal and paroxysmal excessive neural discharges, we should define the network connectivity implied by these activities [2,3]. In fact, several techniques have been proposed and discussed to determine in an accurate way the epileptogenic zone. The first classification made of epileptic phenomena in the 19th century was proposed by the British neurologist Jackson [4], then the International League Against Epilepsy [5] suggested further information to heal this illness. Moreover, several research are based on signal simulation, modeling and electrophysiological signal [6]. Actually, researches are combing electrophysiological signal with l Functional MRI (EEG-fMRI) to study the generalized peak waves (GSW) activity in idiopathic and secondary generalized epilepsy (SGE) [7].

In this work, we will proceed on four linear and non linear analysis techniques: phase plane, electrophysiological time courses, topographic maps and dynamic attractors to compare the healthy from the epileptic rest state. All simulated models are generated on python using the Virtual Brian TVB toolbox. The TVB is a free toolbox for modeling large-scale dynamics [8]. We applied two neuronal models to simulate cerebral activities in order to detect common and differences between healthy and epileptic rest state dynamics. This work is composed of three sections: the materials (normal and epileptic models, phase plan), the second is reserved for the experimental results and the third part depicts conclusion and further perspectives.

2   Methods

In order to understand the human brain functions, several researches proposed preprocessing schemes to decipher different brain states: rest, cognitive (during reflex), illness (epileptic). Other researches are based on exploiting, simulating several models to have a better vision of the brain activities. Since, real activities are delicate with mysterious phenomena which made analysis and interpretation harder, these models may lead to a better diagnosis and description. We proposed, here, two neuronal models of healthy and epileptic rest state: FitzHugh-Nagumo and Epileptor to simulate the behavior of an isolated neuron then a coupled networks neurons. These simulated data would be used to distinguish healthy from epileptic brain activities. This comparison would be studied using a linear and a non linear approach.

2.1   The FHN Model (FitzHugh-Nagumo)

FHN model is a plan neural model based on two variables states: a membrane potential (voltage responsible of the system excitability), and a recovery variable (for feedback purposes) [9]. The FHN model is described in the following two Eqs. (1) and (2):

$$\frac{dv}{dt} = f(v) - w + I = v(v - \alpha)(1 - v) - w + I \tag{1}$$

$$\frac{dw}{dt} = bv - \gamma w \tag{2}$$

With $f(v)$ is a third-degree polynomial, a, b, γ are constant parameters. v, w are two variable states, and I is an external current that acts as a stimulus.

2.2   The Epileptor Model

Epileptor model is a phenomenological neural mass model that would reproduce epileptic seizures dynamics in the same way of the intracranial EEG acquisition [10]. It is a behavioral model composed of six differential equations as indicated below:

$$\dot{x}_1 = y_1 - f_1(x_1, x_2) - Z + I_{ext1}$$
$$\dot{x}_1 = c - dx_1^2 y_1 \tag{3}$$

$$\dot{z} = \begin{cases} r(4(x_1 - x_0) - z - 0.1z^7) \; if \; x < 0 \\ r(4(x_1 - x_0) - z) \qquad\quad if \; x \geq 0 \end{cases} \tag{4}$$

$$\dot{x}_2 = -y_2 + x_2 - x_2^3 + I_{ext2} + 0.002g - 0.3(z - 3.5)$$
$$\dot{y}_2 = 1/\tau(-y_2 + f_2(x2))$$
$$\dot{g} = -0.01(g - 0.1x_1) \tag{5}$$

Or

$$f_1(x_1, x_2) = \begin{cases} ax_1^3 - bx_1^2 & if \; x_1 < 0 \\ -(slope - x_2 + 0.6(z - 4)^2)x_1 \; if \; x_1 \geq 0 \end{cases} \tag{6}$$

And

$$f_2(x_2) = \begin{cases} 0 & if \; x_2 < 0 \\ a_2(x_2 + 0.25) \; if \; x_2 \geq 0 \end{cases} \tag{7}$$

a, b, c and d are the epileptic model coefficients, I_{ext1} and I_{ext2} are the external current input of the first and second populations.

Hence, we studied the phase plans of FHN and epileptor models to simulate an isolated neuron versus a coupled neurons network behavior. We tested these models performances to describe correctly the rest and epileptic state cerebral function. The main goal here is to generate from our models the electrophysiological signal EEG, MEG and IEEG in both states. Then, we proceeded our comparison using the topography maps results of healthy and epileptic state.

- **Non-linear approach**

We applied a Non-linear set oriented approach (NLA) [11] to describe the cerebral dynamic trajectory among higher observation scales. NLA projects the brain activities dynamics (i.e. simulated EEG signals) on a multi-dimensional stochastic system, (same number of sensors). It uses a subdivision method to discretize the measurement space [12]. In each iteration the entropy rate of the transition matrix was calculated

$$h(M^{(k)}) = -\sum_i \pi_i^{(k)} \sum_j p_{ij}^{(k)} log p_{ij}^{(k)}.$$

This approach detects the number of cerebral dynamics attractors in higher level of observation in order to find metastable states (limit cycles or equilibrium state).

- **Linear approach**

We applied a linear approach on simulated epileptic and healthy electrophysiological signals (EEG, MEG and IEEG) to depict the topography map of the transitory activity as in [2,3,13,14]. In fact, the presence of transitory activities have being proved as a hallmark in epileptogenic zone (epileptic state) and also redundant in normal brain behavior.

3   Results and Discussion

3.1   Isolated Neuron Simulation

We simulated the FHN model and the epileptic model (Epileptor) for a single isolated neuron to generate the time series signal describing the Rest State Network. The phase plane results for the two models are presented in Figs. 1 and 2.

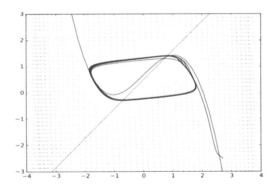

Fig. 1. FHN model Phase plane, $a = -2$, $tau = 1$, $c = -0$, $b = -10$, $I = 0$, $\gamma = 1$.

The Fig. 2 depicts a single isolated neuron epileptic model trajectory.

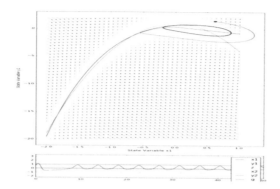

Fig. 2. (Top) epileptic model Phase plane: two nullclines relative to parameters x1 and y1 respectively in green and red, isolated neuron bifurcation in blue. (below) different time series obtained for the 6 model variables.

3.2   A Coupled Network Simulation

To simulate a coupled network, we used 10000 neurons distributed on 76 cortical regions, 38 for the left hemisphere and 38 for the right one. For healthy rest state, we used the stochastic Euler method to simulate the FHN model using a noise Gaussian random process with a standard deviation D = 0.005, we defined the coupling linear function slope a = 0.0042 and the sampling frequency Fe = 256 Hz . We proceed in the same way to generate the epileptic state using a neural epileptor model.

- **Non-linear approach applied to healthy simulated data**

We divided our simulated database (38*100 k points) into two-time series with the same size in order to elevate the computation time.

 We applied the subdivision method, as in [11,12], to discretize the phase plans measurement of our data sets. This discretization will be stopped after the 8th step and hence two transitions matrix are calculated (size $|M| = 250$ and $|M| = 252$ regions with an entropy rate of $h^* = 0.7 * 10^3$ and $h^* = 0.71 * 10^3$ bits per second.). In Fig. 3 we illustrated the first 15 eigenvalues of the M matrix.

 Figure 3 shows a big jump after the first eigenvalue for the two-time series, which proved, in high scale, that the two systems have a unique metastable state whose dynamics, will converge into. These results are also validated in [15] for the study of simulated and real signals in rest state. Thus, we can conclude that the appearance of a single metastable state of this simulated brain dynamics clearly proves existing of a fixed point or a limit cycle in lower observed scale. These results are agreement with the cerebral networks dynamic aspect in rest state.

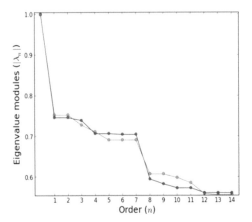

Fig. 3. The largest 15 eigenvalues of the two transition matrix of healthy simulated data.

- **Non-linear approach applied to epileptic data**

We applied the same approach for the epileptic state, discretization procedure is stopped after the 9th iteration, we obtained two matrix of size $|M| = 500$ and $|M| = 502$ regions with an entropy rate of $h^* = 1.3 * 10^3$ bit per second for the first-time series and $h^* = 1.17 * 10^3$ bit per second for the second one. The entropy rate for epileptic data is higher than the healthy one which proves much more disorder in the epileptic brain dynamics. Figure 4 illustrates the first 15 eigenvalues of the two epileptic data matrix.

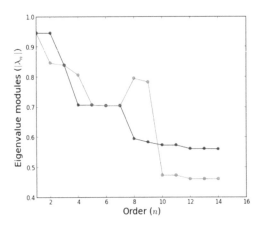

Fig. 4. The largest 15 eigenvalues of the simulated epileptic transition matrix.

In opposite of the healthy data, there is no clear jump among two consecutive eigenvalues, which could be the results of metastable state absence. In the meantime, it is noteworthy that these two dynamics have a chaotic aspect far from the rest state justified by the existence of several epileptogenic zones responsible of excessive discharges and build up seizure.

- **Linear approach**

We used EEG, MEG and IEEG simulated signal by the FHN and epileptor model to detect physiological activities: transitory, transitory with oscillatory and oscillations activities. The EEG time series is illustrated in Fig. 5.

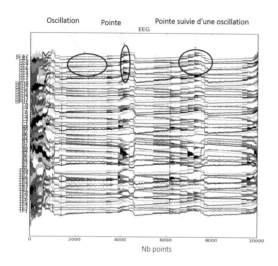

Fig. 5. EEG signal of a coupled epileptic network simulation, depicting three types of activities: transitory, transitory plus oscillatory and oscillations.

Figure 5 demonstrates slow waves, transitory activities, gamma oscillations followed by fast transitory activities, which may have epileptic origins [2, 13, 14]. We also, can observe the existence of bilateral temporal delta waves symmetrical monomorphic with transitory discharges and also intercritical spikes that may represent the ictal state (seizure) [14]. The time series obtained for the IEEG simulation of the epileptic model is illustrated in Fig. 6.

In Fig. 6, we notice the presence of rapid temporal discharge with epileptic spikes (transitory activities). These discharges may involve the temporal epileptogenic zones with further flattening zone by a low frequency rhythm.

The MEG signals of the epileptic simulation are illustrated in Fig. 7.

We noted the absence of temporal signals due to the effect of muscular noise: the presence of myoclonus with post-myoclonic periods of silence.

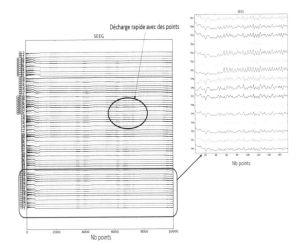

Fig. 6. IEEG time series of an epileptic coupled network.

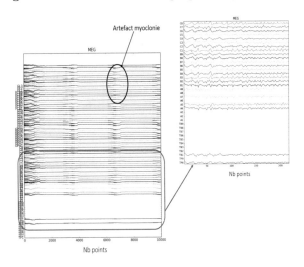

Fig. 7. MEG time series of a coupled epileptic network.

- **Topography of healthy and epileptic signals**

The topographies maps of healthy and epileptic EEG signals are shown in Fig. 8.

The left topography map depicts a dipolar presentation which could represent a normal cerebral transitory activity at rest (alpha rhythm). However the topography map on the right implies a lot of cortical regions which may be explained by the excessive discharges of the epileptic state that propagates in the entire brain and involved further regions. Nevertheless there is a dipolar activity which proves that it is a sickly cerebral activity.

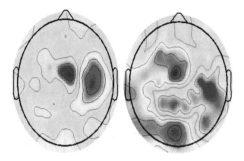

Fig. 8. Topography map of healthy (left) and epileptic (right) EEG signal on transitory activity.

4 Conclusion and Perspective

Our goal, in this work is to simulate and analyze cerebral activity in healthy and epileptic rest state. We modeled the healthy rest and epileptic state, based on the synchronizations hypothesis of the TVB toolbox to generate our analysis on electrical, magnetic and intracerebral brain activities signal. Then, we proceed on the comparison between the phase plane, of the electophysiological time series signals (a nonlinear approach) and the topography maps (a linear approach). For each kind of comparison procedure, we notice differences between cortical activity in healthy and epileptic rest state.

We suggest in further work, to use the MRI magnetic resonance imaging for a 3D visualization of the epileptogenic zone EZ and to define the epileptic networks connectivity in order to determine the affected zone by the excessive discharges using the source localization of either the transitory or the gamma oscillatory activities. Another prominent perspective is the integration of these models on embedded systems to reduce the time consumption since we declared previously that simulating a convenient number of coupled neurons to much the real brain activity is heavy in computation.

References

1. Fisher, R.S., van Emde Boas, W., Blume, W., Elger, C., Genton, P., Lee, P., et al.: Epileptic seizures and epilepsy: definitions proposed by the International League Against Epilepsy (ILAE) and the International Bureau for Epilepsy (IBE). Epilepsia **46**, 470–472 (2005)
2. Jmail, N., Gavaret, M., Bartolomei, F., Chauvel, P., Badier, J.M., Bnar, C.-G.: Comparison of brain networks during interictal oscillations and spikes on magnetoencephalography and intracerebral EEG. Brain Topogr. **29**, 752–765 (2016)
3. Jmail, N., Gavaret, M., Bartolomei, F., Bnar, C.-G.: Despiking SEEG signals reveals dynamics of gamma band preictal activity. Physiol. Meas. **38**, 42–56 (2017)
4. Jackson, J.H.: On the scientific and empirical investigation of epilepsies. Med. Press Circ. (Ed.) **18**, 325–327 (1874)

5. ILAE: Proposal for revised classification of epilepsies and epileptic syndromes. Commission on classification and terminology of the International League Against Epilepsy. Epilepsia **30**, 389–399 (1989)
6. Grouiller, F.: Cartographie fonctionnelle du cerveau pileptique lors des valuations pr chirurgicales. Institut des Neuroscience de Grenoble (2008)
7. Hamandi, K., Salek-Haddadi, A., Laufs, H., Liston, A., Friston, K., Fish, D.R., Duncan, J.S., Lemieux, L.: EEG-fMRI of idiopathic and secondarily generalized epilepsies. Neuroimage **31**, 1700–1710 (2006)
8. Jirsa, V., Jantzen, K., Fuchs, A., Kelso, J.: Spatiotemporal forward solution of the EEG and MEG using network modeling. IEEE Trans. Med. Imag. **21**, 493–504 (2002). https://doi.org/10.1109/TMI.2002.1009385
9. FitzHugh, R.: Impulses and physiological states in theoretical models of nerve membrane. Biophysical J. **1**, 445–466 (1961)
10. Jirsa, V.K., Stacey, W.C., Quilichini, P.P., Ivanov, A.I., Bernard, C.: On the nature of seizure dynamics. Brain J. Neurol. **137**, 1–21 (2014)
11. Hadriche, A., Pezard, L., Nandrino, J., Ghariani, H., Kachouri, A., Jirsa, K.: Mapping the dynamic repertoire of the resting brain. Neuro Image J. **78**, 448–462 (2013)
12. Hadriche, A., Jmail, N., Elleuch, R., Pzard, L.: Different methods for partitioning the phase space of dynamic system. IJCA **93**(15), May 2014
13. Jmail, N., Gavaret, M., Wendling, F., Kachouri, A., Hamadi, G., Badier, J.-M., Bnar, C.-G.: A comparison of methods for separation of transient and oscillatory signals in EEG. J. Neurosci. Methods **199**(2), 273–289 (2011)
14. Jmail, N., Gavaret, M., Bartolomei, F., Bnar, C.-G.: Despikifying SEEG signals using a temporal basis set. In: The International Conference on Intelligent Systems Design and Applications, Marrakech, Moroco (2015)
15. Hadriche, A., Jmail, N., Ghariani, H., Kachouri, A., Pezard, L.: Reduction of continuous neuronal model to discrete binary automata. Int. J. Comput. Appl. **69**(24), 5–10 (2013a). Published by Foundation of Computer Science, New York, USA

Modeling the Effect of Security Measures on Electronic Payment Risks

Marie Ndaw$^{(\boxtimes)}$, Gervais Mendy, and Samuel Ouya

Ecole Superieure Polytechnique, Universite Cheikh Anta Diop, Dakar, Senegal
{marieelisaadam.ndaw,gervais.mendy,samuel.ouya}@ucad.edu.sn
http://www.ucad.edu.sn/lncs

Abstract. Nowadays Banking is very digitized but electronic payment products are vulnerable to different types of attacks. Banks implement many security measures to manage fraud related to the use of debit, credit or prepaid cards. Also, they assess the impact of implemented security measures using a manual method which has some limits including many work sessions, significant level of expertise and estimation error. In this paper, we propose to quantify the impact of security measures on electronic payment risks by defining a mathematical model which is based on FMECA. For testing, the values obtained with the model were compared to reference values given by assessors during different working sessions and the correlation rate is satisfactory. The findings of the study help to assess maturity of implemented security measures and improve security of electronic payment transactions.

Keywords: Payment · Card · Security · Vulnerability
Attack · Risk

1 Introduction

Different types of services are provided using banking cards such as mobile banking which is a service that allows customers to conduct financial transactions remotely using a mobile device and internet banking which is an electronic payment system that enables customers to conduct financial transactions through website. Also, cardholder can use POS (Point Of Sale) which is a point where retail transaction is finalized or ATM (Automated Teller Machine) which is an electronic telecoms device that enables customers to perform financial transactions. Sakharova and Khan in [1] concluded that he use of banking cards may increase the exposure to fraudulent activities. Sullivan in [2] showed that the exiting threats vary with the profile and creativity of the fraudsters who are more and more innovative. As we can see, electronic payments have different types of risks that could lead to financial losses. Arango and Taylor in [3] related that there is potential for the temporary loss of funds or permanent losses. Banks periodically assess the effect of implemented security measures using an actual

© Springer International Publishing AG, part of Springer Nature 2018
A. Abraham et al. (Eds.): IBICA 2017, AISC 735, pp. 255–265, 2018.
https://doi.org/10.1007/978-3-319-76354-5_23

method which is manual, takes time and has some estimation error. So we propose to improve assessment of electronic payment risk security by modeling the residual risks in order to save time on deployment of new security measures.

2   Literature Review

Electronic payment risk management process varies widely across institutions. Fajfar in [4] concluded that identified threats concern an individual who break into an electronic system, steal customers personal data, attack or corrupt data. Different types of fraud are related to electronic payment products. Bhatla et al. in [5] showed that frauds can be classified into three categories related to card, merchant and internet frauds. Payment security has to be negotiated between all partners along the clearing and settlement chains. Users and processors must understand, accept and apply security procedures as related in [6]. Authentication of the cardholder is a fundamental requirement in managing fraud on internet as indicated in [7]. Also, data should be protect while in its custody. Pereira and De Alba in [8] related that the financial, payment and network service providers should implement the appropriate safeguards. Each organization involved put in place strong positive controls. Anderson in [9] showed that various risks have been associated to mobile payments like anti money laundering, credit, liquidity, fraud and Compliance. Burns and Stanley in [10] concluded that the strategic risk with mobile payments is of an attack that makes fraud so easy that a platform or channel becomes unviable.

In [11], it's related that a provider should work closely with merchants and financial institutions to help address fraud risks. The vendor should also be knowledgeable about standards such as the PCI DSS (Payment Card Industry Data Security Standard) and PA DSS (Payment Application Data Security Standard). The provider should provide the tools necessary to support merchant compliance. Also, each stakeholder should address fraud with their EMV(Europay Mastercard Visa) chip migration strategy as showed in [12]. Concerned online transactions security, Conroy in [13] concluded merchants have to stem rising fraud losses, while issuers have to rein in fraud and also maintain consumer confidence. As we can see banks implement several security measures to safe electronic payment transactions and it is necessary to manage all those risks.

Kellogg in [14] related different steps of risk management such as: determining level of inherent risk, monitoring the effectiveness of the risk management practices that are mitigating material risks, determining whether the residual risk is improving, stable or eroding over time. Weese in [15] showed several methods exist for risk assessment. As related in [16], risk management methods include risk assessment, timing of exercise, consistency and implementation. Lipol and Haq in [17] explained FMECA (Failure Modes and Effect Criticality Analysis Method) is a risk analysis method which operate on the whole system; it is most detailed and practical to examine high or small level systems. Bank often evaluate their residual risks by assessing the effect that the current controls have on inherent risks during several working sessions. However, in 2015, we propose a mathematical model which allow a quantification of banking residual risks [18]. After that, we also defined two mathematical models which provide an automatic calculation of internet banking controls maturity [19]. Generally, residual risk are determine by assessing the effect that implemented control has on the overall risk leads [20].

3 Problem

Banking IT security becomes more complex as new information systems are implemented on top of older ones. ATM and internet banking frauds contributed most to frauds in banking sector. Internal control focus a review of all payment activity including mandating dual approval at vulnerable touch points. Most private-sector providers have tools to manage many of these risks. To determine the impact of implemented security measures on identified risks, it is necessary to assess the effect that the security measures has on the overall risk leads. Practicality, residual risks is estimated by assessors during several work sessions using a manual method which has some limits including many work sessions with compromises in case of disagreement, significant level of expertise and time consuming. Furthermore, the impact of security measures are appreciated differently by interlocutors and they are some estimation error rate of residual risks. This may imply unsecured electronic payment transactions and decrease customer confidence and bank profit.

4 Our Contribution

To define our model which quantify the impact security measures on electronic payment risks, we propose the following approach (Fig. 1).

As illustrated above, the proposed approach take into account inherent risks criticality and security measures maturity. For assessment of inherent risks, we will defined scales related to likelihood of occurrence and severity of impact in order to assign inherent value to each risk. For assessment of security measures maturity, we will defined scale related to security measures maturity in order to assign maturity value to each security measure. We will also take into account four types of security measures by defining indexes related to deterrent, preventive, detective and curative security measures. All this parameters will be used to define our model equation.

Fig. 1. Our model approach

4.1   Model Principles

To define our model, we consider the following 7 principles:

- Principle 1: Security measures aim to mitigate electronic payment risks
- Principle 2: Security measures have one maturity and four types
- Principle 3: Deterrent security measures are intended to discourage attacker
- Principle 4: Preventive security measures are intended to minimize an incident occurring
- Principle 5: Detective security measures are intended to identify when an incident has occurred
- Principle 6: Corrective security measures are intended to fix information system components after an incident has occurred
- Principle 7: Mature security measures reduce inherent risks criticality.

4.2   Model Scales

Three scales are defined to propose our model. The scale of security measures maturity has five values at most as indicated below (Table 1).

Table 1. Value of control maturity

Rating	Description	Meaning
1	Not existent	Bank has not even identified the issues to be addressed
2	Initial	Issues exist but the approach to management is disorganized
3	Systematic	The procedures are not sophisticated but they are formalized
4	Managed	Management monitors and measures compliance with procedures
5	Optimized	Processes have been refined to a level of good practice

Each assessed security measure will be assigned one value of maturity using this scale. The scales of likelihood and severity have six values as showed below (Tables 2 and 3).

Table 2. Value of risk likelihood

Rating	Description	Meaning
1	Almost never	It is difficult for threats to exploit vulnerability
2	Unlikely	Threats require significant skills to exploit vulnerability
3	Possible	Threats require moderate skills to exploit vulnerability
4	Highly likely	Threats require minimal skills to exploit vulnerability
5	Almost certain	It is easy for threats to exploit vulnerability
6	Very certain	It is very easy for threats to exploit vulnerability

Each assessed electronic payment risk will be assigned one value of likelihood and one value of severity using the defined scales. The inherent risks criticality is the product of likelihood and severity as related at [21].

Table 3. Value of risk severity

Rating	Description	Meaning
1	Minimal	Any impact on strategic objectives
2	Minor	Impact can be managed within current resources
3	Moderate	Impact can be managed with modest extra resources
4	Significant	Impact cannot be managed without extra resources
5	Severe	Impact cannot be managed without significant extra resources
6	Very severe	Could severely compromise strategic objectives

4.3 The Defined Indexes

We define one index related to maturity of security measures. As indicated below, the maturity index has tree values 0, 1 and 2 (Table 4).

Table 4. Value of maturity index

Rating	Index	Description
1	0	Not existent
2	0	Initial
3	1	Systematic
4	2	Managed
5	2	Optimized

We also define four index related to four types of security measures which have 2 values 0 or 1. Each security measure will be assigned one value of maturity index and one value of type index (Tables 5, 6, 7 and 8).

Table 5. Values of deterrent index

Type of control	Index
Deterrent	1
Not deterrent	0

Table 6. Values of prevention index

Type of control	Index
Preventive	1
Not preventive	0

Table 7. Values of detection index

Type of control	Index
Detective	1
Not detective	0

Table 8. Values of corrective index

Type of control	Index
Corrective	1
Not corrective	0

4.4 New Proposed Model

To define our model, we consider that residual electronic payment risks is the risks after taking into account all types of security measures. We also use the defined 8 principles, 3 scales and 5 indexes. Our model which quantify assessment of security measures effect on electronic banking risks is declined as follows:

$$C_{res} = [P * G] - [(\sum_{e=0}^{r}(a * i)/r) + (\sum_{e=0}^{s}(b * j)/s) + (\sum_{e=0}^{t}(c * k)/t) + (\sum_{e=0}^{u}(d * l)/u)] \quad (1)$$

– Equation (2) provides inherent risk by doing the product likelihood and severity of risk.

$$Equation(2) = P * G \quad (2)$$

– Inherent likelihood: P
– Inherent severity: G
– Equation (3) quantifies the maturity of four type security measures. It takes into account deterrent, preventive, detective and corrective security measures considering the principles 2, 3, 4, 5, 6 and 7.

$$Equation(3) = (\sum_{e=0}^{r}(a * i)/r) + (\sum_{e=0}^{s}(b * j)/s) + (\sum_{e=0}^{t}(c * k)/t)$$

$$+ (\sum_{e=0}^{u}(d * l)/u) \quad (3)$$

– maturity(a), index(i) and number(r) of deterrent security measures
– maturity(b), index(j) and number(s) of preventive security measures
– maturity(c), index(k) and number(t) of detective security measures
– maturity(d), index(k) and number(u) of corrective security measures

Globally, the model is obtained by deducting Eqs. (3) to (2) because security measures are defined to reduce inherent risks criticality considering the principle 1.

5   Tests and Results

5.1   Application of the Model on All Electronic Payment Risks

For testing, we collaborated with banks which offer traditional banking services and electronic payment products like withdrawal, payment or transfer using credit, debit or prepaid cards. This sample is representative because other banks have approximately the same risks considering their similar activity, infrastructure and their dependance to laws and regulations. For reasons of confidentiality, we will not disclose identity of those banks.

5.2   Identification of Electronic Payment Risks

To identify electronic payment risks, it is necessary to take account threats and vulnerabilities because the concerned risk is defined as the exercise of a threat against a vulnerability. After identification sessions, we have collected 117 risks and 205 security measures related to electronic payment.

5.3   Assessment of Inherent Electronic Payment Risks

Assessment of inherent electronic payment risks is the process of determining likelihood of the threat being exercised against the vulnerability and the resulting impact from a successful compromise. Each risk is assigned a value of likelihood and severity using the previous defined scales without take into consideration security measures.

5.4   Assessment of Residual Electronic Payment Risks

Residual risks assessment estimates likelihood of threats which are not avoided by security measures. It was done during working sessions with the concerned assessors by combining inherent risks with security measures assessed using maturity scale and types of index. After that, we obtain an assessment of residual risks. Those values are called reference and they will compared to model values.

5.5   Application of the Model on All Electronic Payment Risks

For testing, we apply the model on each electronic payment risks and compare the obtained values to values given by assessors which are called references values. We remarked that model values are equal, upper or lower than reference values but most model values are equal to reference values. Before approving the model, we apply it on types of transaction, vulnerability and attack.

5.6   Application of the Model on Types of Transaction

We test the model on 4 types of transaction (POS, ATM, internet and mobile) regardless of the number of risks it contains. For this, we classify all electronic payment risks by types of transaction before applying the model. After that, we calculate for each types of transaction, the average of maturity by type of security measures, identify the number of security measures and apply the proposed model in order to obtain model values. The residual values and correlation rate between model and reference are shown in the following Table 9:

Table 9. Comparison between Reference vs Model by types of transaction

Types of transaction	Reference values	Model values	Residual values	Correlation rate
ATM	7	6,63	−0.37	95%
POS	12	12,4	0.4	97%
Internet	9	9.33	0.33	96%
Mobile	8,47	8,13	−0.34	96%

As indicated in the table above, we can see that residual values between model and reference by types of transaction are positive or negative. That means model values are upper or lower than reference values. Also we remark that, correlation rate by type of transactions vary between 94% and 97% with an average which is equal to 96%. That means, model values are approximately equal to reference values. So we can conclude that the results are satisfactory by types of transaction.

5.7   Application of the Model on Types of Vulnerability

We also test the model on 7 types of vulnerability (site, data base, application, network, hardware, software and users) regardless of the number of risks it contains. For this, we classify all electronic payment risks by types of vulnerability before applying the model. After that, we calculate for each type of vulnerabilities, the average of maturity by types of security measure, identify the number of security measures and apply the proposed model in order to obtain model values. The residual values and correlation rate between model and reference are shown in the following Table 10:

Table 10. Comparison between Model and Reference by types of vulnerability

Types of vulnerability	Reference values	Model values	Residual values	Correlation rate
Site	10.17	10	−0.17	98%
Data Base	6	6.4	0.4	94%
Application	7	7.26	0.26	96%
Network	9.58	9.16	−0.42	96%
Hardware	10	10.5	0.5	95%
Software	9	9.33	0.33	96%
Users	7.1	6.75	−0.35	95%

As indicated in the table above, we can see that residual values between model and reference by types of vulnerability are positive or negative. That means model values are upper or lower than reference values. Also we remark that, correlation rate by types of vulnerability vary 94% and 98% with an average which is equal to 96%. That means for each types of vulnerability, model values are approximately equal to reference values. Correlation rate average is the same to the previous test because the tests are done on all electronic payment risks and its value 96% shows that model values are globally equal to reference values. So we can conclude that the results are satisfactory by types of vulnerability.

5.8 Application of the Model on Different Types of Attack

We finally test the model on 9 types of attack (social engineering, spoofing, sniffing, spyware, mobile malware, web attacks, physical attacks, cryptanalysis and others) regardless of the number of risks it contains. For this, we classify all risks by types of attack before applying the model. After that, we calculate for each types of attack, the average of control maturity, identify the number of controls and apply the proposed model in order to obtain model values. The residual values and correlation rate between model and reference are shown in the following Table 11:

Table 11. Comparison between Model and Reference by types of attack

Types of attack	Reference values	Model values	Residual values	Correlation rate
Social engineering	7	6.7	−0.3	96%
Spoofing	6.35	6.11	−0.24	96%
Sniffing	10	9.83	−0.17	98%
Spyware	8	7.8	−0.2	98%
Mobile malware	6	6.28	0.28	96%
Web attacks	6	6.23	0.23	96%
Physical attacks	8.1	7.75	−0.35	96%
Cryptanalysis	9	9.26	0.26	97%
Others	7.58	7.16	−0.42	94%

As indicated in the table above, we can see that residual values between model and reference by types of attack are positive or negative. That means model values are upper or lower than reference values. Correlation rate by types of attack vary 94% and 98% with an average which is equal to 96%. That means for each type of attacks, model values are approximately equal to reference values. Correlation rate average is the same because the tests are done on all electronic payment risks and its value 96% shows that model values are globally equal to reference values. So we can conclude that the results are satisfactory by types of attack.

5.9 Measurement of Time Saving

We finally measure the time saving provided by our model on deployment of new security measures. As showed below in the table below, our model helps to save 500 h for assessment of 100 electronic payment risks by 10 assessors. This may be beneficial for banks because it helps to avoid wasting considerable time during residual risks assessment and fraudsters are more and more innovative. Time saving could be used to deploy new security measures because residual threats must be eliminated by additional security measures in order to reduce electronic payment risks to an acceptable level (Table 12).

Table 12. Time saving provided by our model

Electronic payment risk	Time saving per assessor	Time saving for 10 assessors
1 risk	30 min	50 h
20 risks	10 h	100 h
40 risks	20 h	200 h
60 risks	30 h	300 h
80 risks	40 h	400 h
p100 risks	50 h	500 h

6 Conclusion

Banks identify the various factors determining the electronic payment risks proportion at higher degree and suggest protective measures to minimize risks involved but the residual risk assessment method has many limits. The study was taken with an objective to automatize calculation of security measures effect on electronic payment risks. After model application and comparison with the values given by assessors during work sessions, the obtained correlation rate is 96%. Our model provides time saving on deployment of new security measures considering potential frauds which could lead to important financial losses. Also, it helps to improve electronic payment security and to increase banks profit. In our future works, we will try to reduce model parameters and apply it to others banking risks.

References

1. Sakharova, I., Khan, L.: Payment card fraud: challenges and solutions, Department of Computer Science, The University of Texas at Dallas, Technical report UTDCS3411, November 2011
2. Sullivan, R.J.: Risk management and nonbank participation in the U.S. Retail Payments System, Federal Reserve Bank of Kansas City, Second quarter (2007)

3. Arango, C., Taylor, V.: The Role of Convenience and Risk in Consumers: Means of Payment. Currency Department, Bank of Canada, July 2009
4. Fajfar, M.: Role and security of payment systems in an electronic age remarks, Frank, Harris, Shriver & Jacobson LLP, Prepared for IMF Institute Seminar on current Developments in Monetary and Financial Law, 1 June 2004
5. Bhatla, T.P., Prabhu, V., Dua, A.: Understanding credit card frauds, Tata Consultancy Services. Cards Business Review 1, June 2003
6. E-Payments without frontiers, ECB Conference, 10 November 2004
7. Mobile Payments: Risk, Security and Assurance Issues, IL 60008 USA An ISACA Emerging Technology White Paper, November 2011
8. Pereira, A.L., de Alba, A.M.: Understanding the new payment methods, their risks, and opportunities. LexisNexis, Risk Solutions (2011)
9. Anderson, R.: Risk and Privacy Implications of Consumer Payment Innovation, Cambridge University
10. Burns, P., Stanley, A.: Fraud Management in the Credit Card Industry. Payment Card Center, Federal Reserve Bank of Philadelphia, April 2002
11. Payments 101: Credit and Debit Card Payments, Key Concepts and Industry Issues, Pat McLoughlin, First Data White Paper, October 2010
12. Card-Not-Present Fraud Working Committee White Paper: Near-Term Solutions to Address the Growing Threat of Card-Not-Present Fraud, Biometrics in Banking and Payments. Javelin Strategy & Research, April 2015. www.javelinstrategy.com
13. Conroy, J.: 3D Secure: The Force for CNP Fraud Prevention, January 2016
14. Kellogg, P.: Evolving operational risk management for retail payments. Federal Reserve bank of Chicago, Emerging Payments Occasional Papers Series (2003)
15. Weese, D.: Overview of risk assessment methods and applications, June 2006
16. Manual and Electronic Payment: Security Best Practice Treasury and Trade Solutions. Citi Inc., New York (2016). U.S.A.GRA26733, www.transactionservices.citi.com
17. Lipol, L., Haq, J.: Risk analysis method: FMEA/FMECA in the organizations, University of Boras, Sweden. IJBAS-IJENS, 11(5) (2011)
18. Ndaw, M., Mendy, G., Ouya, S.: A quantification model of internal control impact on banking risks using FMECA. IEEE, WICT, Marakech (2015)
19. Ndaw, M., Mendy, G., Ouya, S.: Quantify the maturity of internet banking security measures in WAEMU Banks. EAI, INTERSOL, Dakar (2017)
20. Risk Assessment Process Information Security, All-of-Government Risk Assessment Process: Information Security, February 2014
21. Dumbrav, V., Maiorescu, T., Iacob, V.S.: Using probability impact matrix in analysis and risk assessment. J. Knowl. Manag. Econ. Inf. Technol. (2013). Special Issue

Modeling an Anomaly-Based Intrusion Prevention System Using Game Theory

El Mehdi Kandoussi[✉], Iman El Mir, Mohamed Hanini, and Abdelkrim Haqiq

Computer, Networks, Mobility and Modeling Laboratory,
Faculty of Sciences and Technology, Hassan 1st University, Settat, Morocco
kandoussi.elmehdi@gmail.com, iman.08.elmir@gmail.com,
haninimohamed@gmail.com, ahaqiq@gmail.com

Abstract. In Cloud Computing environment, the availability, authentication and integrity became a more challenging problem. Indeed, the classical solutions of security based on intrusion detection system and firewalls are easily bypassed by experienced attackers. In addition, the use of different technologies in term of security didn't mitigate the attack considerably. To achieve network system's security with the complexity and the diversity of attack types is too difficult and costly. However, to make them more resistant to attacks, anomaly-based Intrusion Prevention System (IPS) are used. Such systems take into consideration the probability of legitimacy of a packet if it didn't match any signature of malicious packets. In this paper, a competitive normal form game is developed based on the probability of packets' legitimacy and the trust that an IPS has over the owner of the packet. Furthermore, a decision is made about dropping, accepting or testing packet in the network, and different Nash Equilibriums are calculated based on the system's parameters. Our approach demonstrated its feasibility in term of prediction of the cases in which the system could be compromised and the actions that should be performed in case of an intrusion.

Keywords: Cloud computing · Security · Anomaly-based IPS · Game theory
Nash equilibrium

1 Introduction

Cloud computing has recently emerged as a well evolved computer technology area. According to the National Institute of Standards and Technology (NIST) [1] introduces cloud computing as "a model for enabling convenient, on demand network access to a shared pool of configurable computing resources (e.g., network, servers, storage, applications and services) that can be rapidly provisioned and released with minimal management effort or service provider interaction.

In the last few years, Cloud Computing becomes more challenging in term of security. Furthermore, different attacks bypassed easily the static measures of security based on rules of a security policy or a signature database of malicious packets [2]. For this reason different solutions as IPS that are proactive are developed. These types of measures that not only alert in case of an attack but also suspect malicious behavior of

© Springer International Publishing AG, part of Springer Nature 2018
A. Abraham et al. (Eds.): IBICA 2017, AISC 735, pp. 266–276, 2018.
https://doi.org/10.1007/978-3-319-76354-5_24

packet and stop them are called an anomaly-based IPS [3]. Unlike Intrusion Detection System (IDS) that detects known attacks effectively and cannot detect new ones without the corresponding signature in the database [4]. For This reason, different methods based on the legitimacy of a packet are developed. In this paper a game theoretic approach is developed for an anomaly-based IPS in order to enhance network security. In this context, a competitive normal form game with two players is proposed. This model uses different parameters including the estimated loss of a malicious packet, the cost of the attack, and the cost of testing or monitoring packet behavior and also the probability of legitimacy in order to make a decision maximizing the utility of the defender.

In this paper, we propose a game-theoretical model to enhance the Cloud security using anomaly based Intrusion Prevention systems to optimize the system behavior against attack and derive the equilibrium conditions for the both players.

For this purpose, our main contributions are listed as follows:

- We applied the anomaly-based IPS to evaluate the legitimacy of a packet and to determine the appropriate actions to play.
- We developed a non-cooperative game between two competing players (IPS and attacker).
  And we calculated the different Nash Equilibriums so as to define the optimal strategies to improve the defensive process.
- The outcome of the game is the probability distribution over the actions of the defender and attacker with respect to the probability of legitimacy.
- We discussed the effectiveness of the proposed solution through some numerical findings considering multiple system's parameters for decision-making while maximizing the defender's utility.

This paper is organized as follows: In Sect. 2, previous works and research done based on IPS are introduced. In Sect. 3, the security game model is introduced and is resolved by finding the corresponding Nash Equilibriums according to the parameter system's variation. In Sect. 4, numerical results are used to prove the effectiveness of our model. In Sect. 5, we finish by concluding and proposing an extension of our work.

2   Related Work

In the context of this work, various researches have been performed in order to enhance security in a complex network environment based on stochastic modeling [5, 6]. However, stochastic modeling using Markov chain gives only idea about how the system will converge and didn't take into consideration the outside interaction of the system.

For this reason, the use of game theoretic approach gives more flexibility to interact by choosing the appropriate action that maximizes the utility. Consequently, different methods improved security by using normal form game [7, 8]. For example, in [7] a normal form game is used in order to detect the potential attack that an attacker will follow to compromise the system. Moreover, to predict the attack path, the model described in [7] eliminates the inefficient measures of security such as migrating to an insecure server to mitigate attack. In addition, stochastic game is also used in order to model attacker-defender interaction.

The authors in [14] have proposed an effective game theory-based false alarm minimization scheme for signature-based IDS. They have suggested a scheme that involves a game theory-based correlation engine to correlate IDS alarms with network vulnerabilities to reduce the false positive alarm rate of the IDS. They have modeled the attack-defense interaction as a non-cooperative game. They have considered different strategies for both attacker and IDS (defender) to analyze the Nash Equilibrium of the game and construct a selection of the sensitive vulnerability of the network to be eventually correlated with the potential IDS alarms to define the final true positive alarms.

In [15], the authors have suggested to display honeypots into an Advanced Metering Infrastructure where this latter presents an integration of many technologies that offer suitable interactions between the consumers in the smart grid and third party systems. They have presented a honeypot game to tackle DDoS attack and investigate some strategies to find an optimal Bayesian Nash equilibrium between legitimate users and attackers. Through some experiments, they validated the effectiveness of defense using strategic honeypot Game model against DDoS attacks in the smart grid.

In [16], the authors have considered a set of mobile agents perform as the sensors of invalid actions in the cloud environment. They get started a non-cooperative game with the attacker and then compute the Nash equilibrium value and utility in order to distinguish an attack from legitimate requests and make a decision about the intensity of attack and its source. They simulated the proposed defense method and demonstrated its efficiency in attack detection with 86% precision and proved that the usage of mobile agents offers a reduced system overhead and higher detection process.

In [9], a zero-sum stochastic game is developed to predict the adversary's actions, to determine vulnerable network assets and to suggest some measures of security. In general, security games are a quantitative method to model this interaction [10]. In another point of view, deep learning is a new approach that used anomaly-based IPS to detect anomalies and classify them [13].

Even if anomaly-based IPS uses a game theoretic approach, the uncertainty about the packet to be dropped or accepted still exists. In our proposed game, a non-cooperative game model tries to give the probability that the suspected packed will be tested and also the interval of legitimacy in which the packet is in order to compromise the system. Thus, the test environment takes into consideration some facts about the owner of the packet in order to make a decision. In general, this probability to accept the packet in case of a test depends on the previous history of the user. Indeed, for users who manipulate a lot of malicious files they will have more probability that their processed packet will be dropped unlike normal users. The objective behind the two probabilities described above is to identify more precisely the type of the user interacting with the network. In addition, an attacker model, which is more general than normal user model, is used to illustrate the behavior of the types of users.

3   Game Model Description

3.1   Game Formulation

The security game model proposed in this paper is a normal form game. This latter is a non-cooperative game since no prior agreement has been done between the players. The actions are taken simultaneously. Then according to this a payoff is distributed to players. For this reason, a matrix representation is used to show the utilities of players based on the profile played. In our model, two players are considered: an attacker and an IPS. The game is defined as follows:

- The set of players $I = \{Attacker, IPS\}$
- The set of actions for each player
    - $S_{Attacker} = \{Malicious, Legitimate, Unknown\}$
    - $S_{IPS} = \{Drop, Accept, Test\}$
- The utility function for each player is detailed below in Table 1:

Table 1. Matrix representation of the game

IPS	Attacker		
	Malicious	Legitimate	Unknown
Drop	0	$-p$	$-\alpha p$
	$-c$	$-p$	$-\alpha p - (1-\alpha)c$
Accept	$-l$	p	$\alpha p - (1-\alpha)l$
	$l - c$	p	$\alpha p - (1-\alpha)(l-c)$
Test	$-\beta l - C^T$	$-C^T$	$-\beta(1-\alpha)l - C^T$
	$\beta l - c$	βp	$\beta[\alpha p + (1-\alpha)l] - (1-\alpha)c$

For example if the attacker chooses a malicious packet to send and the IPS accepts this latter, the profile played is (*Accept, Malicious*). The corresponding utilities for each player are illustrated above in the intersection of the third line and second column of the matrix. In this case, the attacker has a positive utility equal to $l - c$ in the other hand the IPS will suffer a loss equal to $-l$.

The significance of the mathematical expressions used in utility of each player will be detailed below.

In case when (*Drop, Legitimate*) is played, the malicious user will have a negative payoff equal to $-p$ since his request is dropped. Concerning the IPS utility, it's the same as the attacker since it's a legitimate request that should be routed to a service. In addition, the fact that this latter is negative and not positive is that the IPS is enabling to differentiate between a malicious user and a normal one. Consequently, the action space of the second player named: Attacker can take also into consideration the behavior of a player without attention to compromise.

In reality, p represents the benefit from having a response to the request. The variable $c(resp.l)$ in the other expressions represents the cost of the attack (resp. the loss caused by the attack). The cost of testing a packet in an isolated environment in order to identify its type is denoted C^T. The parameter β is the probability to accept a packet when it is tested. For example, if $\beta = 0(resp.\beta = 1)$ the packet will be automatically dropped (resp. accepted). The parameter α is the probability of the unknown packet to be Legitimate. For example, if $\alpha = 0(resp.\alpha = 1)$ it means that the packet is malicious (resp. Legitimate).

In general, the cost of the attack is lower than the estimated loss caused by this latter. Moreover, the cost of the attack in this model is supposed to be greater than the benefit gained from a legitimate request. In this case, we have: $p < c < l$.

Concerning the cost of testing, it's lower than the loss l. We have: $C^T < p < l$.

To summarize the hypothesis introduced above:

- $0 \leq \alpha \leq 1$ and $0 \leq \beta \leq 1$
- $0 < p < c \ll l$
- $0 < C^T < p < l$
- For future use, the utility of the IPS (resp. attacker) is denoted $U_{IPS}(resp.U_{Attacker})$.

3.2   Game Resolution

To solve a normal form game is to find the set of Nash Equilibrium. The formal definitions of a Nash Equilibrium and a strictly dominated strategy are as follows:

The notation $s = (s_i, s_{-i})$ is used instead of using $s = (s_1, \ldots, s_n)$ and denotes the action profile played by n players and s_{-i} is the profile played without the player i.

- **Definition (*Strictly dominated Strategy and Nash Equilibrium*)**
    - A strategy $s_i \in S_i$ of the player i is strictly dominated by s_i' if:

$$\forall s_{-i} \in S_{-i} : u_i(s_i, s_{-i}) < u_i(s_i', s_{-i}) \tag{1}$$

    - The profile strategy $s^* = \langle s_1^*, \ldots, s_n^* \rangle$ is a Nash Equilibrium if:

$$\forall i \in I, s_i \in S_i : u_i(s_i^*, s_{-i}^*) \geq u_i(s_i, s_{-i}^*) \tag{2}$$

The first inequality (1) signifies that a strictly dominated should never be played by a rational player (which is the majority of the cases). Concerning the second inequality (2), it indicates that given a profile s_{-i}^* the player is enabling to increase his payoff by deviating from s_i^*. It means that s^* is the profile that maximizes the utility of all the players.

In the rest of this section, a pure Nash Equilibrium (resp. Mixed Nash Equilibrium) is denoted by $NE_{pure}(resp.NE_{mixed})$. According to the variables of the game, the resolution of this latter is as follows:

- If $\dfrac{c}{l-p} > 1$: *Malicious* and *Unknown* are two strategies that are strictly dominated by *Legitimate*. As a result, the unique equilibrium with utilities of each player is:

$$NE_{pure} = (Accept, Legitimate) \tag{3}$$

$$\begin{cases} U_{IPS} = p \\ U_{Attacker} = p \end{cases} \tag{4}$$

- If $\dfrac{c}{l-p} \leq 1$: We have two cases.

  - If $0 < \dfrac{l - C^T}{l + p} < \dfrac{l}{l + 2p} < \dfrac{C^T}{p} < \dfrac{l + C^T}{l + p} < 1$:

  - We have a mixed Nash Equilibrium. In this case, the *Attacker* must randomize in order to make the *IPS* indifferent to choosing either strategy. Formally, this approach is translated by the following equation:

$$EU_{IPS}(Drop) = EU_{IPS}(Accept) \Leftrightarrow \begin{cases} x = \dfrac{2p}{l + 2p} \\ y = 1 - x = \dfrac{l}{l + 2p} \end{cases} \tag{5}$$

The same reasoning goes with the *IPS*.

$$EU_{Attacker}(Malicious) = EU_{Attacker}(Legitimate) \Leftrightarrow \begin{cases} x' = \dfrac{l - c - p}{l - 2p} \\ y' = 1 - x' = \dfrac{c - p}{l - 2p} \end{cases} \tag{6}$$

As a result, we have:

$$NE_{mixed} = \left(x'.Drop + y'.Accept, x.Malicious + y.Legitimate \right) \tag{7}$$

$$\begin{cases} U_{IPS} = -x'.(1 - x).p + (1 - x').\left[-l.x + (1 - x).p\right] \\ U_{Attacker} = x.\left[-c.x' + (1 - x').(l - c)\right] + (1 - x).\left[-p.x' + p.(1 - x')\right] \end{cases} \tag{8}$$

-

If $0 < \dfrac{C^T}{p} < \dfrac{l}{l + 2p} < \dfrac{l - C^T}{l + p} < \dfrac{l + C^T}{l + p} < 1$:

For $\forall \alpha \in [0, 1]$ *and* $\beta \in [0, 1] \backslash \left\{ \dfrac{c}{l - p} \right\}$, we have a mixed Nash Equilibrium. The same approach as described above can be used to obtain the explicit expression of this latter as the players' utilities.

- If $\beta = \dfrac{c}{l - p}$ *and* $\left(\alpha \in \left[\dfrac{l}{l + 2p}, \dfrac{(1 - \beta)l - C^T}{((1 - \beta)l + p} \right] \textit{ or } \alpha \in \left[\dfrac{\beta.l + C^T}{\beta.l + p}, \dfrac{(1 - \beta)l - C^T}{((1 - \beta)l + p} \right] \right)$, we have a pure Nash Equilibrium:

$$NE_{pure} = (Test, Unkown) \tag{9}$$

$$\left\{ \begin{array}{l} U_{IPS} = -\beta.(1 - \alpha)l - C^T \\ U_{Attacker} = \beta \big[\alpha p + (1 - \alpha)l \big] - (1 - \alpha)c \end{array} \right. \tag{10}$$

4   Numerical Results

In order to illustrate the behavior of an attacker in front of an IPS based on a test environment and the measure of security that this latter should take, these values are used: The values in (11) are taken in the same size as that in [11, 12].

$$L = 50, \; c = 5, \; p = 4 \textit{ and } Ct = 1. \tag{11}$$

According to the figure above (see Fig. 1), for $\alpha \in [0, 0.68]$ *or* $[0.89, 1]$ the attacker randomizes over the support $\{Malicious, Legitimate\}$ with less probability to invest in malicious packet. This is due to the awareness of the intrusion prevention system of the malicious packets. In the other hand, for values of $\alpha \in [0.68, 0.89]$, an unknown packet is used with less probability to be harmful. The reason is that the packet will be processed by the test environment. In addition, this gives an idea about the legitimacy of a malicious packet in order to be tested than to be accepted. In another point of view, the interval $[0.68, 0.89]$ represents also vulnerability for a successful intrusion.

In the figure above (see Fig. 2), the utility of the attacker is negative for values of $\alpha \in [0, 0.68]$ *or* $[0.89, 1]$. This proves that the attacker will not invest in packet having legitimacy in these two intervals. Consequently, it will try to compromise the system by using a packet representing legitimacy in $[0.68, 0.89]$. Indeed, for these values, its utility is positive.

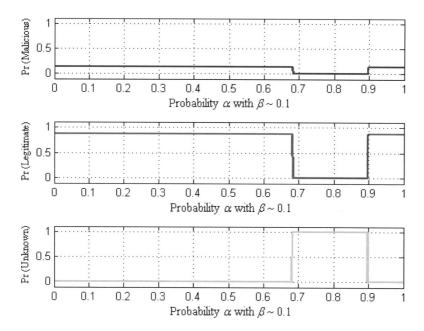

Fig. 1. Probability distribution over attacker's actions with respect to the probability of legitimacy α

According to the figure above (see Fig. 3), for $\alpha \in [0, 0.68]$, the packet will be dropped since it represents less legitimacy. Therefore, for $\alpha \in [0.68, 0.89]$, the packet will be tested. In addition, the value of β, translates in a sense the idea that the IPS has about the attacker. Indeed, since this latter is small all packets that represent a threat will be dropped. In section above, there is a Nash equilibrium that translates the behavior of a normal user. Consequently, the majority of his packets are accepted.

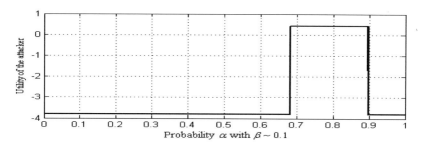

Fig. 2. Utility of the attacker with respect to the probability of legitimacy α

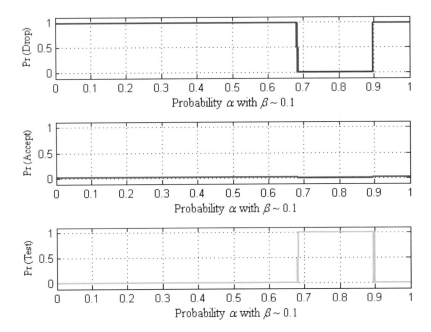

Fig. 3. Probability distribution over defender's actions with respect to the probability of legitimacy α

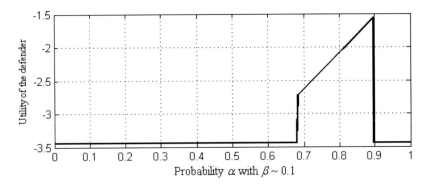

Fig. 4. Utility of the defender with respect to the probability of legitimacy α

As illustrated above (see Fig. 4), the utility of the defender is negative for packet representing less legitimacy below than 0.68. For value of $\alpha \in [0.68, 0.89]$, the utility of the defender increase greatly. The reason is that the IPS uses his test environment in order to accept or drop the received packet in this interval. In general, the IPS without a test environment will have a negative utility equal to -3.5. This proves the contribution of testing in term of security. For value greater than 0.89, we have a negative utility this is due to the attacker that randomizes over: {*Malicious, Legitimate*}. In case of the others IPS without a testing environment as in [9], we will not have a growth of defender's utility for $\alpha \in [0.68, 0.89]$.

5   Conclusion and Future Work

In this paper, an intrusion prevention system based on a test environment is studied using a game theoretic approach. Such a system uses a prior knowledge in order to make a decision in terms of dropping, accepting or testing packet. The introduced model, predicts the behavior of malicious packets that compromise the system even if they behave legitimately. Unlike other static solutions which are based on a database signature so as to drop or accept packets and in the majority of the cases are configured to accept by default unknown ones, the proposed solution demonstrates its effectiveness to detect the cases where the attacker uses malicious packets. As future work, we seek to mitigate attack impact in cloud computing environment by using a dynamic game, which behaves according to the history of the user in such a way as to maximize the probability of detection and facilitates the decision-making.

References

1. Mell, P., Grance, T.: The NIST definition of cloud computing (2011)
2. Hock, F., Kortiš, P.: Commercial and open-source based Intrusion Detection System and Intrusion Prevention System (IDS/IPS) design for an IP networks. In: 13th International Conference on Emerging eLearning Technologies and Applications (ICETA), Slovakia, pp. 1–4. IEEE (2015)
3. Garcia-Teodoro, P., Diaz-Verdejo, J., Macia-Fernandez, G., Vazquez, E.: Anomaly-based network intrusion detection: techniques, systems and challenges. Comput. Secur. **28**(1), 18–28 (2009)
4. Ajay Kumara, M.A, Jaidhar, C.D: Hypervisor and virtual machine dependent intrusion detection and prevention system for virtualized cloud environment. In: 1st International Conference on Telematics and Future Generation Networks (TAFGEN), Malaysia, pp. 28–33. IEEE (2015)
5. El Mir, I., Kim, D.S., Haqiq, A.: Security modeling and analysis of an intrusion tolerant cloud data center. In: Third World Conference Complex Systems (WCCS), Morocco, pp. 1–6. IEEE (2015)
6. El Mir, I., Kim, D.S., Haqiq, A.: Security modeling and analysis of a self-cleansing intrusion tolerance technique. In: 11th International Conference Information Assurance and Security (IAS), Morocco, pp. 111–117. IEEE (2015)
7. El Mir, I., Kandoussi, E.M., Hanini, M., Kim, D.S., Haqiq, A.: A game theoretic approach based virtual machine migration for cloud environment security. Int. J. Commun. Netw. Inf. Secur. (IJCNIS) **3**(9), 345–357 (2017)
8. Wu, Q., Shiva, S., Roy, S., Ellis, C., Datla, V.: On modeling and simulation of game theory based defense mechanisms against Dos and DDoS attacks. In: The spring Simulation Multiconference, pp. 312–318. ACM, San Diego (2010)
9. Ibidunmoye, E.O., Alese, B.K., Ogundele, O.S.: Modeling attacker-defender interaction as a zero-sum stochastic game. J. Comput. Sci. Appl. **2**(1), 27–32 (2013)
10. Alpcan, T., Basar, T.: Network Security: A Decision and Game-Theoretic Approach. Cambridge University Press, New York (2010)
11. Kamhoua, C.A., Kwiat, L., Kwiat, K.A., Park, J.S., Zhao, M., Rodriguez, M.: Game theoretic modeling of security and interdependency in a public cloud. In: 7th International Conference Cloud Computing (CLOUD), USA, pp. 514–521. IEEE (2014)

12. Kwiat, L., Kamhoua, C.A., Kwiat, K.A., Tang, J., Martin, A.: Security-aware virtual machine allocation in the cloud: a game theoretic approach. In: 8th International Conference Cloud Computing (CLOUD), New York, pp. 556–563. IEEE (2015)

13. Van, N.T., Thinh, T.N., Sach, L.T.: An anomaly-based network intrusion detection system using deep learning. In: 2017 International Conference on System Science and Engineering (ICSSE), Ho Chi Minh City, pp. 210–214. IEEE (2017)

14. Subba, B., Biswas, S., Karmakar, S.: False alarm reduction in signature-based IDS: game theory approach. Secur. Commun. Netw. **18**(9), 4863–4881 (2016)

15. Wang, K., Du, M., Maharjan, S., et al.: Strategic honeypot game model for distributed denial of service attacks in the smart grid. IEEE Trans. Smart Grid **5**(8), 2474–2482 (2017)

16. Nezarat, A., Shams, Y.: A game theoretic-based distributed detection method for VM-to-hypervisor attacks in cloud environment. J. Supercomput. **10**(73), 4407–4427 (2017)

Optimized Security as a Service Platforms via Stochastic Modeling and Dynamic Programming

Oussama Mjihil[1]([⊠]), Hamid Taramit[1], Abdelkrim Haqiq[1,3], and Dijiang Huang[2]

[1] Computer, Networks, Mobility and Modeling Laboratory, FST, Hassan 1st University, Settat, Morocco
o.mjihil@uhp.ac.ma, hamidtrmt@gmail.com, ahaqiq@gmail.com
[2] School of Computing, Informatics and Decision Systems Engineering, Arizona State University, Tempe, AZ, USA
dhuang8@asu.edu
[3] e-NGN Research Group, Africa and Middle East, Rabat, Morocco

Abstract. As an important Cloud feature, security assessment has attracted massive attention in the recent years. However, the broad adoption of Cloud computing and its fast growth are raising critical scalability issues for security assessment frameworks. In this work, we addressed the scalability issue from different perspectives. First, we suggested a decentralized multi-tenant security assessment tool, where both the Cloud Service Providers (CSP) and tenants are able to perform security assessment and mitigation. Secondly, the framework uses analytical stochastic modeling altogether with dynamic programming (here, we use value iteration algorithm) to find the optimal policy concerning the routing of security requests between different execution agents. Thirdly, we designed an iterative heuristic algorithm that helps the CSP to improve the system parameters to get better performance. The simulation results have demonstrated the benefit of using our framework in terms of identifying the optimal routing policy of security requests and improving system configuration.

Keywords: Security · Performance evaluation
Dynamic programming

1 Introduction

Cloud commuting models are evolving side by side with the clients' needs in terms of security, scalability, performance, and so on. Recently, multi-tenancy has emerged as an essential feature that is broadly implemented in recent Cloud platforms [12] to handle the growing size and the complexity of the tenants' networked systems. Traditional security assessment tools are designed to operate in a centralized fashion. In other words, a monolithic software platform is

© Springer International Publishing AG, part of Springer Nature 2018
A. Abraham et al. (Eds.): IBICA 2017, AISC 735, pp. 277–287, 2018.
https://doi.org/10.1007/978-3-319-76354-5_25

in place to perform all the security related operations, such as scanning Virtual Machines (VMs), getting the reachability information, constructing attack models, evaluating the system security, and selecting the appropriate countermeasures. In Cloud computing, which is designed to handle multiple tenants and a large amount of resources, monolithic design of security tools seems challenging and not practical.

The aim of this work is to study the performance and optimize the resource consumption of a SECurity as a Service (SECaaS) platform that will be used, simultaneously, by the Cloud Service Provider (CSP) and tenants. The system consists of a pool of security agents having different characteristics and different categories of tenants that generate a workload of security requests. It is practically common, in Cloud computing, that tenants are heterogeneous and their requests may have different priorities. We enhanced our model to take into consideration two categories of tenants (premium and standard), which have different priorities on the queues of two categories of security agents (premium and standard). These agents have different characteristics, such as the performance and the execution costs for each type of requests.

To optimize the security agents' resource exploitation, we added to the system model a set of controllers to route the security requests. The optimal routing policy of requests is obtained dynamically using value iteration algorithm [15]. Furthermore, we have built an algorithm that uses the statistics concerning the optimal policies and budget functions to determine which configuration will give optimal resource exploitation and better performance.

The main contributions of this work are summarized as follows:

- Our framework addresses the security assessment's scalability in multi-tenant Cloud platforms by deploying distributed security agents.
- We used an analytical model to evaluate the performance of the entire security system.
- We considered the resource limitation that security tools may encounter in practice. Thus, we suggested a decision support tool to help the CSPs determine the system configuration that produces the optimal resource usage.

The rest of this paper is organized as follows. After presenting the related work in Sect. 2, we show the system model in Sect. 3, we formulate the problem in Sect. 4, we present the equivalent discrete-time time problem in Sect. 5, then we introduce our algorithm concerning the system configuration enhancement in Sect. 6, followed by the numerical results in Sect. 7, and finally, we conclude our paper in Sect. 8.

2   Related Work

CSPs are aware of the importance of their infrastructures' scalability, response time, and reliability. In multi-tenant Cloud environments, many CSPs provide security tools as a Cloud service.

Security as a Service Models for the Cloud. Cloud computing model is designed to enable resource sharing between different tenants at a large scale. Accordingly, Varadharajan et al. [16] proposed a SECaaS model that CSPs provide to the Infrastructure as a Service (IaaS) tenants. Their work aims to extend the CSPs' control over the tenants' infrastructures in the limit of their privacy guidelines. Moreover, the tenants' preference may vary from the use of their own security tools to the entire reliance on the CSP's security platforms.

Pawar et al. [13] suggested a SECaaS for multi-Cloud and federated Cloud environments to protect data, applications, and hosts. Garfinkel et al. [4] proposed an Intrusion Detection System (IDS) with Virtual Machine Monitor (VMM) to detect attacks carried out on VMs. Unlike host-based IDS, the suggested technique aims to inspect the VMs from the outside. This introspection requires the ability to analyze the software running inside VMs which may not meet the Tenant's privacy requirement or allow an insider attacker to introspect the tenants' VMs. To make any introspection between the Tenants' VMs impossible, Some works [12,19] have addressed multi-tenancy in the Cloud using nested virtualization, so tenants will be completely separated and the security status of a tenant's VM will not impact others.

Butt et al. [2] introduced a self-service Cloud computing to address the tenant's control flexibility over their VMs while ensuring their privacy and security. Instead of using one entrusted domain (e.g., Xen dom0), the new privilege model allows to have an administrative domain per each tenant. The privileged domain separation has been also proposed by Yu et al. [18]. Mjihil et al. [11] addressed the scalability issue in multi-tenant Cloud using decomposition techniques.

In all the previously suggested SECaaS frameworks, and to the best of our knowledge, there is few work addressing the scalability issue and no analytical model is used to evaluate the performance and reliability of the suggested security solutions. Instead, The authors used simulations and repeated real case experiments.

Stochastic Models for the Cloud Performance Analysis. Some recent research works have addressed the Cloud performance using different analytical models. Ghosh et al. [5] have addressed the scalability of the Cloud IaaS using a stochastic analytical model for performance quantification. they considered different pools of resources with different characteristics and a multilevel stochastic IaaS model instead of single-level monolithic model. Accordingly, a later work [3] addressed the same problem as [5] while the authors considered that all the pools are in active state, arriving job will be affected only to the pools with non full queues, and the heterogeneity of resources and workloads. Another distinguished work [1] has addressed the Cloud data center performance and Quality of Service (QoS) using Stochastic Reward Nets (SRNs).

3 System Model

Security assessment process generates an important amount of requests of different types, each type has specific nature and different execution time. A security request can be a scan of a specific host or list of hosts, an investigation about security policies or network configuration, countermeasure application, attack model generation, security evaluation, or security metrics computation. In this work, we consider all these requests as a unified security traffic, which have an average arrival and service rates.

Using this framework, the CSP makes available a shared pool of security agents to serve efficiently the security requests generated by different tenants. As shown in Fig. 1, we consider two types of workloads (premium requests and standard requests) and two categories of agents (main agent and standard agent). These agents have queues of limited sizes and different service rates.

Premium Requests (PRs) arrive from a Poisson process at rate λ_p, and Standard Requests (SRs) arrive from a Poisson process at rate λ_s. The requests are served at Main Agent (MA) and Standard Agent (SA) by exponential servers at rate μ_m and μ_d respectively.

A controller is placed at the system's input to control the arriving requests depending on capacities and states of MA and SA. The premium requests are whether accepted and allocated in MA's queue or they are rejected. The standard requests are whether routed to MA or SA, or rejected.

In main agent, the premium requests have priority over the standard requests. When a premium request arrives to the main agent and finds a standard request in service, it takes its place and puts it in the waiting line.

We maintain the system stable by ensuring that both MA's queue and SA's queue are stable: $\lambda_p + \lambda_s < \mu_m$ and $\lambda_s < \mu_d$.

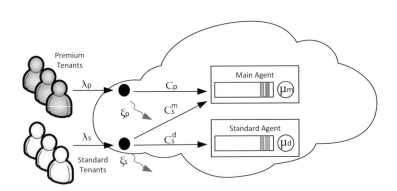

Fig. 1. Shared pool of security agents with priority

4   Problem Formulation

The state of the system at time $t, t \geq 0$, is defined by a stochastic process $(x_t, t \geq 0)$, where $x_t = (x_t^p, x_t^s, x_t^d)$. x_t^p (respectively, x_t^s) is the total number of premium requests (respectively, standard requests) in main agent at time t and x_t^d is the total number of standard requests in standard agent at time t.

The state space is $S = \{0, 1, ..., B_m\}^2 \times \{0, 1, ..., B_s\}$, where B_m is the capacity of the MA, and B_s is the capacity of the SA.

With each state x in S we associate a set of admissible actions $D = \{0, 1\} \times \{0, 1, 2\}$. Thus, an admissible action $a_t(x) = (a_t^p(x), a_t^s(x))$ in state x at time t, with values in D is defined by $a_t = (a_t^p, a_t^s)$ where

$$a_t^p = \begin{cases} 0, & \text{if the arriving PR is rejected;} \\ 1, & \text{if the arriving PR is accepted.} \end{cases}$$

and

$$a_t^s = \begin{cases} 0, & \text{if the arriving SR is rejected;} \\ 1, & \text{if the arriving SR is routed to MA;} \\ 2, & \text{if the arriving SR is routed to SA.} \end{cases}$$

Defining the action space to be the set $A = D^S$, we can now represent an Admissible Policy (AP) as an A-valued stochastic process $\pi = (a_t, t \geq 0)$. Let P denote the set of all AP's. A law of motion corresponding to a policy π is specified by a transition probability $P(y|x, a_t)$, $x, y \in S$, $t \geq 0$, denoting the conditional probability that the system moves to state y at time t^+ when the action $a_t(x)$ is applied to it at time t, while in state x.

At each instant t, an instant cost $C(x_t, a_t)$ is incurred and it is given by:

$$C(x_t, a_t) = c_p x_t^p + \lambda_p \xi_p \mathbb{1}_{\{a_t^p=0\}} + c_s^m(x_t^p + x_t^s) + c_s^d x_t^d + \lambda_s \xi_s \mathbb{1}_{\{a_t^s=0\}}, \quad (1)$$

where c_p is the acceptance cost of the premium requests, c_s^m (respectively, c_s^d) is the cost to pay for each routing of standard request to the main agent (respectively, to the standard agent). ξ_p (respectively, ξ_s) is the rejection cost of premium requests (respectively, standard requests).

$\xi_p > c_p$ and $\xi_s > c_s^m > c_s^d$, and all these coefficients are positive real numbers.

Our objective is to find an optimal policy π minimizing the following discounted cost:

$$\limsup_{T \to \infty} E_\pi \left[\int_0^T e^{-\delta t} C(x_t, a_t)\, dt \right], \quad (2)$$

where $\delta > 0$ is the discount rate.

Remark 1. We assumed that $\xi_p > c_p$ (respectively, $\xi_s > c_s^m$) in order to exclude the situation where the policy that rejects all premium requests (respectively, standard requests) is optimal. And we assumed that $c_s^m > c_s^d$ in order to exclude the situation where the policy that sends all the accepted standard requests to main agent is optimal.

An admissible policy π that assigns to each state x a fixed action $a = \pi_i$ and always uses this action whenever the system is in state x is called a Stationary Policy (SP) [15]. Let $P_s \subset P$ denote the set of all SP's.

We assert from Lippman [10] and Walrand [17] that an optimal policy exists. Moreover, it is stationary.

The inter-arrival and inter-departure times of the security requests are exponentially distributed. Furthermore, the action set D is finite. The assumptions of [10] (Theorem 1, p. 1229) and [17] (Proposition 8.5.3, p. 275) are thereby satisfied, leading to our assertion above.

Hereafter, in view of our previous assertion, we restrict attention to stationary policies and define the δ-discounted cost starting with initial state x associated with the problem (2) by:

$$J^\delta(x) = \min_{\pi \in P_s} E_\pi \left[\int_0^\infty e^{-\delta t} C(x_t, a_t) \, dt \right]. \tag{3}$$

The minimum cost in (3) can be expressed in an alternative form, which facilitates further analysis. Let $0 = t_0 < t_1 < \cdots < t_n \cdots$ be the (random) instants in time denoting transition epochs of the system state $(x_t, t \geq 0)$.

It is convenient to introduce at this point the δ-discounted expected cost over the time-horizon $[0, t_n)$, with initial state x, and following a policy π in P_s, namely,

$$V_n^\delta(x, \pi) = E_\pi \left[\int_0^{t_n} e^{-\delta t} C(x_t, a_t) \, dt \right]. \tag{4}$$

Let

$$J_n^\delta(x) = \min_{\pi \in P_s} V_n^\delta(x, \pi), \quad n = 0, 1, \dots \tag{5}$$

$$J_\infty^\delta(x) = \lim_{n \to \infty} J_n^\delta(x). \tag{6}$$

In the same way as [8], we can prove that the minimum cost in (3) has the alternative expression:

$$J^\delta(x) = J_\infty^\delta(x), \tag{7}$$

for every initial state x.

Let π^* be an optimal stationary policy, T_p the total premium requests accepted under the policy π^*, and T_s (respectively, T_d) the total standard requests routed to MA (respectively, routed to SA) following π^*. The utilizations of MA and SA denoted by ρ_m and ρ_d respectively are defined as follows

$$\rho_m = \frac{T_p \lambda_p + T_s \lambda_s}{(T_p + T_s)\mu_m}, \quad \text{and} \quad \rho_d = \frac{T_d \lambda_s}{T_d \mu_d} = \frac{\lambda_s}{\mu_d}. \tag{8}$$

5   The Equivalent Discrete-Time Problem

In this section, we convert the original continuous-time problem (2) into its discrete-time equivalent by the standard procedure of uniformization [6,7,14].

By suitably introducing dummy departures as in [9,14], the inter-epoch intervals are seen to be i.i.d. random variables with distribution:

$$P[t_{k+1} - t_k > t] = e^{-t(\lambda_p + \lambda_s + \mu_m + \mu_d)}, \quad k = 0, 1, \dots \tag{9}$$

We consider the discrete time system obtained as in [7,14] by sampling the original continuous-time system at its transition epochs. To this end, we introduce the notation $x_k \triangleq x_{t_k}$ and $a_k \triangleq a(x_{t_k})$ and define:

$$\beta = \frac{\lambda_p + \lambda_s + \mu_m + \mu_d}{\lambda_p + \lambda_s + \mu_m + \mu_d + \delta}. \tag{10}$$

We can now conveniently convert the continuous time optimization problem into an equivalent discrete time problem. The β-discounted cost incurred by the n-step discrete time system for the policy π is defined [7,14] as:

$$\widetilde{V}_n^\beta(x, \pi) = E_x^\pi \left[\sum_{k=0}^{n-1} \beta^k C(x_t, a_t) \right]. \tag{11}$$

It then follows, as in [7,14], that

$$V_n^\delta(x, \pi) = \frac{1 - \beta}{\delta} \widetilde{V}_n^\beta(x, \pi). \tag{12}$$

Let

$$\widetilde{V}^\beta(x, \pi) = \lim_{n \to \infty} \widetilde{V}_n^\beta(x, \pi). \tag{13}$$

We can now state the minimization problem (2) in terms of a discrete-time problem of equivalent cost as follows. The minimum β-discounted cost for the n-step and infinite horizon discrete-time systems, respectively, by:

$$\widetilde{J}_n^\beta(x) = \min_{\pi \in P} \widetilde{V}_n^\beta(x, \pi), \tag{14}$$

and

$$\widetilde{J}^\beta(x) = \min_{\pi \in P} \widetilde{V}^\beta(x, \pi). \tag{15}$$

Letting

$$\widetilde{J}_\infty^\beta(x) = \lim_{n \to \infty} \widetilde{J}_n^\beta(x). \tag{16}$$

As in previous section, we have

$$\widetilde{J}_\infty^\beta(x) = \widetilde{J}^\beta(x), \tag{17}$$

for every initial state x.

Finally, the equivalence in the sense of optimal discounted cost, between (2) and the discrete-time formulation above, follows readily from (12) and (16) by noting that:

$$J^\delta(x) = \frac{1 - \beta}{\delta} \widetilde{J}^\beta(x). \tag{18}$$

Thus, we may restrict our attention hereafter to the discrete-time β-discounted cost problem defined by (15). We can now proceed to develop the dynamic programming equation for the problem (15). Walrand [17] (Proposition 8.5.3, p. 275), proves that $\widetilde{J}^{\beta}(x)$ is the unique bounded solution of the following dynamic programming equation:

$$\widetilde{J}^{\beta}(x) = \min_{\pi \in P} \left\{ C(x,a) + \beta \sum_{y \in S} P(y|x,a) \widetilde{J}^{\beta}(y) \right\}, \tag{19}$$

where $C(x,a)$ is the instantaneous cost and $P(y|x,a)$ is the conditional probability (for the discrete-time problem) that the system moves to state y at time $n+1$ when the action $a(x)$ is applied to it at time n while in state x.

6   System Configuration Enhancement

As we have proven, in Sect. 4, the existence of an optimal stationary (time-independent) policy for our problem. This policy is a sequence of actions that solve the dynamic programming Eq. (19). Algorithms, such as Value Iteration Algorithm (VIA), Policy Iteration Algorithm (PIA), and Linear Programming (LP) can all be applied to solve this equation (a.k.a. Bellman's optimality equation). We use the VIA in our work because of its theoretical simplicity and ease of implementation. The VIA yields the optimal stationary policy and corresponding expected total costs for our optimization problem and, hence inserts those results in a table called Lookup Table (LT).

Billing based on the resource consumption is one of the fundamental aspect of Cloud computing. Here, the clients are characterized by their budgets. Following are the budget functions for the premium and the standard categories of tenants, receptively: $B_p = \xi_p - c_p$ and $B_s = (\xi_s - c_s^m) + (\xi_s - c_s^d)$. We consider the constrains mentioned in Remark 1 and that $B_p > B_s$.

Algorithm 1. System Parameters Re-configuration

1: *Initialize_Parameters(Parameters)*
2: **while** *Fixed_Parameters* is not full **do**
3:     **for** all Parameters $\notin$ *Fixed_Parameters* **do**
4:         *Current_Parameter_Range = Budget(Parameters)*
5:         **for** all Values of *Current_Parameter_Range* **do**
6:             *Statistics = Value_Iteration(Parameters)*
7:             $NewPerformance = \left[\frac{B_p'}{B_p' + B_s'} * PR_M \right] + \left[\frac{B_s'}{B_p' + B_s'} * (SR_M + SR_S) \right]$
8:             **if** *NewPerformance > OldPerformance* **then**
9:                 Update *Current_Parameter* and OldPerformance
10:            **end if**
11:        **end for**
12:        **if** The *Current_Parameter* is not updated **then**
13:            add *Current_Parameter* to *Fixed_Parameters*
14:        **end if**
15:    **end for**
16: **end while**

In this section, we suggest an iterative performance improvement procedure, as described in Algorithm 1, to find the best system configuration. In this procedure, the decision routing statistics of the optimal polices and the budget functions are used together to improve, in each iteration, one of the system's parameters (one of the decision costs) in function of the other parameters, which are considered fix.

We measure the efficiency of the system configuration using the statistics produced by the related optimal policy, so a better configuration improves the percentage of the premium requests accepted in MA (PR_M) as well as the standard requests accepted in the MA (SR_M) and SA (SR_S). In this metric, B'_p and B'_s are the budget functions found in the preceding iteration.

7   Numerical Results

We consider in our simulation, fixed values of $\lambda_p = 20$, $\lambda_s = 10$, $\mu_m = 55$, and $\mu_d = 45$. The Algorithm 1 improves the system parameters (c_p, ξ_p, c_s^m, c_s^d, ξ_s). Table 1 summarizes the results obtained from our suggested iterative algorithm.

Table 1. Configuration enhancement

Iteration	Fixed parameters	Optimized parameter	Budgets		Statistics			Performance
			Premium	Standard	SR_M	SR_S	PR_M	
1	$c_p = 2$; $\xi_p = 12$; $c_s^m = 2$; $c_s^d = 0.5$; $\xi_s = 4$	–	10	5.50	30.88%	48.60%	32.09%	43.93
2	$c_p = 2$; $\xi_p = 12$; $c_s^m = 2$; $\xi_s = 4$	$c_s^d = 0.3$	10	5.70	30.48%	54.68%	32.09%	45.35
3	$c_p = 2$; $\xi_p = 12$; $c_s^d = 0.3$; $\xi_s = 4$	$c_s^m = 0.6$	10	7.10	48.06%	51.60%	53.61%	65.12
4	$c_p = 2$; $\xi_p = 12$; $c_s^m = 0.6$; $c_s^d = 0.3$	$\xi_s = 4$	10	7.10	48.06%	51.60%	53.61%	65.12
5	$\xi_p = 12$; $c_s^m = 0.6$; $c_s^d = 0.3$; $\xi_s = 4$	$c_p = 0.5$	11.5	7.10	42.38%	57.49%	98.53%	98.86
6	$c_s^m = 0.6$; $c_s^d = 0.3$; $\xi_s = 4$; $c_p = 0.5$	$\xi_p = 10.5$	10	7.10	42.38%	57.49%	98.53%	98.86

The algorithm was initialized with the configuration presented in the first iteration, which gave low performance (43.93%). The second iteration updates the parameters by optimizing c_s^d while the others are fixed. The new configuration with optimized c_s^d (at value 0.3) gave a policy that increases SR_S, and enhance the performance (45.35%), with a very small increase in the budget of standard tenants. In iteration 3, the parameter c_s^m is optimized at value 0.6, which gives a configuration that increases the amount of accepted requests, however the standard budget increased from 5.70 to 7.10, but the performance raised to 65.12%. In the fourth iteration, we observed that ξ_s is already at its optimized value. Iteration 5 updated the configuration by optimizing c_p at value 0.5, and increased the performance by 33.74%, with PR_M achieving 98.61%. In iteration 6, the new value of ξ_p (10.5) provided a configuration that decreased the premium's budget while keeping the same performance.

We noticed, after just few iterations, that the system performance has increased drastically while the budgets conserved their values or slightly changed.

8   Conclusion

This paper presents a security design framework that can assist the CSPs to build scalable distributed security tools taking into account the resource optimization. The system consists of multiple security agents dedicated to serve the security requests generated by different types of tenants. These requests are routed to the agents using the stationary optimal policy provided by the VIA. Moreover, our framework helps the CSP to find better configuration to optimally use the system resources and suggest appropriate utilization budgets. The simulation results showed two facts. First, Finding the optimal routing policy is not enough for the CSP, because the initialization parameters that can be chosen by an expert may not be accurate, and hence produce poor performance. Secondly, our technique helped improve the system performance and optimizing the tenants' budgets. We envision that the proposed approach can be extended to address further contemporary problems related to the scalability and load balancing.

References

1. Bruneo, D.: A stochastic model to investigate data center performance and QOS in IAAS cloud computing systems. IEEE Trans. Parallel Distrib. Syst. **25**(3), 560–569 (2014)
2. Butt, S., Lagar-Cavilla, H.A., Srivastava, A., Ganapathy, V.: Self-service cloud computing. In: Proceedings of the 2012 ACM Conference on Computer and Communications Security, pp. 253–264. ACM (2012)
3. Chang, X., Xia, R., Muppala, J.K., Trivedi, K.S., Liu, J.: Effective modeling approach for IAAS data center performance analysis under heterogeneous workload. IEEE Trans. Cloud Comput. (2016)
4. Garfinkel, T., Rosenblum, M., et al.: A virtual machine introspection based architecture for intrusion detection. In: NDSS, vol. 3, pp. 191–206 (2003)
5. Ghosh, R., Longo, F., Naik, V.K., Trivedi, K.S.: Modeling and performance analysis of large scale IAAS clouds. Future Gener. Comput. Syst. **29**(5), 1216–1234 (2013)
6. Hajek, B.: Optimal control of two interacting service stations. IEEE Trans. Autom. Control **29**(6), 491–499 (1984)
7. Kumar, P.R., Varaiya, P.: Stochastic systems: estimation, identification, and adaptive control. SIAM (2015). http://epubs.siam.org/doi/book/10.1137/1.9781611974263
8. Lambadaris, I., Narayan, P.: Jointly optimal admission and routing controls at a networks node. Commun. Stat. Stoch. Models **10**(1), 223–252 (1994)
9. Lippman, S.A.: Applying a new device in the optimization of exponential queuing systems. Oper. Res. **23**(4), 687–710 (1975)

10. Lippman, S.A.: On dynamic programming with unbounded rewards. Manag. Sci. **21**(11), 1225–1233 (1975)
11. Mjihil, O., Huang, D., Haqiq, A.: Improving attack graph scalability for the cloud through SDN-based decomposition and parallel processing. In: International Symposium on Ubiquitous Networking, pp. 193–205. Springer (2017)
12. Mjihil, O., Kim, D.S., Haqiq, A.: Security assessment framework for multi-tenant cloud with nested virtualization. J. Inf. Assur. Secur. **11**(5), 283–292 (2016)
13. Pawar, P.S., Sajjad, A., Dimitrakos, T., Chadwick, D.W.: Security-as-a-service in multi-cloud and federated cloud environments. In: IFIP International Conference on Trust Management, pp. 251–261. Springer (2015)
14. Rosberg, Z., Varaiya, P., Walrand, J.: Optimal control of service in tandem queues. IEEE Trans. Autom. Control **27**(3), 600–610 (1982)
15. Tijms, H.C.: A First Course in Stochastic Models. Wiley, Chichester (2003)
16. Varadharajan, V., Tupakula, U.: Security as a service model for cloud environment. IEEE Trans. Netw. Serv. Manag. **11**(1), 60–75 (2014)
17. Walrand, J.: An Introduction to Queueing Networks. Prentice Hall, Englewood Cliffs (1988)
18. Yu, C., Li, L.X., Wang, K., Yu, W.T.: Protecting the security and privacy of the virtual machine through privilege separation. In: Applied Mechanics and Materials, vol. 347, pp. 2488–2494. Trans Tech Publ. (2013)
19. Zhang, F., Chen, J., Chen, H., Zang, B.: Cloudvisor: retrofitting protection of virtual machines in multi-tenant cloud with nested virtualization. In: Proceedings of the Twenty-Third ACM Symposium on Operating Systems Principles, pp. 203–216. ACM (2011)

Multi-view Web Services as a Key Security Layer in Internet of Things Architecture Within a Cloud Infrastructure

Anass Misbah[(✉)] and Ahmed Ettalbi

IMS Team, ADMIR Laboratory, ENSIAS, Rabat IT Center,
Mohammed V University in Rabat, Rabat, Morocco
anassmisbah@gmail.com, ettalbi1000@gmail.com,
a.ettalbi@um5s.net.ma

Abstract. Connected objects are more and more commonly used within different type of organizations in order to collect, summarize and process data of diverse terminals and devices; this concept is called Internet of Things (IoT). One of the major concerns of IoT architectures is the security. In fact, the concept of gathering data from a large variety of objects brought a lot of capabilities and functionalities in terms of decision making, flexibility and improved services. However, it brought also several security issues such as Authentication, Confidentiality and Integrity. In this work we are proposing a new approach of security implementation in IoT architectures; this approach is based on our previous work regarding Multi-view Web services and makes use of the previously proposed mechanisms of dealing with end users needs and access rights. The proposed architecture consists of the introduction of a new security layer that will receive all kinds of messages and requests through the exposed Multi-view Web services, analyze the access rights and messages subscription of the requestor and the object and then relay the request to the appropriate object or service. This architecture can be implemented in a Cloud infrastructure so as to centralize operations, enhance reusability and reach a high level of availability.

Keywords: Internet of Things · Cloud · Multi-view Web services · Security layer
WADL · Architecture · Restful

1   Introduction

Information System architectures are moving fast. Indeed, each day a better solution is proposed to deliver a high quality service to the customer anywhere, anytime and with a responsive design according to the client environment. Therefore, the user experience is improved permanently.

One of these most important concepts is IoT (Internet of Things) [6]. This concept makes it possible for any electrical object to submit data about its identity, state and environment or receive an instruction of state modification or any other such message. This collected data is generally gathered and treated in dedicated Cloud platforms [11] so as to make use of it in different manners and different contexts. However, for the time being, there is still a lack of standardization and security management of IoT objects especially within critical organizations such as: Health, Governments, Industries, Military, Aviation …etc. And here comes a need to address this vulnerability using a control mechanism that covers the whole IoT infrastructure (Objects, servers and applications).

In this work, we are proposing a new approach of handling the security issue of IoT architectures through the Multi-view Web Services [1]. This approach consists of integrating a new security layer between the IoT objects and the Cloud infrastructure. The security layer makes use of Multi-view Web Services as a key element of identifying objects and defining the access rights and messages subscription between Objects and the Cloud applications.

In fact, in the previous work [2], we proposed a Restful implementation of Multi-view Web Services; this implementation brought many advantages such as: standardization, automatic generation, compliance, flexibility, simplicity and portability. We will go ahead in this work and make use of the Restful implementation of Multi-view Web services to put in place the security layer within an IoT infrastructure.

In this work, we will take advantage of the work of [2] WADL implementation to design a new security layer within an IoT – Cloud architecture. This layer provides a security control through WADL Multi-view Web services generated based on the work of [2] approach.

The remainder of this paper is organized as follow: In Sect. 2 we give a brief description of the state of the art as well as a technology overview. Section 3 is dedicated to the description of the problematic and our proposed approach to solve it. Then in Sect. 4 we highlight the advantages of our approach. Eventually, in Sect. 5 we conclude and we give some perspectives.

2 State of the Art

2.1 Multi-view Web Services

The Multi-view concept has been introduced in several previous works, including [3], and has been applied to many domains such as UML (Unified Modeling Language) models and diagrams. This concept came up with a new way to deal with end users needs and access rights since an early step of design and conception. Thus, it brought many advantages mainly regarding security, homogeneity, and integrity as well as redundancy elimination.

The concept of Multi-view relays on two paradigms principally:

- **Views:** A function or method that can be accomplished by at least one user at the same time.
- **Points of view:** A set of views that can be accomplished by one or multiple users.

Web services have became a key component in nowadays Information System architectures, and this is due to the important advantages that they bring, such as interoperability, flexibility and reusability.

The work of [4] proposed an integration of the concept of Multi-view with Web services through the notion of decomposition. It consists of the subdivision of the main

Web services into Sub Web services (Views), and then gathering the Sub Web services in groups of views (Points of view) so as to define for each user the group of views that he is allowed to access.

Furthermore, in the work of [5], an implementation of Multi-view Web services has been projected, it consists of an extension of the standard definition WSDL to a specific new definition called WSDL-Us.

Moreover, in the work of [1] we proposed an improvement of the work [5] implementation based on standard meta-rules of the transformation of any Views and Points of view model to a standard WSDL definition automatically.

Then we went further in the work of [2] and gave a technology overview about Web services and their basic architecture of use, we also highlighted the SOAP-based and Restful-based implementations then we compared and contrasted both implementations particularities.

Besides, in the work of [2] we came up with a new type of Multi-view Web services WADL implementation that relays on Restful architecture style.

2.2   Internet of Things (IoT)

Internet of Things (IoT) [6] can be considered as the extension of the existing interconnection between devices, servers and workstations to a widespread interconnection between objects. An object can be any electrical device capable of transmitting data to indicate information or a state or receiving instructions to modify or adopt a particular behavior.

IoT is used in large amount of professional domains and industries (Domestic, Health, Transportation, Education, Smart Cities, Sensors, Robotics, Manufacturing, Utilities, Logistics, retails, Public services …), and make it possible to handle data with different ways (Machine learning, Big Data, Artificial Intelligence, Expert systems, …) in order to take decisions and optimizing performance. Moreover, there are many existing platforms and projects that put in place a practical approaches making use of IoT mechanisms to create value for businesses and industries.

The concept of IoT brings a lot of advantages in terms of monitoring and managing objects [8]. However, the implemented architectures still relay on manufacturers protocols and messages format. In fact, for the time being, there is a gap in standardization and homogenization [7] of IoT communication means. Thus, the security issues (authentication, confidentiality, privacy and integrity) as well as performance and optimization, remain among the most challenging subjects in IoT architectures [9, 10].

2.3   Security with IoT

Some previous works discussed several models and techniques to handle security over IoT infrastructures. The work of [9] is one of them, in that work a single-sign-on schema has been proposed. However, the proposed model depends on the IoT context and requires a considerable effort to be put in place.

The work of [19] gave an analysis of security issues through several kinds of IoT networks and architectures and this inside different layers of IoT communication solutions. Yet, this work did not propose a concrete new approach to address IoT security issues.

Moreover, the work of [10] analyzed security issues that need consideration and future work. This analyzes cover the IoT technologies, end users and Cloud providers platforms.

In the work of [12], a meta-model supporting security services over the Cloud has been proposed, as well as a discussion of SecaaS (Security as a Service) concept that has emerged with the continuous growing of the Cloud open services and networks. Nevertheless, this approach is Cloud service oriented and do not deal with all kinds of IoT objects.

2.4 The Cloud Computing

The cloud computing is an on-demand service that provides computation, storage and/or software resources through a dedicated infrastructure hosted on the Internet. This service allows different types of organizations to outsource the local IT infrastructure as to get access to the necessary Software/ Hardware without a large amount of investment in Infrastructure, Servers, Applications and administration efforts [11].

In fact, the organizations are more and more using Cloud computing to pay as they use Platforms and Application services. Thus, it became possible to pay as the required need is growing and resize the dedicated capacity accordingly. It is also possible to remotely Design, Build, Deploy and Manage the whole IT infrastructure hosted in the cloud.

Cloud Computing brought many advantages such as simplicity, scalability, flexibility, costless, on-demand, remotely managed, outsourcing and Internet-based.

The Cloud network can be Private (works only for one organization and highly secured), Public (Owned by a public service provider and allows high performance and efficiency of resources) or Hybrid (a combination of a private and public Cloud).

The services provided in the Cloud infrastructure depend on the user's needs. The main service types are the following:

- **IaaS (Infrastructure as a Service):** This architecture provides computing and storage power such as CPU, Memory and Networks capabilities. In this case a complete access is given to the environment and Operating Systems; however, an additional effort should be done to maintain and manage this environment.
- **PaaS (Platform as a Service):** This model provides a runtime environment to develop, deploy and run homemade or acquired applications. The maintenance of the applications is attributed to the customer.
- **SaaS (Software as a Service):** This model provides a set of ready to use applications, the customer can reach the application wherever he is and take advantage of the provider's software updates and maintenance without dealing with the infrastructure or the platform management. The customer will still need to manage the application users and parameter settings.

3   Problematic and the Proposed Solution

3.1   The Problematic

As organizations are becoming aware about the importance of the digital transformation of the IT infrastructure with the aim of reducing costs, improve quality, deliver new time to market services and enhance customer satisfaction, IoT has become one of the widespread concepts deployed broadly in different kind of organizations.

Hence, instead of hosting the whole IT infrastructure of IoT platform, organizations resort to some services providers (such as Microsoft [18], Google [17], Amazon [13], Heroku [14], …) to design, develop, deploy and maintain their IoT object's monitoring applications over the service provider's Cloud Web platforms.

Whereof, a pressing must of security emerge with the need of controlling messages transmitted between IoT objects and applications within the Cloud infrastructure.

In this work we propose to deal with this issue through the injection of a new security layer (Fig. 1). This security layer is composed of Multi-view Web services engine which is basically a set of separated bricks interacting with each other to control the access rights in between IoT Objects and Cloud Applications. These bricks are presented in the following subsections.

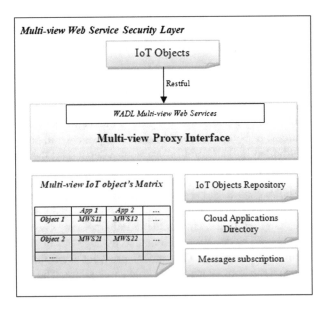

Fig. 1.   Architecture schema of the Multi-view Web service security layer

3.2   Multi-view Proxy Interface

This bloc represents the interface between all IoT objects and the Cloud objects (which can be an actual application or another object). Actually, this part contains one or more Multi-view Web services (we recommend to define one Multi-view Web service per

Object so as to simplify the implementation) that will receive all IoT object's requests, check the eligibility of each object according to the "Multi-view Objects matrix" and relay the request to the Cloud application with the corresponding access rights or deny the request in case insufficient rights.

The multi-view Web services are generated automatically and implemented with a WADL definition by the means of the work of [2]. In fact, in this previous work [2] we proposed automatic rules to generate Multi-view Web services WADL definition according to a set of Views and Points of view.

In this case, IoT objects are considered as Users, then each IoT object has a Point of view (a set of application's functionalities that it has access to). Also, every application is considered as a main Web service with one or more functionalities (Sub Web services) that can be reached by one or a group of users (IoT objects). These Functionalities constitute Views.

Therefore, all communications are achieved using Restful protocol, this protocol have many advantages as highlighted in the work of [2], in the proposed architecture (Fig. 2), we take advantage of the simplicity, lightweight, portability and standardization

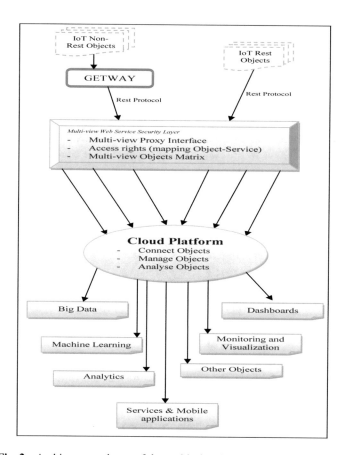

Fig. 2. Architecture schema of the multi-view IoT – Cloud security layer.

of Restful architecture in addition to the fact that a lot of IoT objects can natively "talk" Restful messages.

In case of none-restful objects, we are proposing to implement a Getway (Fig. 2) that will catch the original message format (such as MQTT [6]) and then transform it to Restful format before send it back to the "Multi-view Proxy Interface".

In this interface we define all Application's signatures (input and output parameters) as well as the server binding information.

- **IoT Objects repository:** This bloc is administrated by a super user and it is dedicated to define the total list of IoT objects as well as their specifications (identification, states, description, context, usage…).
- **Cloud Applications Directory:** This bloc contains the overall set of Cloud Applications that can be reached by objects or that can send instructions to objects. For each application we may define information such as: Identity, usability, environment, context of invocation, functionalities, server properties, methods, signatures.
- **Message subscription:** This section is used to dematerialize an affiliation of an application or an object to a stream of information of another application or object. This affiliation might be temporary or permanent and can be managed by the super user. This kind of streaming is very useful in case of monitoring and supervising objects using centralized dashboards.
- **Multi-view Object's matrix:** Table 1 presented below is generated using the information of "Object Repository", "Cloud Applications Directory" and "Message subscription" and allows to define for each IoT object or Cloud application, which Multi-view Web service to be used to communicate with another object or application and with which parameters. Depending on the case, this communication can be streaming information, an instruction, a warning, an alert, a notification or any kind of such notices. These Multi-view Web services will be used by the "Multi-view Proxy Interface" bloc to control the access rights and relay the request to the correspondent Cloud application server or IoT object accordingly.

Table 1. Multi-view Object's matrix

IoT objects	Cloud applications or objects			
	Application 1	Application 2	IoT object 6	…
Object 1	MVWS 11	MVWS 12	xxx	…
Object 2	MVWS 21	MVWS 22	MVWS 26	…
…				

Each Object has a Point of view (a set of Multi-view Web services allowing access to application's functionalities), and each application has one or more Views (functionalities that can be accessed by one or a group of users). A Point of view may be the same for a group of objects.

Each functionality has a specific signature. If one functionality has different behaviors depending on the IoT object, for every IoT object a Multi-view web service will be created.

3.3   The IoT – Cloud Architecture Schema

The architecture illustrated below (Fig. 2) demonstrates how the "Multi-view Web Service Security Layer" interacts with the IoT objects and the Cloud Platform applications. In fact, the "Multi-view Web services Security Layer" plays a central role in terms of communication flow control and orchestration between IoT objects and the Cloud applications.

This layer will handle all types of messages and instructions that could transit in the middle of the overall architecture. The objects can be of any nature and of any activity type such as: sensors, cars, cameras, airplanes, microcontrollers, lighting, radars, fridges, TVs, wash machines…

The overall architecture is composed of the following elements:

- **Non-Restful objects:** IoT objects that are not capable of exchanging Restful messages and can only communicate over a native protocol.
- **Restful objects:** Any device that can send and receive Restful protocol messages so as to adopt a particular behavior, change the current state or send specific information to drive a change to another IoT object.
- **Getway:** An intermediate bloc that represents a bridge for Non-Restful devices allowing the transformation of the original messages to a Restful messages and vice-versa.
- **Cloud Platform:** The centralized infrastructure which allows connecting, managing and analyzing IoT Objects. Several free and paying solutions exist and propose different kind of services with multiple SLA (Service Level Agreement), for instance: Amazon Web Services [13], Heroku [14], ThingSpeak [15], Oracle IoT [16], Google Analytics [17], Microsoft Azure [18], …

Applications that consume IoT object's messages can be one of the following:

- **Big Data:** Analyze large amounts of data to find tendencies.
- **Machine Learning:** Artificial Intelligence bots that are capable of learning from data so as to be able to simulate the human being intelligence and take decisions on him behalf.
- **Analytics:** Reporting about the main indicators of the activity, generally it concerns the KPIs (Key Performance Indicators) of the organization.
- **Services:** Any kind of services that needs an IoT object to provide accurate information (for example a public weather Web service may need thermo sensors to provide the actual information about temperature).
- **Service and Mobile applications:** All kind of mobile phone or tablet applications. One of the most common recent usages is Chat bots and Script bots.
- **Dashboards:** Business Intelligence indicators related to the business strategy, for instance financial, schedule or quality forecasts.
- **Monitoring and Visualization:** All kind of supervision equipments, audio or video streaming terminals (screens, voice alerts …).
- **Other objects:** IoT objects may need to send information or an instruction to another IoT object.

4   The Advantages of the Proposed Architecture

In this work we proposed a security layer based on Multi-view Web services model. This layer is used to secure communication between IoT objects and applications within a Cloud environment.

This approach brings several enhancements in terms of messages exchanging control and regulation. We can cite the followings among others:

- As our approach improve the security within IoT – Cloud architecture, this may encourage more organizations to resort to IoT deployment and then take advantage of all gains and benefits that this concept allows.
- Since the communication is Restful-based, and Rest is widely used, the deployment of the proposed architecture will be easy to implement inside the existing IoT and Cloud platforms, also Restful messages are lightweight and then do not require a large bandwidth.
- The proposed schema covers the overall communications between IoT objects and Cloud application, and then they will be an optimization of the traffic. In fact, every application will subscribe only to the required information instead of receiving all messages broadcasts and then sort data according to the needs.
- The Cloud infrastructure will make it possible to develop and manage application remotely without considering the platform or networking maintenance and then improve cost effectiveness.
- The Cloud application's community is very active and proposes each day noticeable improvements in the existing applications or new business processes applications. Then we can make use of this synergy in order to deploy easily and securely more applications and functionalities.
- The Multi-view Web services bring flexibility, reusability, portability, compatibility, interoperability, security, standardization and self-validated as long as they are generated automatically.

5   Conclusion and Future Work

We proposed in this work a new way to secure IoT objects exchanges within a Cloud platform using the Multi-view Web Services security layer. This architecture combines IoT objects and Cloud platforms and make use of Restful implementation of Multi-view Web services to control data transitions as well as the objects access rights. Thus, the proposed approach allowed more security, flexibility, performance and centralized control.

As a future work, we will go ahead and prototype IoT object's interactions with the Cloud applications through the Multi-view Web services security layer. We will also take some study cases as a PoC (Proof of Concept) and PoV (Proof of Value) of the proposed architecture.

References

1. Misbah, A., Ettalbi, A.: Towards a standard WSDL implementation of Multiview web services. In: 5th International Conference on Multimedia Computing and Systems (ICMCS 2016) – IEEE Conference, Marrakech, Morocco, 29 September–1 October 2016
2. Misbah, A., Ettalbi, A.: Towards a standard Restful WADL implementation of Multiview web services. IJCSNS Int. J. Comput. Sci. Netw. Secur. **17**(4), 315–320 (2017)
3. Kriouile, A.: VBOOM, object-oriented analysis and design method by points of view. Ph.D. thesis, Mohammed V University, Rabat, Morocco (1995)
4. Boukour, R., Ettalbi, A., Nassar, M.: Multiview web service: the integration of the notion of view and point of view in the web services. Int. J. Comput. Sci. Netw. Secur. (IJCSNS) **14**(2), 31–36 (2014)
5. Boukour, R., Ettalbi, A., Nassar, M.: Multiview web service: the description multiview WSDL of web services. In: International Symposium on Signal, Image, Video and Communications (ISIVC 2014), Marrakech, Morocco, 19–21 November 2014
6. Vermesan, O., Friess, P.: Internet of Things: Converging Technologies for Smart Environments and Integrated Ecosystems. River Publishers, Aalborg (2013). ISBN 978-87-92982-96-4
7. Wood, A.: The Internet of Things is revolutionizing our lives, but standards are a must. The Guardian, 31 March 2015
8. Mitchell, S., Villa, N., Stewart-Weeks, M., Lange, A.: The internet of everything for cities: connecting people, process, data, and things to improve the 'Livability' of cities and communities. Cisco Systems. Accessed 10 July 2014
9. Witkovski, A.: An IdM and key-based authentication method for providing single sign-on in IoT. In: Proceedings of the IEEE GLOBECOM, p. 1 (2015). ISBN 978-1-4799-5952-5. https://doi.org/10.1109/glocom.2015.7417597
10. Singh, J., Pasquier, T., Bacon, J., Ko, H., Eyers, D.: Twenty Cloud security considerations for supporting the Internet of Things. IEEE Internet Things J. **3**(3), 1 (2015). https://doi.org/10.1109/jiot.2015.2460333
11. Hassan, Q.: Demystifying Cloud computing. J. Def. Softw. Eng. **1**, 16–21 (2011). Accessed 11 Dec 2014
12. Furfaro, A., Garro, A., Tundis, A.: Towards security as a service (SecaaS): on the modeling of security services for Cloud computing. In: 2014 International Carnahan Conference on Security Technology (ICCST), pp. 1–6, 01 October 2014. https://doi.org/10.1109/ccst.2014.6986995
13. Amazon, official Website. https://aws.amazon.com/fr/documentation/
14. Heruko official Web development online center. https://devcenter.heroku.com/
15. ThingSpeak community tutorials. http://community.thingspeak.com/tutorials/
16. Oracle IoT portal. https://www.oracle.com/fr/solutions/internet-of-things/index.html
17. Google Analytics solutions Website. https://www.google.com/analytics/
18. Microsoft IoT products and solutions. https://www.microsoft.com/en-us/internet-of-things/
19. Jing, Q., Vasilakos, A.V., Wan, J., Lu, J., Qiu, D.: Security of the Internet of Things: perspectives and challenges. Wirel. Netw. https://doi.org/10.1007/s11276-014-0761-7

Access Domain-Based Approach
for Anomaly Detection and Resolution
in XACML Policies

Maryem Ait El Hadj[1(✉)], Yahya Benkaouz[2], Ahmed Khoumsi[3],
and Mohammed Erradi[1]

[1] Networking and Distributed Systems Research Group, ITM Team, ENSIAS,
Mohammed V University in Rabat, Rabat, Morocco
maryem_aitelhadj@um5.ac.ma, mohamed.erradi@gmail.com
[2] Conception and Systems Laboratory FSR, Mohammed V University in Rabat,
Rabat, Morocco
y.benkaouz@um5s.net.ma
[3] Department of Electrical and Computer Engineering, University of Sherbrooke,
Sherbrooke, Canada
ahmed.khoumsi@usherbrooke.ca

Abstract. Access control protects systems' resources against unauthorized access via a set of policy rules. In distributed environments, access control policies might be aggregated from multiple tenants and could be managed by more than one administrator. Therefore, errors in the rules definitions may compromise the system security by leading to unauthorized access or denying authorized access. This may result into anomalies, i.e. conflicting rules and redundant rules. In this paper, we propose an approach to detect and resolve anomalies in XACML (eXtensible Access Control Markup Language) policies. We introduce the concept of a rule access domain, which is used to accurately identify and resolve policy anomalies.

Keywords: XACML policies · Clustering · Anomaly detection
Anomaly resolution · Access domain

1 Introduction

Distributed computing environments (e.g. Cloud Computing) offer various benefits to users, such as scalability, elasticity and flexibility. However, due to the current big data exponential growth, the regulation of resources managed by such environments raises multiple security challenges. Hence, users need authorization systems to help them share their resources, data and applications with a large number of users without compromising security and privacy. Access control models have been suggested to deal with such requirements. Access control models are concerned with determining the allowed activities of legitimate users, mediating attempt by a user to access a resource in a given system.

© Springer International Publishing AG, part of Springer Nature 2018
A. Abraham et al. (Eds.): IBICA 2017, AISC 735, pp. 298–308, 2018.
https://doi.org/10.1007/978-3-319-76354-5_27

In large collaborative platforms, access control policies might be aggregated from multiple tenants and could be managed by more than one administrator. Therefore, errors in the rules definitions may compromise the system security by leading to unauthorized access or denying authorized access. This may result into anomalies, i.e. conflicting and redundant rules. Detecting and resolving automatically such anomalies in large sets of complex policies is of fundamental importance.

The authors of [1] proposed a method to detect anomalies within XACML (eXtensible Access Control Markup Language) policies. The proposed method is based on segmenting the policy into clusters before detecting anomalies within each cluster. Given an XACML policy, they proceed as follows: (1) extract the rules of the XACML policy, (2) compute a similarity score for each pair of rules, (3) regroup similar rules into the same cluster. Finally, (4) detect anomalies within each cluster [1]. This approach is justified by the fact that the probabilities of anomalies between similar rules are much lower than the probabilities of anomalies between non-similar rules. The results obtained in [1] demonstrate the correctness of this approach.

In the present paper, we first reformulate in Sect. 4 the definition of anomalies by using the concept of rule access domain, and then in Sect. 5 we improve [1] by proposing a method that not only detects anomalies, but also resolves them.

The rest of the paper is organized as follows: Sect. 2 presents related work. In Sect. 3 we overview XACML policies. In Sect. 4, anomalies are formulated by using the concept of rule access domain, and then Sect. 5 presents an iterative procedure to detect and resolve anomalies in a cluster. Finally, the conclusion and expected future work are drawn in Sect. 6.

2 Related Work

Various conflict detection strategies have been proposed [5,7,8]. They are based on different techniques: Answer Set Programming, UML, and coq, to represent the policy and then detect different types of anomalies such as conflicting rules, redundancy, inconsistency, etc. However, these proposed strategies do not resolve the detected anomalies. The resolution of such anomalies was already handled by XACML itself. In fact, XACML offers a set of rule combining algorithms (RCA) [9] to overcome the issue of conflicting rules. The RCA are: Deny-Overrides, Permit-Overrides, First Applicable and Only-One-Applicable. The first two algorithms are trivial. For instance, permit-overrides returns permit if one of the conflicting policies evaluates to permit. Otherwise, the result is deny. First applicable returns the result of the first policy that matches the request. Only-one applicable obligates the system to have one and only one policy that matches the request, otherwise the system returns Not-applicable. However, the RCAs need to be defined manually by the policy administration. Moreover, only one RCA can be applied to all kinds of identified anomalies. Therefore, this technique remains static and cannot be applied to distributed and dynamic systems.

In the other hand, some XACML policy evaluation engines, such as Sun PDP [6] and XEngine [4], have been developed to handle the request/policy evaluation. These implementations can detect conflicts by checking if a request matches multiple rules with different effects, then conflicts are resolved by one predefined XACMLs Policy Combining Algorithms (PCA) which allows the policy administrator to apply one strategy for all the identified conflicts. In contrast, XAnalyzer [3] provides policy analysis at policy design time and gives the designer the ability to choose the adequate PCA for each specific policy. However, these systems are based on the intervention of the designer or the administrator to set up the adequate combining algorithm.

The authors of [1] present a method to detect anomalies within XACML policies, that takes into account a large set of rules and attributes. It proposes an anomaly detection method performed in each cluster of rules, instead of the whole policy set, which implies less processing time. In the present paper, we generalize [1] by proposing a method that not only detects anomalies, but also resolves them.

3   Preliminaries

3.1   XACML: Overview

XACML defines a policy language using attributes of requester, resources, and environment. It provides an authorization architecture which supports Attribute-Based Access Control [10]. An XACML policy is composed of a Target, that identifies the capabilities that should be exposed by the requester (the targeted resources for example), and a set Rules. The Targets values are compared against the XACML request values. If the request attributes match the Targets attributes, the policy will be further evaluated using the Rules component, else the XACML engine decides that the request is not applicable to the policy.

3.2   Expression and Semantics of Rules

In an XACML policy, each rule has three categories of attributes: subject, resource and environment. A rule is specified by a condition and an action decision, the latter being noted X_{act}, where X is a decision Permit or Deny, and act is a set of actions. $Permit_{read}$ and $Deny_{write}$ are two examples of action decisions.

The condition (or profile) of a rule is specified by one or several assignments $att \in V_{att}$, where att identifies an attribute and V_{att} is a set of possible values of att. There is at most one assignment for each attribute. The absence of assignment for an attribute att means the implicit existence of the assignment $att \in ALL_{att}$, where ALL_{att} denotes the set of all possible values of att. The assignments corresponding to the same category are separated by a comma ",", while a semicolon ";" means the passing to the next category. An access request is defined by attribute values (at most one value for each attribute) and one action.

We say that a value v of an attribute att satisfies an assignment $att \in V_{att}$ of a rule r_i, if v is an element of V_{att}. We say that an access request R matches a rule r_i (we can also say r_i matches R) if every attribute value of R satisfies the corresponding assignment of r_i.

Consider a rule r_i whose action decision is X_{act}. The semantics of r_i is as follows: for any access request R that matches r_i and whose action a belongs to act, the decision X is made for the action a. The formal expression of a rule profile is therefore as follows:

$$X_{act}(att_1 \in V_{att_1}, att_2 \in V_{att_2}, ..., att_n \in V_{att_n})$$

Example 1: $Permit_{read}$ (position $\in$ {Doctor}, specialist $\in$ {Generalist}, team $\in$ {Oncology}, experience $\in$ {+10}, grade $\in$ {Registrar}, department $\in$ {oncology}; type $\in$ {PR/CAT}, formatType $\in$ {AST}, degreeOfConfidentiality $\in$ {Secret}; organisation $\in$ {EMS}, time $\in$ [8:00, 12:00]). The attributes of the *Subject* category are: position, specialist, team, experience, grade, department. The attributes of the *Resource* category are: type, formatType, degreeOfConfidentiality. The attributes of the *Environment* category are: organisation, time.

4   Formal Definitions of the Considered Anomalies

Anomalies are patterns in data that do not conform to a well-defined notion of normal behavior [2]. In XACML policy, an anomaly in a policy P is defined as the existence of access request matching several rules of P. We have considered two types of anomalies:

- **A redundancy** occurs when there exist useless (or redundant) rules, i.e. rules whose removal does not modify the behavior of the policy. We consider redundancy as an anomaly, because it may affect the performance of a policy, since verifying if an access request respects a policy depends on the size of the policy.
- **A conflict** occurs when there exist two or more rules that generates contradictory decisions on an access request, e.g. $Permit_{read}$ and $Deny_{read}$.

4.1   Access Domain and Compatible Rules

We have seen that a rule is expressed in the form $X_{act}(att_1 \in V_{att_1}, att_2 \in V_{att_2}, ..., att_n \in V_{att_n})$. For the purpose to prepare a clear approach of anomaly resolution in Sect. 5, we reformulate equivalently a rule in the following form: $X_{act}((att_1, att_2, ..., att_n) \in V_{att_1} \times V_{att_2} \times ... \times V_{att_n})$. That is, instead of specifying separately a set of values for each attribute att_i, we specify a unique a set of values for the n-tuple $(att_1, att_2, ..., att_n)$ of all attributes. Such a set is called access domain of r and noted AD_r. So a rule r can be simply expressed in the form $r = X_{act}((att_1, att_2, ..., att_n) \in AD_r)$. For simplicity, we sometimes note it $r = X_{act}(AD_r)$.

4.2   Redundancy

Consider two rules $r_i = X_a(AD_{r_i})$ and $r_j = X_b(AD_{r_j})$. r_i is redundant to r_j iff:

 – $AD_{r_i} \subseteq AD_{r_j}$, intuitively r_j matches all requests matched by r_i, and
 – $a \subseteq b$.

Intuitively, every decision taken by r_i on any request is also taken by r_j. Therefore, r_i is useless and hence can be removed from the policy.

Example 2: Consider the following rules r_1 and r_2:

 – r_1: $Permit_{\{read,write\}}$((position; fileType; time) $\in$ {Doctor, Nurse} $\times$ {Source, Documentation} $\times$ [8:00, 18:00]).
 – r_2: $Permit_{\{read\}}$((position; fileType; time) $\in$ {Nurse} $\times$ {Documentation} $\times$ [8:00, 18:00]).

r_2 is redundant to r_1, because ({Nurse} $\times$ {Documentation} $\times$ [8:00, 18:00]) $\subset$ ({Doctor, Nurse} $\times$ {Source, Documentation} $\times$ [8:00, 18:00]) and {read} $\subset$ {read, write}.

4.3   Conflict

Consider two rules $r_i = X_a(AD_{r_i})$ and $r_j = Y_b(AD_{r_j})$. r_i and r_j present a conflict (or are conflicting) iff:

 – $AD_{r_i} \cap AD_{r_j} \neq \emptyset$
 – $X \neq Y$, and
 – $a \cap b \neq \emptyset$ (i.e. $a \cap b$ contains actions that are at the same time permitted by r_i and forbidden by r_j).

Example 3: Consider the following rules r_1 and r_2:

 – r_1: $Deny_{\{read\}}$((position; fileType; time) $\in$ {Doctor, Nurse} $\times$ {Source, Documentation} $\times$ [8:00, 18:00]).
 – r_2: $Permit_{\{read,write\}}$((position; fileType; time) $\in$ {Nurse} $\times$ {Documentation} $\times$ [8:00, 18:00]).

r_1 and r_2 are conflicting, because their access domains are not disjoint, while the action *read* is permitted by r_2 and forbidden by r_1.

5   Anomaly Detection and Resolution

In this section, we describe how to detect and resolve the two types of anomalies mentioned in the previous section.

5.1   Detection and Resolution of Redundancy Between Two Rules

The response time of a policy to an access request depends on the number of rules to be parsed in the policy. So redundancy (i.e. existence of useless rules) may affect the performance of a policy, and hence is treated as an anomaly. Thus, removing redundancies is considered as one of the effective solutions for optimizing XACML policies and improving the performance in policy decision time.

Given two rules $r_i = X_a(AD_{r_i})$ and $r_j = Y_b(AD_{r_j})$, r_i is detected to be redundant to r_j if the two conditions of Sect. 4.2 are satisfied. The resolution of that anomaly consists in removing the redundant rule.

Example 4: Consider the previous Example 2 where we have shown that r_2 is redundant to r_1:

- r_1: $Permit_{\{read,write\}}((\text{position}; \text{fileType}; \text{time}) \in \{\text{Doctor}, \text{Nurse}\} \times \{\text{Source}, \text{Documentation}\} \times [8{:}00, 18{:}00])$.
- r_2: $Permit_{\{read\}}((\text{position}; \text{fileType}; \text{time}) \in \{\text{Nurse}\} \times \{\text{Documentation}\} \times [8{:}00, 18{:}00])$.

Hence, resolution of this redundancy consists in removing r_2.

5.2   Detection and Resolution of Conflict Between Two Rules

Let us first define the following notations, given two rules, r_1 and r_2, whose actions decision are X_a and Y_b respectively:

- **Common domain:** $CD = AD_{r_1} \cap AD_{r_2}$
- **Common actions:** $CA = a \cap b$

Given two rules $r_i = X_a(AD_{r_i})$ and $r_j = Y_b(AD_{r_j})$, a conflict is detected if the three conditions of Sect. 4.3 are satisfied. The resolution of the detected conflict is realized as follows:

Permissive strategy: r_1 is not modified and r_2 is replaced by the following two rules:

- $r_2' = Deny_b(AD_{r_j} \setminus CD)$
- $r_2'' = Deny_{b \setminus CA}(CD)$

Intuitively, the unique modification that has been done is not denying the common actions of r_1 and r_2 for requests matching both r_1 and r_2. It is easy to check that r_1, r_2' and r_2'' are conflict-free with each other.

Restrictive strategy: r_2 is not modified and r_1 is replaced by the following two rules:

- $r_1' = Permit_a(AD_{r_i} \setminus CD)$
- $r_1'' = Permit_{a \setminus CA}(CD)$

Intuitively, the unique modification that has been done is not permitting the common actions of r_1 and r_2 for requests matching both r_1 and r_2. As in the permissive strategy, it is easy to check that r'_1, r''_1 and r_2 are conflict-free with each other.

Example 5: Consider the following rules r_1 and r_2:

- r_1: $Permit_{\{read,write\}}((position;$ fileType;   time$) \in \{Doctor,$ Nurse$\} \times$ Documentation$\} \times 8{:}00, 18{:}00])$.
- r_2: $Deny_{\{read,create\}}((position;$ fileType;   time$) \in \{Nurse\} \times \{Documentation\} \times [8{:}00, 18{:}00])$.

The access domains of r_1 and r_2 are:

- $AD_{r_1} = \{Doctor, Nurse\} \times \{Documentation\} \times [8{:}00, 18{:}00]$.
- $AD_{r_2} = \{Nurse\} \times \{Documentation\} \times [8{:}00, 18{:}00]$.

Common access domain: $CD = AD_{r_1} \cap AD_{r_2} = AD_{r_2}$

The resolution is as follows:

Permissive strategy: r_1 is not modified and r_2 is replaced by:

- $r'_2 = Deny_{\{read,create\}}((position; fileType; time) \in \emptyset)$, so this rule is not considered since its access domain is empty;
- $r''_2 = Deny_{\{create\}}((position; fileType; time) \in \{Nurse\} \times \{Documentation\} \times [8{:}00, 18{:}00])$.

Restrictive strategy: r_2 is not modified and r_1 is replaced by:

- $r'_1 = Permit_{\{read,write\}}((position; fileType; time) \in \{Doctor\} \times \{Documentation\} \times [8{:}00, 18{:}00])$.
- $r''_1 = Permit_{\{write\}}((position; fileType; time) \in \{Nurse\} \times \{Documentation\} \times [8{:}00, 18{:}00])$.

5.3   Anomaly Detection and Resolution in a Cluster of Rules

Anomaly detection and resolution in a cluster is an iterative process that consists in verifying the existence of anomalies and, if any, in modifying the rules of the cluster until all the obtained rules are anomaly-free. The approach consists in first constructing a graph (Algorithm 1) where each node represents a rule, and each node is connected to all other nodes. A link between two nodes r_1 and r_2 means that we have to verify if there is an anomaly between r_1 and r_2, and resolve it, if any. Then, the anomaly detection and resolution (Algorithm 2) will verify and modify iteratively the graph until we obtain a graph of nodes without any link, which means that we have obtained an anomaly-free set of rules. At each iteration of Algorithm 2, the anomaly resolution is applied to a pair of linked nodes r_1 and r_2 as explained below.

Algorithm 1. Graph Construction

Input: Cluster of rules (C_k)
Output: Graph G(N) N: Set of nodes
1: **procedure** GRAPHCONSTRUCT(C_k)
2:     **for** $r \in C_k$ **do**
3:         n.rule=r; N.add(n) ▷ create a node for each rule.
4:     **end for**
5:     **for** a pair $(n_i, n_j) \in N$ **do**
6:         n_i.linkTo(n_j) ▷ link all the created nodes.
7:     **end for**
8:     **return** $G(N)$
9: **end procedure**

If we detect that r_1 is redundant to r_2, then resolution consists in removing the node of r_1 from the graph (lines 4–8 of Algorithm 2).

If a conflict between r_1 and r_2 is detected, we have seen in Sect. 5.2 that there are two strategies:

In the **permissive strategy** (lines 15–17 of Algorithm 2), r_1 is not modified and r_2 is replaced by r_2' and r_2''. The graph is then updated as follows:

1 - In the node of r_2, replace the AD and act of r_2 by the AD and act of r_2'.
2 - Create a new node that contains the AD and act of r_2'', and link this node to all the nodes of the graph, except those of r_1 and r_2.
3 - Remove the link between the nodes of r_1 and r_2, because there is no anomaly between them (after the modification of r_2 in Point 1).

In the **restrictive strategy** (lines 18–20 of Algorithm 2), r_2 is not modified and r_1 is replaced by r_1' and r_1''. The graph is then updated as follows:

1 - In the node of r_1, replace the AD and act of r_1 by the AD and act of r_1'.
2 - Create a new node that contains the AD and act of r_1'', and link this node to all the nodes of the graph, except those of r_1 and r_2.
3 - Remove the link between the nodes of r_1 and r_2, because there is no anomaly between them (after the modification of r_1 in Point 1).

If no anomaly is detected between a pair of linked nodes r_1 and r_2, the resolution algorithm simply removes the link between r_1 and r_2 (lines 29–30 of Algorithm 2). Also, remove any node whose rule has an empty access domain or empty set of actions (lines 22–26 of Algorithm 2).

Example 6: Consider a cluster consisting of the following four rules:

- $r_1 = Permit_{\{read,write\}}$ ((position; fileType; time) $\in \{$Doctor, Nurse$\} \times \{$Documentation$\} \times [8:00, 18:00])$.
- $r_2 = Permit_{\{read\}}$ ((position; fileType; time) $\in \{$Nurse$\} \times \{$Documentation$\} \times [8:00, 18:00])$.
- $r_3 = Deny_{\{read,delete\}}$ ((position; fileType; time) $\in \{$Nurse$\} \times \{$Source, Documentation$\} \times [8:00, 18:00])$.

Algorithm 2. Anomaly Detection and Resolution

Input: Cluster of rules $(G(N))$
 1: **procedure** ANOMALYRESOLUTION$(G(N))$
 2:     **while** G contains a pair (n_1, n_2) of linked nodes **do**
 3:         **if** n_1.rule.decision $= n_2$.rule.decision **then**
 4:             **if** n_1.rule.setOfActions $\subseteq n_2$.rule.setOfActions **then**   ▷ REDUNDANCY
 5:                 N.remove(n_1)
 6:             **elseif** n_2.rule.setOfActions $\subseteq n_1$.rule.setOfActions
 7:                 N.remove(n_2)
 8:             **end if**
 9:         **elseif** n_1.rule.decision $\neq n_2$.rule.decision and   ▷ CONFLICT
10:                     n_1.rule.setOfActions $\cap n_2$.rule.setOfActions $\neq \emptyset$
11:             $np =$ the node in $\{n_1, n_2\}$; np.rule.decision $=$ Permit
12:             $nd =$ the node in $\{n_1, n_2\}$; nd.rule.decision $=$ Deny
13:             $CD = np$.rule.accessDomain $\cap nd$.rule.accessDomain
14:             $CA = np$.rule.setOfActions $\cap nd$rule.setOfActions
15:             **if** Permissive strategy **then**
16:                 nd.rule.accesDomain $= nd$.rule.accesDomain$\backslash CD$ ▷ r_2'
17:                 add new node nn; nn.rule $=$ (Deny, nd.rule.setOfActions$\backslash CA$, CD)   ▷ $r_2^{\bar{n}}$
18:             **elseif** Restrictive strategy
19:                 np.rule.accesDomain $= np$.rule.accesDomain$\backslash CD$ ▷ r_1'
20:                 add new node nn; nn.rule $=$ (Permit, np.rule.setOfActions$\backslash CA$, CD)   ▷ $r_1^{\bar{n}}$
21:             **end if**
22:             **for** every node $n \in \{np, nd, nn\}$ **do**
23:                 **if** n.rule.accesssDomain$=\emptyset$ or n.rule.setOfActions$\emptyset$ **then**
24:                     N.remove(n)
25:                 **end if**
26:             **end for**
27:             **if** nn has not been removed **then**
28:                 link it to all the other nodes of the graph
29:             **elseif** np and nd have not been removed
30:                 remove the link between them
31:             **end if**
32:         **end if**
33:     **end while**
34:     **return** $G(N)$ ▷ Returns the modified graph G(N)
35: **end procedure**

- $r_4 = Deny_{\{write,create\}}$ ((position; fileType; time) $\in \{$Nurse$\} \times \{$Source, Documentation$\} \times [8:00, 18:00])$.

Since we consider a cluster of four rules, we get the 4-node and 6-edge Graph 1 of Fig. 1.

First iteration: Let us consider the pair (r_1, r_2) of Graph 1. Since $AD_{r_2} \subseteq AD_{r_1}$ and $\{$read$\} \subseteq \{$read, write$\}$ thus, r_2 is redundant to r_1. The resolution procedure

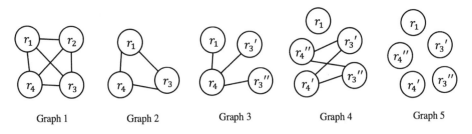

Graph 1   Graph 2   Graph 3   Graph 4   Graph 5

Fig. 1. Constructed graphs from Algorithm 2

will remove r_2 and its edges from Graph 1. We obtain Graph 2 of Fig. 1.

Second iteration: Let us consider the pair (r_1, r_3) of Graph 2.

Since $AD_{r_1} \cap AD_{r_3} = \{\text{Nurse}\} \times \{\text{Documentation}\} \times [8\text{:}00, 18\text{:}00] \neq \emptyset$, a conflict is detected between r_1 and r_3. If we use the permissive strategy, we keep r_1 and r_3 is replaced by:

- $r_3' = Deny_{\{read,delete\}}((\text{position}; \text{fileType}; \text{time}) \in \{\text{Nurse}\} \times \{\text{Source}\} \times [8\text{:}00, 18\text{:}00])$;
- $r_3'' = Deny_{\{delete\}}((\text{position}; \text{fileType}; \text{time}) \in \{\text{Nurse}\} \times \{\text{Documentation}\} \times [8\text{:}00, 18\text{:}00])$.

We then obtain Graph 3 of Fig. 1.

Third iteration: Let us consider the pair (r_1, r_4) of Graph 3.

Since $AD_{r_1} \cap AD_{r_4} = \{\text{Nurse}\} \times \{\text{Documentation}\} \times [8\text{:}00, 18\text{:}00]) \neq \emptyset$, a conflict is detected between r_1 and r_4. If we use the permissive strategy, we keep r_1 and r_4 is replaced by:

- $r_4' = Deny_{\{write,create\}}((\text{position}; \text{fileType}; \text{time}) \in \{\text{Nurse}\} \times \{\text{Source}\} \times 8\text{:}00, 18\text{:}00])$.
- $r_4'' = Deny_{\{create\}}((\text{position}; \text{fileType}; \text{time}) \in \{\text{Nurse}\} \times \{\text{Documentation}\} \times [8\text{:}00, 18\text{:}00])$.

We obtain Graph 4 of Fig. 1.

Iterations 4 to 7: The connected pairs of Graph 4 are (r_3', r_4''), (r_3'', r_4'), (r_3', r_4') and (r_3'', r_4''). For each of these pairs, the intersection of access domains or the intersection of sets of actions is empty. Therefore, their four links are removed through the four iterations 4 to 7. We obtain Graph 5 of Fig. 1, that has no link. Therefore, the algorithm terminates.

6   Conclusion

Access control policies might be aggregated from multiple parties and could be managed by several administrators. Therefore, the definition of the policy rules might contain several anomalies, which leads to high implementation complexity, as well as affects the performance of the policy execution. In this direction, we propose an approach to detect and resolve anomalies in XACML policies. We introduce the concept of policy rule access domain, which is used to identify and resolve effectively policy anomalies. As future work, we aim to integrate the proposed anomalies resolution method within our existing implementation of anomaly detection.

References

1. Ait El Hadj, M., Ayache, M., Benkaouz, Y., Khoumsi, A., Erradi, M.: Clustering-based approach for anomaly detection in xacml policies. In: the 14th International Joint Conference on e-Business and Telecommunications (ICETE 2017) - Volume 4: SECRYPT, pp. 548–553 (2017)
2. Chandola, V., Banerjee, A., Kumar, V.: Anomaly detection: a survey. ACM Comput. Surv. (CSUR) **41**(3), 15 (2009)
3. Hu, H., Ahn, G.-J., Kulkarni, K.: Anomaly discovery and resolution in web access control policies. In: Proceedings of the 16th ACM Symposium on Access Control Models and Technologies, pp. 165–174. ACM (2011)
4. Liu, A.X., Chen, F., Hwang, J., Xie, T.: Xengine: a fast and scalable XACML policy evaluation engine. ACM SIGMETRICS Perform. Eval. Rev. **36**, 265–276 (2008)
5. Mourad, A., Tout, H., Talhi, C., Otrok, H., Yahyaoui, H.: From model-driven specification to design-level set-based analysis of XACML policies. Comput. Electr. Eng. **52**, 65–79 (2015)
6. Proctor, S., et al.: Suns XACML implementation (2004). sunxacml.sourceforge.net
7. Ramli, C.D.P.K.: Detecting incompleteness, conflicting and unreachability XACML policies using answer set programming. arXiv preprint arXiv:1503.02732 (2015)
8. St-Martin, M., Felty, A.P.: A verified algorithm for detecting conflicts in XACML access control rules. In: Proceedings of the 5th ACM SIGPLAN Conference on Certified Programs and Proofs, pp. 166–175. ACM (2016)
9. OASIS Standard: extensible access control markup language (XACML) version 2.0 (2005)
10. Yuan, E., Tong, J.: Attributed based access control (ABAC) for web services. In: IEEE International Conference on Web Services (ICWS 2005), p. 569, July 2005

Biometric Template Privacy Using Visual Cryptography

Sana Ibjaoun[1,2(✉)], Anas Abou El Kalam[1(✉)], Vincent Poirriez[2(✉)],
and Abdellah Ait Ouahman[1(✉)]

[1] University Cadi Ayyad, Marrakesh, Morocco
ibjaoun.sanaa@gmail.com, elkalam@hotmail.fr, aitouahman@yahoo.fr
[2] University of Valenciennes and Hainaut Cambrésis, Valenciennes, France
Vincent.Poirriez@univ-valenciennes.fr

Abstract. One of the critical steps in designing a secure biometric system is protecting the templates of the users. If a biometric template is compromised, it leads to serious security and privacy threats because unlike passwords, it is not possible for a legitimate user to revoke his biometric identifiers and switch to another set of uncompromised identifiers. This work propose a new cancelable biometric template using visual cryptography (VC) for preserving privacy to biometric data such as finger-vein. The finger vein template is divided into two noisy images, known as shares, using VC, where one share is stored in the system database and the other is kept with the user on a smart card. The private image can be revealed only when both shares are simultaneously available at the same time. The individual share images do not reveal the identity of the private image. We experimentally evaluate performance of the proposed scheme. The evaluation is performed based on the two metrics: (i) False Acceptance Rate (FAR), and (ii) False Rejection Rate (FRR).

Keywords: Biometric · Authentication · Template protection · Privacy
Visual cryptography · Finger vein

1 Introduction

Using biometric authentication compared to traditional authentication, like password or token, has the advantage of security: they in fact (i) cannot be lost or forgotten. (ii) are very difficult to copy or share. (iii) are extremely hard to forge or distribute. (iv) cannot be guessed easily. Unfortunately recently, protecting biometric templates has become an issue. In a client/server-type biometric authentication system, biometric templates are stored in a database on the authentication server. In this case, it is difficult to prevent internal fraud by server's administrator, such as taking out biometric templates from the server. Besides, it is impossible to revoke biometric unlike password or token, and therefore if biometric is leaked out once and threat of forgery has occurred, the user cannot securely use his biometric anymore. Thus, biometric template protection schemes are required to protect the biometric data/feature and at the same time, maintain capability to identify and verify identity [1]. Cancelable Biometrics [2] is a biometric

© Springer International Publishing AG, part of Springer Nature 2018
A. Abraham et al. (Eds.): IBICA 2017, AISC 735, pp. 309–317, 2018.
https://doi.org/10.1007/978-3-319-76354-5_28

verification scheme which was introduced to address this problem. This scheme enables the system to store and match templates while keeping them secret. This preserves user's privacy and enhances security since it is impossible to recover the original biometric from the transformed version. A compromised template can be revoked using another transformation.

In order to realize Cancelable Biometrics it is important to preserve the accuracy. Secondly, it is required to prevent the attacker from recovering the original biometric feature from the transformed feature. Ideally, the transformed feature itself does not leak any information about the original one. Nevertheless, none of the existing methods meets both requirements at the same time.

In this paper, we propose a novel method of Cancelable Biometrics which meets both requirements at the same time. We use visual cryptography to preserve the privacy of biometric template by decomposing the original image into two images in such a way that the original image can be revealed only when both images are simultaneously available, further, the individual component images do not reveal any information about the original image.

The rest of this paper is organized as follows. Biometric template protection techniques are introduced in Sect. 2. Section 3 presents brief review of visual cryptography. Then, in Sect. 4, we describe our proposed method to protect the biometric template. After that, in Sect. 5, we present the security analysis. Tests and performance evaluation are presented in Sect. 6. Finally, in Sect. 7 we draw up our conclusions.

2   Biometric Template Protection

There are different levels at which a biometric template can be secured, e.g., hardware level and software level. The hardware-based approach involves designing a closed recognition system, where the biometric template never leaves a physically secure module such as a smart card or a hand-held device [3].

In the software based techniques, the biometric data is usually combined with some external key, such as a password or a system generated random number and the resultant data is stored in the system database instead of the original biometric template. It is expected that the protected template reveals little information about the original template. Based on the way in which the matching is performed, the software based template protection techniques can be divided into three main categories: Encryption, Biometric cryptosystems, and Template transformation.

2.1   Encryption

In encryption based techniques, the biometric template is encrypted using an encryption key, possibly derived from a password, during enrolment. During authentication, the stored data is decrypted using the corresponding decryption key and is matched with the captured query. One of the main limitations of encryption based techniques is insecure key management since the decryption key is exposed to the system during each attempt to authenticate and thus can be easily stolen by the adversary. The advantage, however is that any sophisticated matching procedure can be employed thereby preserving the matching accuracy.

2.2 Biometric Cryptosystems

Biometric cryptosystems [4, 5] are similar to password based key generation systems as they are used to secure cryptographic key or to directly generate cryptographic key from biometric features. Since the biometric measurements obtained during enrollment and authentications are different, these features cannot be used directly for the generation of cryptographic key generation. To facilitate key generation helper data or secure sketch of the biometric features are stored during enrollment. Therefore, biometric cryptosystems are also known as helper data systems. The main advantage of biometric cryptosystem is that exact recovery of original biometric data allows its use as an encryption key in another cryptosystem. Biometric cryptosystems are classified as key release, key binding and key generation systems depending on how the secure sketch is obtained. Secure sketch is public information about biometric features stored in databases during enrollment. Fuzzy vault [6, 7], fuzzy commitment [8] and secure sketches [9, 10] are the most popular techniques used for constructing biometric cryptosystems.

2.3 Template Transformation

Template transformation techniques [11–17] transform the biometric template based on parameters derived from external information such as user passwords or keys. During authentication, the same transformation function is applied to the query and matched with the stored template in the transformed domain. Usually, geometric transformations involving projection onto a new space determined by the password are applied to the biometric features. The main advantage of template transformation techniques is that if the user transforms his biometric on a separate personal device and sends only the transformed template to the biometric system, the original biometric is never revealed in the system.

Note that the three techniques discussed above are independent in nature and can be used in any combination. For example, the templates protected using either the template transformation or biometric cryptosystem can be further encrypted and a transformed template can be secured using biometric cryptosystem which in turn can also be encrypted.

3 Visual Cryptography Approaches for Securing Biometric Template

Naor and Shamir [18] introduced the visual cryptography scheme (VCS) as a simple and secure way to allow the secret sharing of images without any cryptographic computations. VCS is a cryptographic technique that allows for the encryption of visual information such that decryption can be performed using the human visual system. The basic scheme is referred to as the k-out-of-n VCS, which is denoted as (k, n) VCS [18]. Given an original binary image T, it is encrypted in images, such that:

$$T = S_{h1} \oplus S_{h2} \oplus \dots \oplus S_{hk}$$

where $\oplus$ is a Boolean operation, S_{hi}, $hi \in 1, 2, \dots, k$ is an image which appears as white noise, $k \leq n$ and n is the number of noisy images. It is difficult to decipher the secret image using individual's [18]. The encryption is undertaken in such a way that k or more out n of the generated images are necessary for reconstructing the original image T. However, no combination of $k - 1$ shares can reveal the secret. Figure 1 shows an example of *2-outof-2* VC on a binary image.

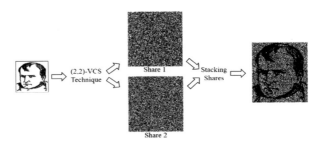

Fig. 1. An example of 2-out of-2 VCS

The scheme proposed by Naor and Shamir defines methods to create shares for binary images only. However, new approaches have been proposed that extend to gray-scale and color images too. Biometric template protection using VC schemes have been introduced for various modalities like fingerprint [19], face [20, 21], palmprint [22], and iris [23–25].

Any VC scheme is characterized by two parameters:

- Pixel expansion: Each pixel of the original image is transformed into n shares, one for each transparency/share image. This involves encoding of a pixel into more than one sub-pixel. The number of sub-pixels required to encode a pixel determines the pixel expansion.
- Contrast: Contrast refers to the quality of the reconstructed image after the decryption.

4   Proposed Method

In this section, we propose a new method of Cancelable Biometrics using VC to protect the finger vein template from attack in system database. There are two main phases: Enrollment phase and Authentication phase.

4.1   Enrollment Phase

During the enrollment process, user extracts his biometric template and performs image scrambling. Image scrambling is used to make images visually unrecognizable such that unauthorized users have difficulty decoding the scrambled image to access the original image. The scrambled image is then sent to a trusted third-party entity. Once the trusted entity receives it, the scrambled image is decomposed into two noisy images and the original data is discarded. The first share is stored in the database and the other is stored on the user smart card. This ends the enrollment phase. Figure 2 shows the block diagram that illustrates the enrollment phase of the proposed approach.

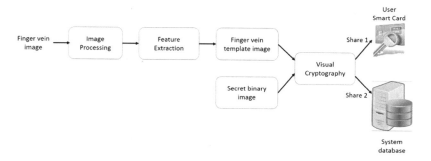

Fig. 2. Enrollment phase of the proposed scheme

4.2 Authentication Phase

For this phase, the user will provide his biometric information on the specific device, here we will extract his finger vein template. The user will provide also his smart card on which the share is stored. At the same time, the system will find his corresponding second share from the database. These both shares will be superimposed in order to reconstruct the scrambled image. An inverse permutation sequence is obtained by using the same key, and applies this sequence to the scrambled image in-order to reconstruct the original image.

If both the provided template and the constructed one match then the user is given access to the service. But, if they don't match then access is denied. The block diagram of proposed system is shown in Fig. 3.

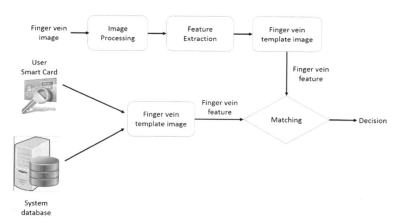

Fig. 3. Authentication phase of the proposed scheme

5   Analysis

5.1   Preserving Accuracy

Basic VC leads to the degradation in the quality of the decoded images, where the white background of the original image becomes gray in the decrypted image. Therefore, an error in computing the score would occur due to applying the VC and the matching accuracy may degrade compared to the version without transformation [11]. It is important to reduce the accuracy degradation and to preserve the accuracy. The overlaying or superimposing operation in visual cryptography is computationally modeled as the binary OR operation which causes the contrast level of the target image to be lowered. In our scheme we use the XOR operator instead of OR operator. In this way the loss in contrast in target images is addressed. Furthermore, the target image can be down-sampled by reconstructing just one pixel from every block. Thus, the reconstructed image will be visually appealing while requiring less storage space. Hence, the matching score is not affected with the quality of the image when applying our proposed method, thus the accuracy is preserved.

5.2   Protection Criteria

In our scheme, the biometric template is encrypted into two shares which are distributed amongst the user and database server. The four biometric template protection criteria as specified by Jain et al. [26] can also be fulfilled by our scheme. These are: diversity, revocability, security and performance.

Revocability. In our scheme VC use a randomly selected image while generating shares. These shares can be suspended or replaced time to time to generate revocable template.

Diversity. The same for different applications template can be decomposed into different constituting shares, by using different secret binary image.

Security. It is computationally hard to obtain the secret image by any individual share. Also, by storing the shares on two different spaces, the chances of obtaining the secret image is minimized. Data stored on the distributed server prevents unauthorized modification and inaccurate updates.

Performance. The performance of recognition systems does not degrade when the original biometric template is reconstructed from its constituting shares using our approach as the contrast (quality) is optimally preserved while reconstructing.

6   Performance Evaluation

In our experiment, we applied the proposed method to finger-vein pattern matching and examine the verification performance. The performance of a biometric system is essential to determine whether the system has a potential to be applied in real life situations.

These performances are measured by two key performance metrics: (i) False Acceptance Rate (FAR), the probability that an unauthorized person is accepted as an authorized person, and (ii) False Rejection Rate (FRR), the probability that an authorized person is rejected as an unauthorized person.

The system has been trained using a number of 72 finger images. Figure 4 shows an example of infrared finger image. The modified Hausdorff distance was calculated between each image and the other ones. It means that a number of 2628 attempts occurred, from which 144 was genuine attempts and the rest of 2484 was impostor attempts.

Fig. 4. Example of infrared finger image

Figure 5 presents the distribution of scores we obtained using this biometric dataset (legitimate scores are represented in blue continuous line, impostor ones in red dotted line). We decided to fix the threshold at 15. According with selected threshold, we obtained a FAR value of 3% and a FRR value of 5%.

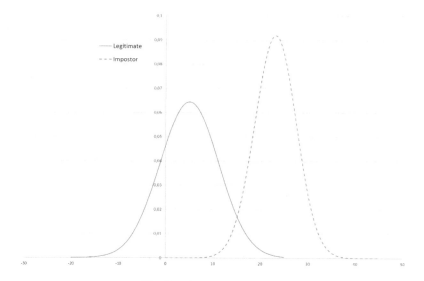

Fig. 5. Distribution of scores

The resulting ROC curve is given in Fig. 6. The EER (Equal Error Rate) performance point is obtained at the intersection of the ROC curve (blue dotted) and the line FAR = FRR (red continuous). This value has not yet made any practical use since it is not generally desired that the FAR and the FRR be the same but it is an indicator of the accuracy of the biometric device. In other words, the lower the EER, the more efficient the system. For our system we have an EER with a value of 0.04 (4%).

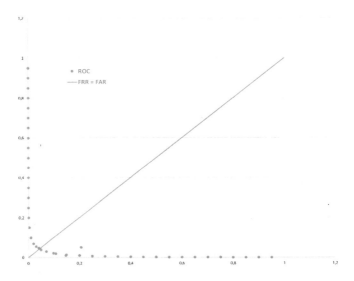

Fig. 6. ROC curve and EER

7   Conclusion

In this paper, we proposed a novel method of Cancelable Biometrics using visual cryptography to ensure the user privacy. The template is decomposed into two noise-like images using (2, 2) VCS. The XOR operator is used to superimpose the two noisy images and fully recover the original template. Thus, the matching accuracy is not affected with the quality of the image when applying our proposed method. Moreover, the two shares do not leak any information of the original image, in other words, our proposed method has perfect secrecy. Additionally, we applied our proposed method to finger-vein template and experimentally obtained very high authentication performance.

References

1. Stavroulakis, P., Stamp, M.: Handbook of Information and Communication Security. Springer, Heidelberg (2010)
2. Ratha, N.K., Connell, J.H., Bolle, R.M.: Enhancing security and privacy in biometric-based authentication systems. IBM Syst. J. **40**(3), 614–634 (2001)
3. Maltoni, D., Maio, D., Jain, A.K., Prabhakar, S.: Handbook of Fingerprint Recognition. Springer, London (2009)
4. Soutar, C., Roberge, D., Stoianov, A., Gilroy, R., Kumar, B.V.: Biometric encryption using image processing. In: Photonics West 1998 Electronic Imaging, International Society for Optics and Photonics, pp. 178–188 (1998)

5. Uludag, U., Pankanti, S., Prabhakar, S., Jain, A.K.: Biometric cryptosystems: issues and challenges. Proc. IEEE **92**(6), 948–960 (2004)
6. Juels, A., Sudan, M.: A fuzzy vault scheme. In: Proceedings of IEEE International Symposium on Information Theory, Lausanne, Switzerland, p. 408 (2002)
7. Nandakumar, K., Jain, A.K., Pankanti, S.: Fingerprint-based fuzzy vault: implementation and performance. IEEE Trans. Inf. Forensics Secur. **2**(4), 744–757 (2007)
8. Juels, A., Wattenberg, M.: A fuzzy commitment scheme. In: Proceedings of the 6th ACM Conference on Computer and Communications Security, pp. 28–36. ACM (1999)
9. Li, Q., Chang, E.C.: Robust, short and sensitive authentication tags using secure sketch. In: Proceedings of the 8th Workshop on Multimedia and Security, pp. 56–61. ACM (2006)
10. Sutcu, Y., Li, Q., Memon, N.: Protecting biometric templates with sketch: theory and practice. IEEE Trans. Inf. Forensics Secur. **2**(3), 503–512 (2007)
11. Ratha, N.K., Connell, J.H., Bolle, R.M., Chikkerur, S.: Cancelable biometrics: a case study in fingerprints. In: 18th International Conference on Pattern Recognition (ICPR 2006), vol. 4, pp. 370–373 (2006)
12. Ratha, N.K., Chikkerur, S., Connell, J.H., Bolle, R.M.: Generating cancelable fingerprint templates. IEEE Trans. Pattern Anal. Mach. Intell. **29**(4), 561–572 (2007)
13. Maiorana, E., Campisi, P., Fierrez, J., Ortega-Garcia, J., Neri, A.: Cancelable templates for sequence based biometrics with application to on-line signature recognition. IEEE Trans. Syst. Man Cybern. Part A Syst. Hum. **40**(3), 525–538 (2010)
14. Connie, T., Teoh, A., Goh, M., Ngo, D.: Palmhashing: a novel approach for cancelable biometrics. Inf. Process. Lett. **93**(1), 1–5 (2005)
15. Teoh, A.B.J., Ngo, D.C.L., Goh, A.: Biohashing: two factor authentication featuring fingerprint data and tokenised random number. Pattern Recogn. **37**(11), 2245–2255 (2004)
16. Hirata, S., Takahashi, K.: Cancelable biometrics with perfect secrecy for correlation-based matching. In: Tistarelli, M., Nixon, M.S. (eds.) ICB 2009. LNCS, vol. 5558, pp. 868–878. Springer, Heidelberg (2009)
17. Chen, X., Bai, X., Tao, X., Pan, X.: Chaotic random projection for cancelable biometric key generation. In: Intelligent Science and Intelligent Data Engineering, pp. 605–612. Springer, Heidelberg (2013)
18. Naor, M., Shamir, A.: Visual cryptography. In: Advances in Cryptology EUROCRYPT 1994, pp. 1–12. Springer, Heidelberg (1995)
19. Monoth, T., Anto, P.B.: Tamperproof transmission of fingerprints using visual cryptography schemes. Procedia Comput. Sci. **2**, 143–148 (2010)
20. Ross, A., Othman, A.: Visual cryptography for biometric privacy. IEEE Trans. Inf. Forensics Secur. **6**(1), 70–81 (2011)
21. Ross, A., Othman, A.A.: Visual cryptography for face privacy. In: SPIE Defense, Security, and Sensing, International Society for Optics and Photonics, p. 76,670B (2010)
22. Divya, C., Surya, E.: Visual cryptography using palm print based on DCT algorithm. Int. J. Emerg. Technol. Adv. Eng. **2**(12), 2250–2459 (2012)
23. Revenkar, P., Anjum, A., Gandhare, W.: Secure iris authentication using visual cryptography. arXiv preprint arXiv:10041748 (2010)
24. Chin, C.S., Jin, A.T.B., Ling, D.N.C.: High security iris verification system based on random secret integration. Comput. Vis. Image Underst. **102**(2), 169–177 (2006)
25. Sinduja, R., Sathiya, R., Vaithiyanathan, V.: Sheltered iris attestation by means of visual cryptography (SIA-VC). In: 2012 International Conference on Advances in Engineering, Science and Management (ICAESM), pp. 650–655 (2012)
26. Jain, A.K., Nandakumar, K., Nagar, A., et al.: Biometric template security. EURASIP J. Adv. Sig. Process. **2008**(113), 1–17 (2008)

Scalable and Dynamic Network Intrusion Detection and Prevention System

Safaa Mahrach[1(✉)], Oussama Mjihil[1], and Abdelkrim Haqiq[1,2]

[1] Computer, Networks, Mobility and Modeling Laboratory, FST,
Hassan 1st University, Settat, Morocco
mahrachsafaa@gmail.com, o.mjihil@uhp.ac.ma, ahaqiq@gmail.com
[2] e-NGN Research Group, Africa and Middle East, Rabat, Morocco

Abstract. Network Intrusion Detection and Prevention Systems (NIDPS) are widely used to detect and thwart malicious activities and attacks. However, the existing NIDPS are monolithic/centralized, and hence they are very limited in terms of scalability and responsiveness. In this work, we address how to mitigate SYN Flooding attacks that can occur in the management network (OpenFlow) as well as in the production network taking into account the network scalability. Our suggested framework is a distributed and dynamic NIDPS that uses the Programming Protocol independent Packet Processors (P4) to process the network packets at the switch level and perform two main functions. First, it detects the SYN flooding attacks based on the SYN packets' rate and threshold. Secondly, our system uses a reviewed way to activate the SYN cookies in order to block/drop illegitimate packets. Our framework takes advantage of the switch programmability (i.e., using P4 language), distributed packet processing, and centralized Software Defined Networking (SDN) control, to provide an efficient and extensible NIDPS.

Keywords: SDN · Network security · DDOS · IDS · IPS
P4 language

1 Introduction

Network Security is attracting increasingly more attention due to the increasing number of attacks and their sophistication. Attacks, such as resource abuse (i.e., the usage of network resources for illegal purposes) and malicious insiders (i.e., malware injection attacks), are among the most severe threats. Attackers may spam, perform malicious code execution, and exploit vulnerabilities to form botnets which will be used to launch advanced attacks, such as Distributed Denial-of-Service (DDoS).

Network Intrusion Detection and Prevention Systems (NIDPS) [12] are considered as a good security solution to detect and react to malicious attacks. They have been widely deployed to enhance network security. However, the traditional NIDPS expose the network to important issues (i.e., consuming significant resources and increasing the response time), and also they are still limited

A. Abraham et al. (Eds.): IBICA 2017, AISC 735, pp. 318–328, 2018.
https://doi.org/10.1007/978-3-319-76354-5_29

to control and secure a scalable network. In this sense, we propose a new scalable and flexible NIDPS based on Software Defined Networking (SDN) [11], which is a new architecture that enables the automatic management of network services from a high abstraction level. This feature is achieved by decoupling the control plane from the data plane. SDN use OpenFlow [9] as a standard protocol, which is defined to implement packet-forwarding rules in switches. Few works have been proposed [19, 20] in this area. However, these studies focused on developing specific high-level SDN applications to provide improved security solutions. As results, the path between the data and control planes quickly becomes a bottleneck, which hinders the network performance and limits its scalability and responsiveness.

For enabling practical SDN-based security applications, the data plane needs to support more advanced functionalities. In this sense, a new high-level language has been proposed: Programming Protocol independent Packet Processors (P4) [5]. P4 language rises the abstraction level of the network programming by indicating to the switch how to operate, rather than being limited by the switch provider's design. This opportunity meets our objective to activate intrusion detection and prevention roles in a programmable switch. So, this latter can analyze quickly the local traffic, detect and react to malicious activities, and notify the controller.

The main contribution of this work is the design of a scalable and dynamic NIDPS that enables the data plane to perform advanced detection and reaction functions at an early stage of the network communication. With this framework, we activate a dynamic mitigation mechanism against SYN Flooding attacks while enabling the switch to: (1) count the SYN packets' rate and detect the SYN flooding attacks based on the SYN packets' rate and threshold, (2) react automatically using the SYN cookies to block/drop illegitimate packets. Moreover, the switch will have the ability to notify the SDN controller, which will be able to manage and distribute the appropriate decisions in large networks.

In this paper, we present the design of a scalable and dynamic network intrusion detection and prevention system. The rest of the paper is organized as follows. Section 2 presents a general background about Intrusion Detection and Prevention Systems and P4 Language. In Sect. 3, we discuss related work. We present the dynamic defense mechanism for the SYN Flooding attacks, in Sect. 4. Finally, we give our conclusion in Sect. 5.

2 Background

2.1 Network Intrusion Detection and Prevention Systems (NIDPS)

NIDPS are proactive and reactive systems able to actively prevent and block network intrusions that are detected. NIDPS have been widely deployed to enhance network security, but they are still limited to manage and secure the scalable networks. According to some existing articles [19, 20], the traditional NIDPS expose the network to the following major limitations:

– **Latency:** They require inspection and blocking actions on each network packet, which consume resources and also increase detection latency and response time.
– **Resource consumption:** NIDPS services usually consume significant resources (e.g., SPAN port mirroring technology).
– **Inflexible network reconfiguration:** NIDPS are static and not able to automatically reconfigure the networking system and offer pointed traffic inspection and control.

2.2   P4 Language

Programming Protocol independent Packet Processors (P4) is a programming language for describing how packets are processed by the data plane [1]. It's also considered as a protocol between the controller and the network devices. P4 program is based upon an abstract forwarding model consisting of a parser and a set of match+action table resources, divided between ingress and egress (Fig. 1). The parser defines the headers present in each incoming packet. Each match+action table identifies the type of lookup (i.e., rules) to perform, the input fields to use and the actions that may be applied [1].

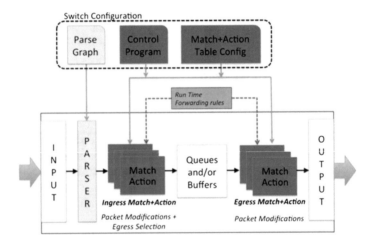

Fig. 1. Abstract forwarding model [1]

– **Counters:** P4 maintains information across packets using stateful memories: Counters, Meters and Registers [1]. In this work, we use Counters to perform different measurements, such as the number of packets or bytes generated by each host that can enable the detection of eventual worm propagation or Denial-of-service (DOS) attacks. Counters are categorized in two types: direct and indirect.

```
Counter ip_pkts_by_dest{
   type:   packets;
   direct: ip_host_table;}
```

In the example above we use direct access for ip-host-table, which allocates one counter for each table entry. The current counter gives to the controller the ability to read the number of packets sent to each host.

- **Tables:** As P4 specification, tables are the fundamental units of the match action pipeline. Tables define rules to perform the input fields to use, and the actions that may be applied. Actions in P4 are declared as functions (i.e., compound actions) which are built from primitive actions (i.e., basic actions).
- **Control Program:** The control program organizes the layout of tables within ingress and egress pipeline, and the packet flow through the pipeline. It may-be expressed with an imperative program (i.e., main program in P4 language) which may apply tables, call other control flow functions, or test conditions.

3 Related Work

3.1 Practical SDN for Security Solutions

Programming Protocol independent Packet Processors (SDN) is designed to improve the network management while separating the control decision from the data plane. It can also be useful for enhancing the agility of network security applications. For instance, the authors of [21] implemented practical security functions (e.g., Firewall, NIDS, etc.) in the Floodlight platform in order to encourage researchers and practitioners to use and develop SDN-based network security services. Particularly for IDPS, the authors of [20] presented a new IPS architecture called SDNIPS, based on SDN and Open Virtual Switch (OVS), which provides a dynamic defensive mechanism for clouds. However, these studies focus on developing specific high-level SDN applications to provide improved security solutions. As results, the path between the data and control planes quickly becomes a bottleneck, which hinders the network performance and limits its scalability (because there are usually a small number of controllers than switches in a network). In this sense, the SDN community declared the requirement of programmable switch, which can support more advanced functionalities. There were few works that fundamentally address flow management at SDN switch level. Avant Guard [14] activated the data plane with new security extensions which reduce the threats of the saturation attack, detect, and response to unusual traffic volume within the data plane. OpenFlow Extension Framework (OFX) [16] enabled practical SDN security applications within unmodified OpenFlow infrastructure. OFX allows applications to dynamically load software modules directly into an existing OpenFlow switches where application process (e.g., monitoring, detection) can execute closer to the data plane.

3.2 Stateful SDN Data Plane Applications

As discussed above, the SDN community has already recognized the need to implement some specific functions directly in the switch itself. For example, the authors of [6] made a survey of the recently proposed approaches for the stateful SDN data plane [4,10]. In addition, they analyzed the vulnerabilities of these stateful proposals and list some potential attacks that can exploit these vulnerabilities. OpenState [4] and FAST (Flow-level State Transitions) [10] were among the first platforms designed to handle the flow states inside the switch using finite state machines as an extension of OpenFlow protocol. This extension allows the switch to locally handle stateful rules, and hence provide more control and security extents.

3.3 Packet Processing Language

P4 is a new high-level language which supports configurable switches. It aims to provide a description of customized packet processing which gives the possibility to activate different network functions (e.g., Monitoring, Security, and so on) on dynamic switches. In [18] Vörös et al. studied how P4 can be used as a security middle-ware programming language. They demonstrated how to implement a stateful firewall on switches using P4 language. The authors of [2] offer a mitigation system against network spoofing attacks while implementing a collection of antispoofing techniques in OpenFlow 1.5 and P4 match and action rules. They also developed algorithms to automatically redistribute the required rules over the network switches.

4 System Design

DDOS (Distributed Denial of Service) is one of the oldest and the most popular attacks that is growing in size and frequency in traditional networks. Accordingly, it is considered among the major threats that exploit the SDN vulnerabilities [3,13]. We distinguish three types of DDOS attacks: the volumetric network based attacks (e.g., UDP flood, ICMP flood), the protocol attacks (e.g., SYN flood, Ping of death), and the application level attacks (e.g., HTTP flood, DNS flood). Our approach addresses SYN flooding attacks which remain very popular attacks [22]. The data plane provides interesting capabilities, such as stateful memories (e.g., counters and registers in P4 program) that can be used to detect abnormal behavior of network. In addition, it has a powerful high-throughput packet processing and filtering capabilities that can be useful for mitigating denial of service attacks. This opportunity meets our aim to enable a scalable and dynamic NIDPS at the switch level using P4 language. As a use case, we activate a dynamic mitigation mechanism against the SYN Flooding attacks for both the data center (i.e., production network) and the SDN architecture (i.e., communication between switches and controllers).

4.1   Selected Methods

In our defensive mechanism we select SYN cookie technique and the method used in [2] to modify the SYN packet to become SYN-ACK packet.

– **SYN cookie** [8] is used in our system as state-less technique to prevent the memory consumption caused by the half open SYN attacks (SYN food attacks) (see Fig. 4). The mitigating system (i.e., server or switch in our case) enabling SYN cookie method responds to the client SYN request by a SYN-ACK packet with pre-generated cookie (i.e., cookie or the Initial Sequence Number (ISN) is created while hashing details about the initial SYN packet and its TCP options). When the ACK of client is received (with pre-generated cookie +1) the system can validate it by checking the TCP sequence number against the cookie value that was encoded in the SYN-ACK packet.
  We select **TCP-reset** [17] as SYN cookie technique. When the client is authenticated (i.e., using SYN cookie) and classified as legitimate client. In this case the mitigating system send back a TCP-reset packet (i.e., with source IP of the original server) to the sender in order to enable him to re-establish the connection directly with the server.
  Inspiring by the Linux kernel method, which automatically enables the SYN cookies only when the SYN queue is full, we activate the SYN cookie only when the rate of SYN requests reached the defined threshold, rather than enabling them constantly.

To realize this, the method performs the following primitive steps on the SYN packets that are received and not yet authenticated (Table 1):
  The selected methods will be implemented using the high-level P4 language (see Sect. 2.2), which allowed us to indicate to the switch how to operate and process the received packets while creating specific parsers, tables, actions, and control flow.

4.2   Design Architecture Description

In this part, we describe the operation of our detection and prevention system implemented on p4 switch. We are going through the different steps of our framework (Figs. 3 and 4):

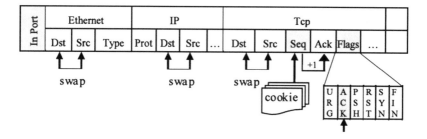

Fig. 2. Modifying SYN packet to become a SYN-ACK packet [2]

Table 1. Steps to modify SYN packet to become a SYN-ACK packet.

1	Swap the source and destination IP addresses
2	Exchange the source and destination Ethernet addresses
3	Swap the source and destination TCP ports
4	Set the ACK bit in the TCP flags (to convert the SYN packet into a SYN-ACK packet)
5	Increment the client-Seq number field by one
6	Copy the incremented Seq number field to the ACK number field
7	Write a pre-generated random cookie to the Seq number field
8	Recalculate the IP/TCP checksum values (in P4)
9	Send back the SYN-ACK packet on the incoming port (on which the SYN packet was received)

1. We define the header fields (i.e., IPv4, TCP) to use and then we start parsing packets and extracting the header field values.
2. We use registers and counters to count the SYN packets' rate (i.e., list of connections for which a SYN has been received and an ACK has not yet been received) while defining specific metrics, such as IP source, destination, protocol TCP source, destination, Flags-SYN.
3. We create our control program in P4, which checks if the SYN requests rate attains the defined threshold.
4. If the SYN packets rate reaches the defined threshold, the SYN cookie technique will be activated. For this end, our program starts intercepting the received SYN packets and sends back a SYN/ACK (i.e., SYN/ACK packet is generated from SYN request (see Sect. 4.1) with a pre-generated cookie (see Sect. 4.2)) (Fig. 4).
5. If the switch receive an ACK with validated cookie (i.e., the defined random cookie+1) from the client, the connection is authenticated and a RST packet (the ACK packet is transformed into a RST packet using the same primitive steps used to transform a SYN into a SYN/ACK) is sent back to the client in order to enable it to re-establish the connection directly with the server (Fig. 4).
6. If the third ACK contains an incorrect cookie (i.e., not validated ACK), it will be dropped.

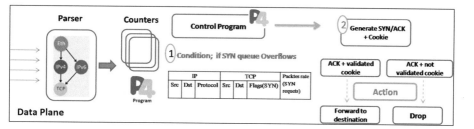

Fig. 3. Architecture of dynamic defense mechanism (SYN flooding attacks)

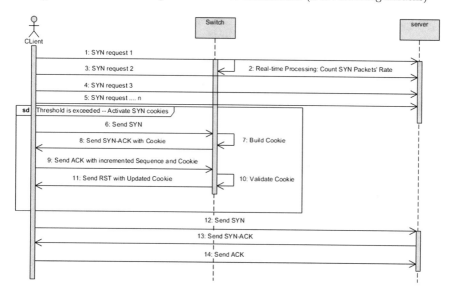

Fig. 4. Dynamic defense mechanism against SYN flooding attacks

a. Pre-Generated Cookie: According to [15], the implementation of the SYN cookies must fulfill the following basic requirements:

- Cookies should contain some details of the initial SYN packet and its TCP options.
- Cookies should be unpredictable by attackers. It is recommended to use a cryptographic hashing function in order to make the decoding of the cookie more complicated. For this end, we select the recommended Linux SYN cookies method for generating and validating cookies [7].

Cookie generation:

$$H_1 = hash(K_1, IP_s, IP_d, Port_s, Port_d); \tag{1}$$

$$H_2 = hash(K_2, count, IP_s, IP_d, Port_s, Port_d); \tag{2}$$

$$ISN_d(cookie) = H_1 + ISN_s + (count * 2^{24}) + (H_2 + MSS) \mod 2^{24}; \tag{3}$$

Table 2. Parameters of the Linux implementation

Parameter	Description
K_1, K_2	Secret keys
IP_s, IP_d	Source and destination IP addresses
$Port_s, Port_d$	Source and destination ports
ISN_s, ISN_d	Source and destination initial sequence numbers
ACK	Acknowledgement number
SEQ	Sequence number
MSS	2 bit index of the client's Maximum Segment Size
Count	32 bit minute counter
Hash()	32 bit cryptographic hash

Cookie validation:

$$ISN_d = ACK - 1; \tag{4}$$

$$ISN_s = SEQ - 1; \tag{5}$$

$$count(cookie) = (ISN_d - H_1 - ISN_s)/2^{24}; \tag{6}$$

$$MSS(cookie) = (ISN_d - H_1 - ISN_s) \mod 2^{24} - H2 \mod 2^{24}; \tag{7}$$

As we can see above, we calculate the two hash values H1 and H2 (based on TCP options, secret keys k1, k2 and count) then we use them with ISNs and MSS to generate the cookie (ISNd), as it is shown in (3). For the cookie validation, there are 2 integrity controls (count(cookie) and MSS(cookie)). The first one checks the age of the cookie. The second evaluates whether the value of the MSS is within the 2 bit range (0–3). If the cookie meets both integrity controls, it is considered valid, and the connection can be accepted Table 2.

P4 language gives us the possibility to generate the hash values (H1 and H2) using the bellow function and to perform the equations (cookies generation and validation) as arithmetic operations.

```
field_list_calculation hash_value_name {
    input {  fields;}
    algorithm : hash_algo;
```

5   Conclusion

With the increasing networks size, it is very challenging to implement a distributed and dynamic NIPDS that can handle all the traffic, in real-time, without becoming the bottleneck of the network. Our solution consists of offloading some control functions to the switch, so they can support a fast and early stage detection and prevention techniques. The switch will keep the state information of the network traffic, detect anomalies, and deploy countermeasures. As a use case, we addressed syn-flooding attacks, for this end, we used a P4 program that is counting the SYN packets' rate to detect abnormal communication patterns, and then the program activates syn-cookies as a prevention technique in order to block/drop illegitimate packets.

References

1. P4 language specification. https://p4lang.github.io/p4-spec/. Accessed 12 Sep 2017
2. Afek, Y., Bremler-Barr, A., Shafir, L.: Network anti-spoofing with SDN data plane. In: IEEE Conference on Computer Communications, IEEE INFOCOM 2017, pp. 1–9. IEEE (2017)
3. Benton, K., Camp, L.J., Small, C.: Openflow vulnerability assessment. In: Proceedings of the Second ACM SIGCOMM Workshop on Hot Topics in Software Defined Networking, pp. 151–152. ACM (2013)
4. Bianchi, G., Bonola, M., Capone, A., Cascone, C.: Openstate: programming platform-independent stateful openflow applications inside the switch. ACM SIGCOMM Comput. Commun. Rev. 44(2), 44–51 (2014)
5. Bosshart, P., Daly, D., Gibb, G., Izzard, M., McKeown, N., Rexford, J., Schlesinger, C., Talayco, D., Vahdat, A., Varghese, G., et al.: P4: programming protocol-independent packet processors. ACM SIGCOMM Comput. Commun. Rev. 44(3), 87–95 (2014)
6. Dargahi, T., Caponi, A., Ambrosin, M., Bianchi, G., Conti, M.: A survey on the security of stateful SDN data planes. IEEE Commun. Surv. Tutor. 19, 1701–1725 (2017)
7. Echevarria, J.J., Garaizar, P., Legarda, J.: An experimental study on the applicability of SYN cookies to networked constrained devices. Softw. Pract. Exp.
8. Fontes, S.M., Hind, J.R., Narten, T., Stockton, M.L.: Blended syn cookies, 6 June 2006. US Patent 7,058,718
9. McKeown, N., Anderson, T., Balakrishnan, H., Parulkar, G., Peterson, L., Rexford, J., Shenker, S., Turner, J.: Openflow: enabling innovation in campus networks. ACM SIGCOMM Comp. Commun. Rev. 38(2), 69–74 (2008)
10. Moshref, M., Bhargava, A., Gupta, A., Yu, M., Govindan, R.: Flow-level state transition as a new switch primitive for SDN. In: Proceedings of the Third Workshop on Hot Topics in Software Defined Networking, pp. 61–66. ACM (2014)
11. Nunes, B.A.A., Mendonca, M., Nguyen, X.N., Obraczka, K., Turletti, T.: A survey of software-defined networking: past, present, and future of programmable networks. IEEE Commun. Surv. Tutor. 16(3), 1617–1634 (2014)
12. Scarfone, K., Mell, P.: Guide to intrusion detection and prevention systems (IDPS). NIST special publication 800(2007), 94 (2007)
13. Shin, S., Gu, G.: Attacking software-defined networks: a first feasibility study. In: Proceedings of the Second ACM SIGCOMM Workshop on Hot Topics in Software Defined Networking, pp. 165–166. ACM (2013)
14. Shin, S., Yegneswaran, V., Porras, P., Gu, G.: Avant-guard: scalable and vigilant switch flow management in software-defined networks. In: Proceedings of the 2013 ACM SIGSAC Conference on Computer & Communications Security, pp. 413–424. ACM (2013)
15. Simpson, W.A.: TCP cookie transactions (TCPCT) (2011)

16. Sonchack, J., Smith, J.M., Aviv, A.J., Keller, E.: Enabling practical software-defined networking security applications with OFX. In: NDSS, vol. 16, pp. 1–15 (2016)
17. Touitou, D., Pazi, G., Shtein, Y., Tzadikario, R.: Using TCP to authenticate IP source addresses, 12 July 2011. US Patent 7,979,694
18. Vörös, P., Kiss, A.: Security middleware programming using P4. In: International Conference on Human Aspects of Information Security, Privacy, and Trust, pp. 277–287. Springer (2016)
19. Xing, T., Huang, D., Xu, L., Chung, C.J., Khatkar, P.: Snortflow: a openflow-based intrusion prevention system in cloud environment. In: 2013 Second GENI Research and Educational Experiment Workshop (GREE), pp. 89–92. IEEE (2013)
20. Xing, T., Xiong, Z., Huang, D., Medhi, D.: SDNIPS: enabling software-defined networking based intrusion prevention system in clouds. In: 2014 10th International Conference on Network and Service Management (CNSM), pp. 308–311. IEEE (2014)
21. Yoon, C., Park, T., Lee, S., Kang, H., Shin, S., Zhang, Z.: Enabling security functions with SDN: a feasibility study. Comput. Netw. **85**, 19–35 (2015)
22. Zargar, S.T., Joshi, J., Tipper, D.: A survey of defense mechanisms against distributed denial of service (DDoS) flooding attacks. IEEE Commun. Surv. Tutor. **15**(4), 2046–2069 (2013)

A Hybrid Feature Selection for MRI Brain Tumor Classification

Ahmed Kharrat[(✉)] and Mahmoud Neji

MIRACL-Laboratory: "Multimedia, InfoRmation systems and Advanced Computing Laboratory", FSEG, University of Sfax, BP1088, 3018 Sfax, Tunisia
Ahmed.kharrat@isims.usf.tn, mahmoud.neji@fsegs.rnu.tn

Abstract. Because a great number of features affects the performance of classification systems, a growing emphasis is placed on the feature selection. This work seeks to obtain an optimal feature subset through a hybrid algorithm of Simulated Annealing-Genetic Algorithms (SA-GA). Our proposed approach mutually avoids being stuck in a local simulated annealing minimum with the very high convergence rate of the genetic algorithm crossover operator and thus guarantee a high computational efficiency of support vector machine. To evaluate the proposed approach, a real dataset of brain tumor Magnetic Resonance Images was used. The proposed approach was compared to the methods of simulated annealing and a genetic algorithm used separately. The obtained results showed that SA-GA outperforms simulated annealing and genetic algorithms when they are applied in isolation, in terms of accuracy and computing time.

Keywords: Simulated annealing · Genetic algorithms · Feature selection
Computing time

1 Introduction

An increasing mortality rate among different age categories in the world is caused by a brain tumor. This illness is manifested through the growth of abnormal cells inside or around the brain [1]. There exist various types of brain tumors. Some of them are noncancerous and others are. The National Brain Tumor Foundation (NBTF) reported that during the last three decades, the total of humans who have developed brain tumors and died from them has almost tripled [2]. Early detection of the brain tumor is of great importance and the major challenge for further studies. Thus, computer and image processing devices are used in analyzing the tumor and the tumor area.

To create a Computer-Aided Detection (CAD) system, various image processing techniques such as image segmentation, feature extraction and selection, and classification are essentially integrated. In fact, feature selection represents an active research domain in pattern recognition [3], machine learning and data mining [4]. Irrelevant and redundant features invite further search as they make patterns less detectable and rules necessary for forecasting or classification less evident, in addition to the high overfitting risk. The selection of feature subsets requires determining the appropriate feature to maximize the accuracy of prediction or classification. A major aim of the available

© Springer International Publishing AG, part of Springer Nature 2018
A. Abraham et al. (Eds.): IBICA 2017, AISC 735, pp. 329–338, 2018.
https://doi.org/10.1007/978-3-319-76354-5_30

research works was to determine an optimal feature subset. Selecting features is usually based on the parameters of computational time and the quality of the generated feature subset solutions. In fact, fast and accurate classification, using the minimum number of features is often opted for. This can obviously be obtained through feature selection. In this paper, we suggest a novel hybrid algorithm for an optimal selection of feature subsets. The proposed algorithm generates better feature subsets as compared to other algorithms at a lesser execution time. Moreover, the quality of selected subsets is further improved as the algorithm is run.

The remainder of this paper is structured as follows. Section 2 reviews relevant literature on the issue of feature selection. Section 3 then handles the proposed approach. Experimental results are presented and compared with other existing approaches in Sect. 4. Conclusions and recommendations for future research are finally drawn in Sect. 5.

2   Review of Existing Techniques

Recently, several hybrid approaches have been proposed. For instance, a feature selection algorithm based on correlation and a genetic algorithm [5], *t*-statistics and a genetic algorithm [6], mutual information and a genetic algorithm [7], principal component analysis and an ACO algorithm [8], artificial bee colony algorithm [15], chi-square approach and a multi-objective optimization algorithm [9] rely on filter and wrapper methods. Interestingly, these approaches first apply filter methods to select a feature pool, then the wrapper method to obtain the optimal feature subset from the selected feature pool. Thus, feature selection becomes faster as in the filter method, the effective number of considered features is rapidly reduced. Despite the low probability of good predictor elimination by filter methods, hybrids of filter and wrapper methods can be little accurate because an isolated relevant feature can be as discriminating as an irrelevant one in the presence of feature interactions.

Wrapper schemes use the K-Nearest Neighbor (K-nn) as a learning algorithm. In this approach, feature selection is "wrapped" in a learning algorithm. K-nn can be used for numerous training sets and provide accurate information about distance, weighted average and pixels. Meanwhile, an accurate K-nn algorithm depends on the presence of noisy or irrelevant features, or feature scales inconsistent with their importance. Moreover, the choice of k affects the K-nn algorithm. Empirical evidence suggests that its memory is intensive while its classification is slow [10].

3   Proposed Algorithm

The SA-GA hybrid algorithm is proposed here to efficiently select optimal feature subsets. It depends on a simulated annealing [16], a genetic algorithm [17], a support vector machine and a greedy search algorithm. Our hybrid approach is characterized by the avoidance of being trapped in a local minimum of SA, a high GA crossover operator convergence. Meanwhile, it guarantees a strong local search greedy algorithm and a support vector machine (SVM) with a high computational efficiency.

Interestingly, the SA-GA approach selects feature sets without recourse to the filter steps. It combines the mutation-based search SA algorithm with good global searchability and the GA capacity to implement both crossover and mutation operations. Thus, GA overcomes the convergence issue, but the crossover and low fixed mutation rate combination often trap the search in a local minimum. Moreover, both SA and GA have a weak local search capability. Meanwhile, the local search ability of greedy algorithms is good, but their global search ability is weak.

SA-GA performs three search stages.

Stage 1: SA-GA employs an SA to avoid the risk of bad random choice of an initial population as shown in Fig. 1. This figure illustrates the difficulties of bad and random initialization of generating initial population by GA. The first case demonstrates the risk of deviating from the desired optimum because of bad random choice of initial population. By contrast, in the second case, the initialization is better thanks to SA. Hence, SA generates an initial population for the GA better than the population generated randomly by GA. This process results in a better exploration and exploitation of search space. As SA is a global search algorithm, it guarantees the convergence to a global optimum.

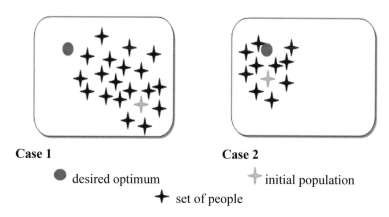

Case 1   **Case 2**

● desired optimum   ✦ initial population

✦ set of people

Fig. 1.   One kernel Exploration of research space by genetic algorithm.

Due to very high temperature, SA tolerates new solutions; which brings about a near random search through the search space. However, at a low temperature close to zero, it only accepts improvements.

Stage 2: Our proposed algorithm (SA-GA) performs optimizations by means of a GA. We set the GA population at 100. The best solutions are detected by SA and make up the initial population. Crossover in GA is aimed at forming new and hopefully better solutions by exchanging information between pairs of good solutions. Rapid convergence to a good solution is facilitated by the crossover operator. In fact, thanks to the mutation operator, new genes are introduced into the population and genetic diversity is retained.

***Stage* 3:** The greedy algorithm locally searches the k-best solutions provided by SA and GA and chooses the best neighbors which are defined in terms of the fitness function. As computational efficiency is essential, we employ a fast and robust supervised learning algorithm (SVM) to analyze data, recognize patterns and assess candidate solutions.

In order for SVM to perform effectively, the feature selection method should be reliable in terms of discarding noisy, irrelevant and redundant data and preserve the discriminating power of data. Without feature selection, SVM input space is large and disturbed; which lowers the SVM performance. SVM accuracy rate depends on the quality of the feature's dataset, and other factors including the kernel function and the two parameters C and λ as well.

SVM provides the optimal solution depending on several kernel functions. Our study employs the RBF kernel function to discover the optimal solution.

In RBF, C and γ should be appropriately set. The C parameter refers to the penalty cost. The value of C influences the classification outcome.

The γ parameter affects the outcome more than C, as the partitioning outcome in the feature space depends on its value. If the parameter values are not set properly, then the classification outcomes are inappropriate [11].

Hence, good global search capability, rapid convergence to a near optimal solution, along with good local search ability and high computational efficiency are achieved through our proposed algorithm (SA-GA).

The estimation of all the features is based on the fitness function in Eq. (2). A fitness value is used to measure the 'fitness' of a feature to a population. The initial genetic process population encompasses the best solutions detected by SA. Low fitness features are discarded by GA and high fitness ones are preserved. In our algorithm (SA-GA), SVM evaluates candidate feature subset solutions. Prior to this step, each feature is scaled between 0 and 1 for normalization purposes. A 5-fold cross validation was necessary to estimate the SVM classifier testing accuracy. Solution fitting evolves in parallel with the accuracy. In case of equal accuracy rate of two solutions, the solution that relies on fewer features wins.

$$Accuracy = \frac{(TP + TN)}{(TP + TN + FP + FN)} \times 100\%. \tag{1}$$

$$fitness = W_A \times Accuracy + W_{nb} \times \frac{1}{N}. \tag{2}$$

Where W_A is the weight of accuracy and W_{nb} is the weight of N feature contributing in classification, where $N \neq 0$.

4   Performance Evaluation

4.1   Data

Sone benchmark images were freely downloaded from the Harvard Medical School brain atlas [12]. We tested our classification algorithm for several normal brains and pathological brain MR images. All benchmark axial images were three weighted (enhanced T1, proton density (PD) and T2). These images were acquired at several positions of the transaxial planes as 256 × 256 sizes. The subject's left was conventionally on the right. The images in the dataset were shown in a side view, i.e. in the sagittal image. Our case study considered a total of 83 transaxial images (29 normal and 54 pathological brains, suffering from a low-grade glioma, Meningioma, bronchogenic carcinoma, Glioblastoma multiform, Sarcoma or Grade IV tumors) in several brain locations.

4.2 Experimental Results

We proposed to classify human brain MR images according to a methodology that consists of three steps: feature extraction, feature selection and classification.

For each image, we extract 26 features by means of WT-SGLDM. To check the performance of the proposed method, nine additional features were extracted. A total of 44 features was thus obtained. The SA-GA parameters served to reduce the number of extracted features as described in Table 1.

Table 1. Contingency table.

Parameters	Value/Method
Initial temperature	T0 = 75
Temperature change	Ti + 1 = 0.09 × Ti
Number of iterations for each temperature	50
Selectivity function	Fitness
SA Stop condition	Tstop = 0,01
Generation number	100
Size of initial population	Size of the solution obtained by SA
Selection Method	Tournoi
Probability of crossover	Pc = 0,9
Mutation probability	Pm = 0,1
Crossing method	Crossing to a point

The best SA-GA selected features during the execution are illustrated in Table 2. The classification performance of **95.65%** was obtained with 4 of the whole available features, thus classifying normal and pathological brains by means of the least features and reducing the classifier cost.

Using only four features: mean of contrast, mean of homogeneity, Mean of energy and range of correlation, classification was obtained at an accuracy rate of 95.65. Actually, the SA-GA algorithm selected features according to the appearance of images of the tumors database. The area of the tumor in abnormal brain images was brighter, with a regular color distribution. The selection of the contrast and the correlation features as descriptive characteristics of the tumor was thus significant.

Table 2. Results of feature selection performed by SA/GA.

Feature selection	Feature set	Classifier accuracy
SA/GA	**7 features:** Mean of contrast (M.CON), Mean for Information measure of correlation I (M.IMC I), Mean dissimilarity (M.DISS), Range of correlation (R.CORR), Range of variance difference (R.DVAR), Range Information measure of correlation I (R.IMC I), Range Information measure of correlation II (R.IMC II)	95.65%
	5 features: Mean of contrast (M.CON), Mean dissimilarity (M.DISS), Mean of homogeneity (M.HOMO), Range of correlation (R.CORR), Range Information measure of correlation II (R.IMC II)	95.65%
	4 features: Mean of contrast (M.CON), Mean of energy mat (M.ENER MAT), Mean of homogeneity (M.HOMO), Range of correlation (R.CORR)	95.65%

Profile regularity and repetition in a signal can be detected thanks to the correlation. Furthermore, color distribution at a tumor area was regular as the values are fairly close. Homogeneity is then chosen as a descriptive characteristic of the tumor according to these aspects. In particular, the homogeneity has an opposite behavior of the contrast. In fact, the homogeneity characteristic is related to the texture homogeneous regions. As for energy characteristic, it is responsible for extracting image regular contour.

This comparison gives importance to three features more than others because they are selected several times by simulated annealing, genetic algorithm and the SA-GA process (Table 3). These three features are Mean of contrast (M.CON), Mean of homogeneity (M.HOMO) Range of correlation (R.CORR). They lead us to extract the abnormal areas of a brain MRI image and specifically distinguish tumors from noise in the image, which facilitates and optimizes the classification and segmentation system. Thus, contrast, homogeneity and correlation feature present the most distinctive features of a tumor, since they combine light distribution (correlation and contrast) with extracting textures homogeneous zones (homogeneity). Therefore, the SA-GA approach achieves better results than SA and GA in isolation in terms of reducing effective and reliable data.

Figure 2 shows that the simulated annealing reached the maximum accuracy (99.99%) by 10, 80 and 100 generations. Meanwhile, this maximum precision is achieved by the genetic algorithm for more generations: 10, 30, 40, 70 and 100. On the other hand, our proposed SA-GA approach achieves a high degree of accuracy almost similar to that achieved by SA. We conclude that this is due to the precision of the initial search zone of the proposed SA-GA approach.

The results of convergence to an optimal solution for the three implemented methods indicate that the genetic algorithm achieves 99.99% accuracy by producing different numbers of generations. However, more stability is obtained by simulated annealing, as the maximum accuracy obtained for its solutions varies between only two values regardless of the number of generations. By adding simulated annealing before the execution

Table 3. Comparison of results obtained by the genetic algorithm, simulated annealing and SA-GA algorithm.

Feature set	Genetic Algorithms [13]			Simulated Annealing [14]			SA-GA		
	* features	6 features	5 features	9 features	3 features	* features	* features	5 features	4 features
Mean of energy (M.ASM)									
Mean of contrast (M.CON)		X	X			X	X	X	X
Mean of correlation (M.CORR)	X								
Mean of variance (M.VAR)									
Mean of inverse difference moment (M.IDM)				X					
Mean of entropy (M.ENT)									
Mean of sum average (M.SAVG)			X						
Mean of sum variance (M.SVAR)			X		X				
Mean of sum entropy (M.SENT)					X				
Mean of difference variance (M.DVAR)	X								
Mean of difference entropy (M.DENT)									
Mean of Information measure of correlation I (M.IMC I)	X	X					X		
Mean of Information measure of correlation II (M.IMC II)									
Mean of maximal correlation coefficient (M.MAX CORR)				X					
Mean of correlation max (M.CORR MAT)					X				
Mean of cluster Prominence (M.CP)				X					
Mean of cluster Shade (M.CS)									
Mean of dissimilarity (M.DISS)							X	X	
Mean of energy max (M.ENER MAT)					X	X			X
Mean of homogeneity (M.HOMO)		X	X	X	X	X		X	X
Mean of Maximum probability (M.MAX PROB)	X								
Mean of inverse difference moment (M.IDM)	X	X							
Range of energy (R.ASM)									
Range of contrast (R.CON)	X				X				
Range of correlation (R.CORR)		X	X			X	X	X	X
Range of variance (R.VAR)									
Range of inverse difference moment (R.IDM)				X					
Range of entropy (R.ENT)									
Range of average sum (R.SAVG)									
Range of sum variance (R.SVAR)				X					
Range of sum entropy (R.SENT)									
Range of difference variance (R.DVAR)				X	X		X		
Range of difference entropy (R.DENT)				X					
Range of Information measure of correlation I (R.IMC I)					X		X		
Range of Information measure of correlation II (R.IMC II)							X	X	
Range of maximal correlation coefficient (R.MAX CORR)				X					
Range of correlation mat (R.CORR MAT)				X	X				
Range of cluster Prominence (R.CP)									
Range of cluster Shade (R.CS)									
Range of dissimilarity (R.DISS)									
Range of energy max (R.ENER MAT)				X					
Range of homogeneity (R.HOMO)	X	X	X						
Range of Maximum probability (R.MAX PROB)									
Range of inverse normalized difference (R.IDN)									

of the genetic algorithm, we note an important influence on the result of the proposed SA-GA approach revealed by the stabilization of the results obtained.

According to Fig. 2, we note that the highest accuracy achieved by simulated annealing, genetic algorithm and our proposed approach SA-GA does not depend on the number of generations produced since we can have an accuracy of 99.99% by the three methods even with only 10 generations. We can see then that the calculation accuracy depends only on the fitness function. For this, the choice of the selectivity function is very important for optimal solutions.

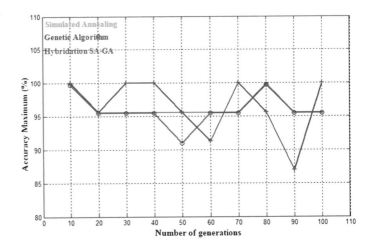

Fig. 2. Accuracy evolution according to the number of generations through the implementation of SA, GA and SA-GA.

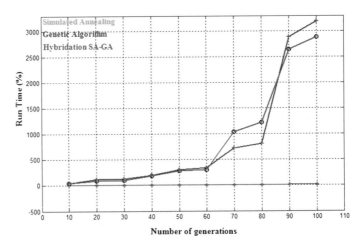

Fig. 3. Computing time depending on the number of generations through the implementation of SA, GA and SA-GA.

According to Fig. 3, the three algorithms: simulated annealing, genetic algorithm and our hybrid SA-GA algorithm need more computing time which increases the number of generations. The difference in terms of computing time between simulated annealing and genetic algorithm is greatly remarkable. The latter spends too much computing time. Thus, the necessary time to produce a solution by genetic algorithm for 10 generations is three times greater than the time required by simulated annealing for 100 generations. We also note that SA-GA pace computing time is almost near the pace of the genetic algorithm according to the number of generations. Thus, the influence of the genetic algorithm remains significant since the difference between the two curves which show these two algorithms is considered low (genetic algorithm needs further 162 min than our proposed approach SA-GA for performing 100 generations).

Simulated annealing however slightly decreased the computing time curve achieved by SA-GA compared to the genetic algorithm curve thanks to the precision and the reduction of the search area in the initialization phase. Therefore, optimization of the genetic algorithm for the selection of the most relevant features by simulated annealing is clear in Fig. 3.

In the selection phase of the most relevant features of the brain images, simulated annealing, genetic algorithm and SA-GA showed that the execution speed is highly dependent on the number of generations produced to have an optimal solution.

5   Conclusion

This present study suggests a hybrid algorithm SA-GA that benefits from the combination of merits of a number of existing algorithms devoted to the selection of optimal feature subsets from a large number of features. The proposed hybrid algorithm included a simulated annealing, a genetic algorithm, a support vector machine and a greedy search algorithm. It avoids being trapped in a local minimum of SA, a high GA crossover operator convergence. On the other hand, it guarantees a strong local search greedy algorithm and a support vector machine (SVM) with a high computational efficiency.

References

1. Gopi, K., Ramashri, T.: A GUI based hybrid clustering technique for brain tumor segmentation and validation in MRI images. Aust. J. Basic Appl. Sci. **10**(8), 84–96 (2016)
2. Shah, S., Chauhan, N.C.: Techniques for detection and analysis of tumours from brain MRI images: a review. J. Biomed. Eng. Med. Imaging **3**(1), 9–20 (2016)
3. Zhou, X., Gao, X., Wang, J., Yu, H., Wang, Z., Chi, Z.: Eye tracking data guided feature selection for image classification. Pattern Recogn. **63**, 56–70 (2017)
4. Bischl, B., Lang, M., Kotthoff, L., Schiffner, J., Richter, J., Studerus, E., Casalicchio, G., Jones, Z.M.: mlr: machine learning in R. J. Mach. Learn. Res. **17**(170), 1–5 (2016)
5. Ronen, M., Jacob, Z.: Using simulated annealing to optimize feature selection problem in marketing applications. Eur. J. Oper. Res. **171**, 842–858 (2006)
6. Tan, F., Fu, X., Wang, H., Zhang, Y., Bourgeois, A.: A hybrid feature selection approach for microarray gene expression data. In: Alexandrov, V.N., van Albada, G.D., Sloot, P.M.A., Dongarra, J. (eds.) Computational Science – ICCS 2006. Lecture Notes in Computer Science, vol. 3992, pp. 678–685. Springer, Heidelberg (2006)
7. Huda, S., Yearwood, J., Jelinek, H.F., Hassan, M.M., Fortino, G., Buckland, M.: A hybrid feature selection with ensemble classification for imbalanced healthcare data: a case study for brain tumor diagnosis. IEEE Access **4**, 9145–9154 (2016)
8. Fatourechi, M., Birch, G., Ward, R.K.: Application of a hybrid wavelet feature selection method in the design of a self-paced brain interface system. J. Neuro Eng. Rehabil. **4**, 11 (2007)
9. Huang, J., Cai, Y., Xu, X.: A wrapper for feature selection based on mutual information. In: 18th International Conference on Pattern Recognition, vol. 2, pp. 618–621 (2006)

10. Buttrey, S.E., Karo, C.: Using k-nearest-neighbor classification in the leaves of a tree. Comput. Stat. Data Anal. **40**, 27–37 (2002)
11. Pardo, M., Sberveglieri, G.: Classification of electronic nose data with support vector machines. Sens. Actuators B Chem. **107**, 730–737 (2005)
12. Harvard Medical School (1999). http://www.med.harvard.edu/aanlib/home.html
13. Gasmi, K., Kharrat, A., Messaoud, M.B., Abid, M.: Automated segmentation of brain tumor using optimal texture features and support vector machine classifier. In: Campilho, A., Kamel, M. (eds.) Image Analysis and Recognition - ICIAR 2012. Lecture Notes in Computer Science, vol. 7325, pp. 230–239. Springer, Heidelberg (2012)
14. Kharrat, A., Ben Halima, M., Ben Ayed, M.: MRI brain tumor classification using support vector machines and meta-heuristic method. In: 15th IEEE International Conference on Intelligent Systems Design and Applications (ISDA), pp. 446–451 (2015)
15. Li, X., Yang, G.: Artificial bee colony algorithm with memory. Appl. Soft Comput. **41**, 362–372 (2016)
16. El-Shamir Ezugwu, A., Oluyinka Adewumi, A., Eduard Frîncu, M.: Simulated annealing based symbiotic organisms search optimization algorithm for traveling salesman problem. Expert Syst. Appl. **77**, 189–210 (2017)
17. Chandra, G.R., Rao, K.R.H.: Tumor detection in brain using genetic algorithm. Procedia Comput. Sci. **79**, 449–457 (2016)

A Statistical Analysis for High-Speed Stream Ciphers

Youssef Harmouch$^{(\boxtimes)}$ and Rachid El Kouch

Department of Mathematics, Computing and Networks,
National Institute for Post and Telecommunication, Rabat, Morocco
{harmouch,elkouch}@inpt.ac.ma

Abstract. The current article statistically analyzes several high-speed stream ciphers. The study focuses on frequency cryptanalysis and the goodness-of-fit test. The purpose of this work is to show if there is a signature left by these stream ciphers in each of the encrypted streams. In addition, the work compares these ciphers to indicate which is the safest against statistical cryptanalysis.

Keywords: Stream cipher · Statistical cryptanalysis · Frequency analysis
Goodness-of-fit

1 Introduction

The common symmetric ciphers used today are in fact block ciphers which uses iterations of a deterministic algorithm that operates on plaintext bits of fixed length - known as block - at a time. Each iteration is called round and uses a different subkey created from the primary key of encryption. There have been numerous operating modes developed for block cipher in order to allow authenticity and confidentiality while some modes also provides the padding for the plaintext block. Padding the plaintext block is simply adding bits to the plaintext block in cases where plaintext bits are shorter than the block size. It should be noted that block ciphers have also been used in Pseudo-Random Number Generators "PRNG" and universal hash functions. Today's most Famous Block Ciphers can be found gathered in [1].

On the other hand, a stream cipher, takes plaintext characters (1 bit or byte) at a time and XOR them with the pseudo-random bits to get the output. The infinite pseudo-random bits actually refer to the key that is known as the keystream. The keystream is normally created using the initial encryption key - called the seed - by the PRNGs. To remain secure, PRNGs should be unpredictable in stream cipher. In addition, stream ciphers should not use the same keystream twice, otherwise the cipher may be broken. The aim of designing stream cipher was to approach the idealistic cipher, known as the One-Time Pad.

The One-Time Pad, which is supposed to use a purely random key that is longer than plaintext, can potentially reach "the perfect secret", that is, the total safe against brute force attacks. Nevertheless, such a cipher would be too impractical to use, because if someone likes to encrypt and send a one-minute full HD video file to

© Springer International Publishing AG, part of Springer Nature 2018
A. Abraham et al. (Eds.): IBICA 2017, AISC 735, pp. 339–349, 2018.
https://doi.org/10.1007/978-3-319-76354-5_31

another, he would need a key of at least 144 megabytes in size (this size is calculated), that is to say, a key with 1,125 Gigabits of length.

As appears the impractical use of a key that is longer than the plaintext, stream ciphers are far from the perfect secret. However, the key change per use makes them difficult to break. In fact, if the keystream sequence is at least as long as the plaintext being encrypted, i.e. the keystream is used once when encrypting the plaintext, before the seed is changed, then the block cipher in counter mode cannot be any more secure than the stream cipher. The vulnerabilities of the two is then the same, according to Albert Manfredi- Principal Engineer at Boeing Defense Systems.

Statistical cryptanalysis is an important tool in a cryptanalysis study. Even if a stream or bloc cipher is protected against today's cryptanalysis attacks [2], the enormous amount of encrypted data can be seriously dangerous for any cipher if statistical bias has occurred.

In today's communication, stream ciphers are widely used. Each human communication-language contains a signature. The signature is actually the occurrence number of letter and the occurrence number of word used in the language. This signature can be easily computed due to quantization [3], i.e. the binary vector representing of the analog signal for human voice is not infinite. Hence, the huge importance of a statistical analysis for stream ciphers.

Consequently, this paper investigates the behavior of several stream ciphers on statistical cryptanalysis attacks such as frequency analysis and the goodness-of-fit. Moreover, this work do contain statistically comparison between stream ciphers based on some experiment and result discussions, so as to deduct the presence or the absence of weakness and safety level obtainable thru each cipher.

2 Overview of Existing Stream Ciphers

Stream ciphers create successive characters of keystreams based on their internal state. They are two types of stream ciphers:

Synchronous stream cipher [4]: the status is updated regardless of plaintext and ciphertext. The keystream bits are subsequently combined with plaintext for encryption or with ciphertext for decryption, which implies that the encryption and decryption machines respectively, Bob and Alice must use the same information, hence synchronization.

If synchronization is lost during the process, some approaches are applied to resynchronize the two machines. Among these approaches, we have: the systematic use of various offsets until achieving the synchronization, or tagging the ciphertext with markers at set points.

In this schema, if synchronization is lost, i.e. there is a corrupted bit in the data stream transmitted between Bob and Alice, then a single bit will be corrupted in the recovered plaintext and the error does not affect the rest of the data stream. Consequently, this mode is very useful in case of high error rate in a communication. However, this scheme can be very susceptible to active attacks if a malicious attacker has access to the stream.

Self-synchronizing stream ciphers [5]: the status is updated based on the previous ciphertext, i.e. the previous X (X is a number) ciphtertext bits help to generate the keystream. This allows Bob and Alice to retrieve data more easily if there are bits added, deleted, or altered in the stream data. In this scheme, the error of one bit will be limited in the overall effect.

As example of self-synchronizing stream cipher, we have RC4 [6] or a block cipher that operates in CFB mode [7] (cipher feedback).

Other ciphers that use this technique are A5/1 [8], A5/2 [9], Helix [10], ISAAC [11], MUGI [12], Phelix [13]...

Stream cipher differs from the block cipher design. This causes a difference in use. For instance, Block ciphers require more memory to save the master key, subkeys, plaintext block and often more data from previous blocks depending on the encryption mode [7], which can also associate confidentiality to the key integrity check. Whereas stream ciphers only work on a few bits at a time, they have relatively low memory requirements, i.e. more suitable to embedded devices, firmware...

Block ciphers are more susceptible to transmission noise, that is, if a bit is corrupted in the ciphertext, the rest of the block is probably unrecoverable. While stream ciphers encrypt bytes independently without connection to each other.

Stream ciphers are usually faster than block ciphers, but they are often less secure and subject to weaknesses based on usage, because of the very strict requirements for the keystream.

Stream ciphers do not provide integrity nor authentication, whereas some block ciphers can provide integrity in addition to confidentiality (depending on encryption mode).

Because of all the above, stream ciphers are typically best for cases where the amount of data is either unknown, or continuous, such as network streams, radio mobile communication... While Block ciphers are more suitable when the amount of data is known or high secret, such as a file sharing, top-secret communication...

In this work, several high-speed stream ciphers are examined to observe the presence or absence of potential vulnerabilities to statistical analysis. The studied stream ciphers are ChaCha8/12/20 [14], HC128/256 [15, 16], Panama [17], Rabbit [18], RC4, Sosemanuk [19], Salsa20/XSalsa20 [20], SEAL [21], WAKE [22]. Despite that RC4, SEAL and WAKE have been broken and are no longer secure, they are still used (e.g. RC4 is widely used in web encryption).

3   Statistical Cryptanalysis

Studying cryptanalysis cannot go past over statistical analysis [23, 24], because even if there is no connection between plaintext and ciphertext, statistical analysis and more specific frequency analysis can show to attacker important information. In this section, we examine our tested stream ciphers to observe the presence or the absence of a signature into theirs ciphertext that can lead to some useful data in plaintext. In this test, we analyze the frequency of character (letter) in English as language. Then, we evaluate each stream cipher PRNG from a statistical point of view and finally, we study the distribution of the encrypted characters based on the Chi-square statistical test, in order to have a global idea about the statistical resistance of each stream cipher.

3.1 Frequency Analysis

With a long enough plaintext, each character occurs with a characteristic frequency. The most frequently used character in English is the 'E' with a frequency of 12.7% followed by 'T' with a frequency of 9.1% [25] (see Table 1).

Table 1. The frequency of letters in English (L denote the Letter and F denote the Frequency in %)

L	F	L	F	L	F	L	F
A	08,2	H	6,1	O	7,5	V	1,0
B	01,5	I	7,0	P	1,9	W	2,4
C	02,8	J	0,2	Q	0,1	X	0,2
D	04,3	K	0,8	R	6,0	Y	2,0
E	12,7	L	4,0	S	6,3	Z	0,1
F	02,2	M	2,4	T	9,1		
G	02,0	N	6,7	U	2,8		

The frequency study will lead to apply a guess attack because it is normal to suggest that the character with a higher frequency in the ciphertext has more probabilities of being the character with a higher frequency in the plaintext. As a result, we define the probability of success of guessing attacks for each stream cipher as the ratio equal to the number of good guessed character divided by the total number of characters. This ratio will help to compare the resistance of stream ciphers to these types of attacks. It should also be mentioned that the probabilities of the guessing attack is also related to the number of obtained ciphertext. The higher the number of obtained ciphertext the higher the guessing attack success probability.

Typically, stream ciphers are mono-alphabetic ciphers, but the keystream changes continuously allows them to act as polyalphabetic [26]. This complicate the frequency cryptanalysis study because the search of the possible mono-decrypted character become a search of the possible poly-decrypted character.

Furthermore, the probability of success of the guessing attack is also related to the probability of searching for the character in the ciphertext from the plaintext. We call this, the mission candidate i.e. we define the group of poly-decrypted character as the possible candidate for each plaintext character. The candidate assignment is based on a binomial distribution, in this way, every candidate of the encrypted character has a probability of being the corresponding character [27]:

$$P(character = x) = \binom{n}{x}(P_{character})^{x}(1 - P_{character})^{n-x} \tag{1}$$

Where $P_{character}$ is the probability of the searching character from ciphertext ($P_{character}$ is equal to the character frequency into plaintext), n is the total number of all ciphertext character and x is the occurrence of the encrypted character that we calculate its probability.

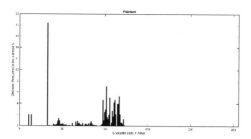

Fig. 1. The percentage of each character in the plaintext

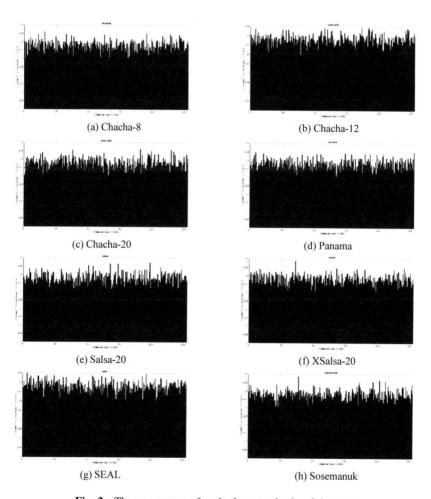

(a) Chacha-8	(b) Chacha-12
(c) Chacha-20	(d) Panama
(e) Salsa-20	(f) XSalsa-20
(g) SEAL	(h) Sosemanuk

Fig. 2. The percentage of each character in the ciphertext

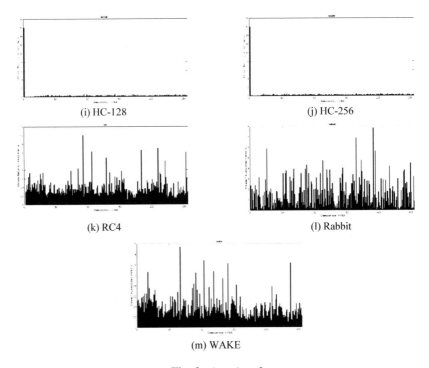

(i) HC-128   (j) HC-256

(k) RC4   (l) Rabbit

(m) WAKE

Fig. 2. (*continued*)

The first observation taken from Fig. 2 is that some ciphertext has a near pseudo-uniform distribution and the character frequency seem bounded between zero and 0,47%. This pseudo-uniform distribution of information into ciphertext attests the huge difficulty of extracting information from ciphertext to determine the plaintext even with clear character frequency (see Fig. 1). Moreover, Fig. 2(i) and (j) showed that one character appeared 16% in the ciphertext while the rest of characters has a frequency seem bounded between zero and 0,746%. For HC128 and HC256, our guessing attacks has showed 0,037934% and 0,048203% as probability of success for guessing one character. However, it is still difficult to apply frequency analysis further because the rest of encrypted data with HC128/256 has a near pseudo-uniform distribution. As for RC4, Rabbit and WAKE in Fig. 2(k), (l) and (m) respectively, the frequency analysis showed a big bad character distribution into ciphertext, which gives attractive information to break the cipher. In this work, our attempt in guessing attacks showed 0,081989%, 0,10249% and 0,10044% as probability of success for guessing one character for Rabbit, RC4 and WAKE respectively. Of note, our work here is not trying to break those ciphers, but to show that with only few attempt, the guessing attacks succeed in guessing one character with a probability of success near to 0.1%. Therefore, we deduce that RC4, Rabbit and Wake are potentially vulnerable to frequency analysis.

In general, the strength of a stream cipher is based on the unpredictability of its PRNG used. Statistically, a good PRNG is linked to the uniform distribution of the character from the set domain to the codomain. For instance, if Bob sends a message to Alice and this message contains only a duplicate of a character, then it will be bad if the encrypted message contains a bias. It is not necessary that all the characters encrypted have the same frequency of appearance, what is bad for a cipher is to find an encrypted character with a frequency of appearance higher than the others.

For that reason, we test the PRNGs of our studied stream ciphers by applying a frequency analysis for a duplication of a random character in the plaintext. The result is illustrated in the following Fig. 3.

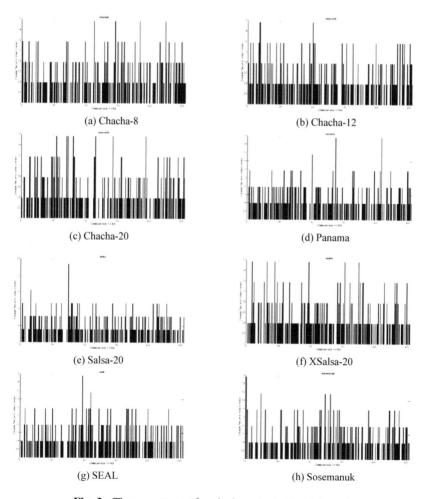

(a) Chacha-8   (b) Chacha-12

(c) Chacha-20   (d) Panama

(e) Salsa-20   (f) XSalsa-20

(g) SEAL   (h) Sosemanuk

Fig. 3. The percentage of each character in the ciphertext

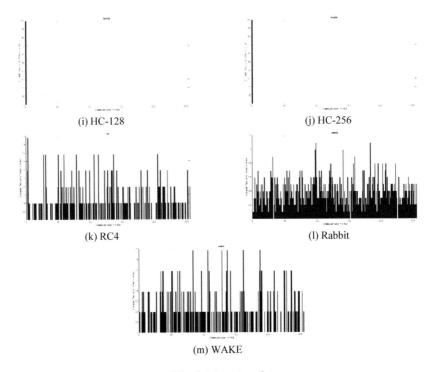

(i) HC-128   (j) HC-256

(k) RC4   (l) Rabbit

(m) WAKE

Fig. 3. (*continued*)

Figure 3 shows the probability of occurrence of each ciphertext character for one plaintext character. We notice that Chacha8/12/20, HC128/256, Panama, Rabbit, Wake and XSalsa showed a good distribution of encrypted characters, which leads to good diffusion and confusion by their PRNGs. On the other hand, RC4, Salsa, SEAL, and Sosemanuk had a bad statistical distribution (e.g., an encrypted character with Salsa has a probability of occurrence equal to 2.4%, which is approximately seven times bigger than it should be). Therefore, we deduce that their PRNGs are not statistically strong.

3.2   Chi-Square Goodness-of-Fit Test

The Chi-squared statistic [28] is a measure of similar degree for two categorical probability distributions. If the two distributions are matching, the chi-squared statistic is zero, if the distributions are very different, some higher numbers will result. The formula for the chi-squared statistic is:

$$\chi^2(P, C) = \sum_i \frac{(P_i - C_i)^2}{C_i} \qquad (2)$$

where P_i is the frequency of a character in the source file, and C_i is the frequency in the corresponding encrypted file. The χ^2 test can be used by cybercriminal to guess the key used into encryption, even if he must try in the worst case all possible key which is

Table 2. Pearson's Chi-Square test

Stream cipher	Expected χ^2	Tested χ^2
Chacha8	717607	703094
Chacha12	717607	720545
Chacha20	717607	682797
HC128	886823	1273949
HC256	892714	1466110
Panama	717607	679200
Rabbit	288098	2743095
RC4	717607	3443095
Salsa	717607	673304
SEAL	717607	812491
Sosemanuk	717607	728598
WAKE	717607	1246379
XSalsa	717607	749167

similar to brute force attack, he can in the best case catch the key in less operation by comparing χ^2 for every key and pick-up the minimum.

To resist the test χ^2, the cipher must distribute the encrypted characters uniformly. Uniformity caused by a number may be quantitatively justified by the Pearson chi-square test [29]. The χ^2 distribution is used to compare the goodness-of-fit of the observed frequencies of a sample measurement with the corresponding expected frequencies of the hypothesized distribution.

The chi-square test value for the same ciphertext used in the frequency analysis is listed in Table 2. It is found that for Chacha8/20, Panama and Salsa the real χ^2 is smaller than the estimated χ^2 implying that the null hypothesis is not rejected and the distribution of the ciphertext is uniform. Contrary to Chacha12, HC128/256, Rabbit, RC4, SEAL, Sosemanuk, WAKE and Xsalsa, the founded χ^2 is bigger than the estimated χ^2 implying that the null hypothesis is rejected and the distribution of the ciphertext is not uniform.

4   Conclusion

This article contains a statistical cryptanalysis study for several high-speed stream ciphers that are Chacha8/12/20, HC128/256, Panama, Rabbit, RC4, Salsa, SEAL, Sosemanuk, WAKE and XSalsa.

The aim of this work is to show either the presence or the absence of any statistical signature in the ciphertext by the ciphers studied. It has been found that RC4, Rabbit and WAKE have a lot of attractive information in the ciphertext, which provides a frequency statistic that can be used in order to reduce the brute-force attack against them. Our guessing attack applied in this paper managed to link a character of plaintext to some characters to the ciphertext with a probability of success equal to 0,1%. This number means that, starting from n characters in the ciphertext, the attack succeeds in cracking an encrypted character $n \times 0{,}001 \times P_{character}$ times. In addition, the study of PRNGs showed that RC4, Salsa, SEAL and Sosemanuk had a poor statistical distribution for keystream generation. As example, the constant input encryption for Salsa has shown that an encrypted character has a probability of occurrence seven times greater than it should be in uniform distribution. In addition, the chi-square test showed that Chacha12, HC128/256, Rabbit, RC4, SEAL, Sosemanuk, WAKE and Xsalsa have a non-uniform distribution in ciphertext.

According to these tests, we deduce from our statistical analysis that Chacha8/20 and Panama are the best stream ciphers among those studied in this paper in terms of hidden statistical information in the ciphertext, followed by Chacha12, HC128/256, Salsa and Xsalsa while it seems that RC4 is the worst of them.

References

1. Harmouch, Y., El Kouch, R.: A fair comparison between several ciphers in characteristics, safety and speed test. In: Rocha, Á., Serrhini, M., Felgueiras, C. (eds.) Europe and MENA Cooperation Advances in Information and Communication Technologies. Advances in Intelligent Systems and Computing, vol. 520. Springer, Cham (2017)
2. Banegas, G.: Attacks in stream ciphers: a survey. http://eprint.iacr.org/2014/677.pdf
3. Gersho, A., Gray, R.M.: Vector Quantization and Signal Compression. Springer, New York (1991). ISBN 978-0-7923-9181-4
4. Wu, H., Preneel, B.: Cryptanalysis and Design of Stream Ciphers, Ph.D. thesis, Katholieke Universiteit Leuven (Belgium) (2008)
5. Daneshgar, A., Mohebbipoor, F.: A Secure Self-synchronized Stream Cipher (2017). CoRR, abs/1709.08613
6. Rivest, R.: The RC4 encryption algorithm. RSA Data Security (1992)
7. NIST Computer Security Division's (CSD) Security Technology Group (STG). "Block cipher modes". Cryptographic Toolkit. NIST. Accessed 12 Apr 2013
8. Maximov, A., Johansson, T., Babbage, S.: An improved correlation attack on A5/1. In: Selected Areas in Cryptography 2004, pp. 1–18 (2004)
9. Goldberg, I., Wagner, D., Green, L.: The (Real-Time) Cryptanalysis of A5/2. Rump session of Crypto 1999 (1999)
10. Ferguson, N., Whiting, D., Schneier, B., Kelsey, J., Lucks, S., Kohno, T.: Helix: fast encryption and authentication in a single cryptographic primitive. In: Fast Software Encryption – FSE 2003. LNCS, vol. 2887, pp. 330–346, Springer, Heidelberg (2003)
11. Jenkins Jr., R.J.: ISAAC. Fast Software Encryption, pp. 41–49 (1996)
12. Watanabe, D., Furuya, S., Takaragi, K., Preneel, B.: A new keystream generator MUGI (PDF). In: 9th International Workshop on Fast Software Encryption (FSE 2002), pp. 179–194. Springer, Leuven. Accessed 07 July 2007
13. Whiting, D., Schneier, B., Lucks, S., Muller, F.: Phelix-fast encryption and authentication in a single cryptographic primitive. In: ECRYPT-Network of Excellence in Cryptology, Call for stream Cipher Primitives - Phase 2-(2005). http://www.ecrypt.eu.org/stream/
14. Bernstein, D.J.: ChaCha, a variant of Salsa20. In: Workshop Record of SASC 2008: The State of the Art of Stream Ciphers. http://cr.yp.to/chacha/chacha-20080128.pdf
15. Wu, H.: The stream cipher HC-128. In: Robshaw, M., Billet, O. (eds.) New Stream Cipher Designs. Lecture Notes in Computer Science, vol 4986. Springer, Heidelberg (2008)

16. Wu, H.: A new stream cipher HC-256. In: Fast Software Encryption (FSE 2004). LNCS 3017, pp. 226–244. http://eprint.iacr.org/2004/092.pdf
17. Daemen, J., Clapp, C.: Fast hashing and stream encryption with panama. In: Vaudenay, S. (ed.) Fast Software Encryption, FSE 1998. Lecture Notes in Computer Science, vol 1372. Springer, Heidelberg (1998)
18. Boesgaard, M., Vesterager, M., Pedersen, T., Christiansen, J., Scavenius, O.: Rabbit: a new high-performance stream cipher. In: Johansson, T. (ed.) FSE 2003. LNCS, vol. 2887, pp. 307–329. Springer, Heidelberg (2003)
19. Berbain, C., et al.: Sosemanuk, a fast software-oriented stream cipher. In: Robshaw, M., Billet, O. (eds.) New Stream Cipher Designs. Lecture Notes in Computer Science, vol 4986. Springer, Heidelberg (2008)
20. Bernstein, D.J.: The Salsa20 stream cipher, slides of talk. In: ECRYPT STVL Workshop on Symmetric Key Encryption (2005). http://cr.yp.to/talks.html#2005.05.26
21. Rogaway, P., Coppersmith, D.: A software-optimized encryption algorithm. In: Anderson, R.J. (ed.) FSE 1994. Lecture Notes in Computer Science, vol. 809, pp. 56–63. Springer, Berlin (1994)
22. Wheeler, D.J., A bulk data encryption algorithm. In: Anderson, R. (ed.) Fast Software Encryption. LNCS, vol. 809, pp. 127–134. Springer (1994)
23. Gérard, B.: Cryptanalyses statistiques des algorithmes de chiffrement la clef secrète. Other [cs.OH]. Université Pierre et Marie Curie - Paris VI (2010). French
24. Junod, P.: Statistical Cryptanalysis of Block Ciphers. École Polytechnique Fédérale de Lausanne
25. Beker, H., Piper, F.: Cipher Systems: The Protection of Communication. John Wiley & Sons (1983) Beutelsbacher, Albrecht: "Kryptologie", Vieweg, Teubner (1993)
26. Klima, R., Sigmon, N.P., Stitzinger, E.: Applications of Abstract Algebra with Maple and MATLAB, Second Edition (2007)
27. Cochran, D.: For Whose Eyes Only? Cryptanalysis and Frequency Analysis. http://www.westpoint.edu/math/Military%20Math%20Modeling/PS5.pdf
28. Ganesan, R., Sherman, A.T.: Statistical Technique for Language Recognition: An Introduction and Guide for Cryptanalysts, 25 February 1993
29. L'ecuyer, P., Simard, R.: TestU01: a C library for empirical testing of random number generators. ACM Trans. Math. Softw. **33**(4), 1–40 (2007)

Weighted Access Control Policies Cohabitation in Distributed Systems

Asmaa El Kandoussi$^{(\boxtimes)}$ and Hanan El Bakkali

ENSIAS, Mohammed V University, Rabat, Morocco
{asmaa.elkandoussi,h.elbakkali}@um5s.net.ma

Abstract. Collaboration between distributed domains has become an emerging demand that allows organization to share resources and services. In order to ensure secure collaboration between them, authorization specification is required. Thus, a global access control policy should be defined. However, the combination of the collaborator's access control policies may create authorization conflicts. In this paper, we propose a new approach based on organization's weight α_i in order to resolve potential detected policy conflicts, also we define how to calculate α_i accordingly and we propose a new algorithm to resolve the detected conflicts.

Keywords: Access control policy · Distributed systems
Conflict resolution

1 Introduction

Actually, more and more organizations collaborate in order to enhance their services and activities. Thus, collaborators have to specify their authorization policy that regulates how others can access to shared resources and services. In this field, many access control models exist, such as Role-based access control (RBAC) [1]. In RBAC, users are assigned role which has greatly facilitated the access control administration in companies. However, RBAC is static and cannot satisfy the dynamic aspect of distributed systems. Many other works have been proposed in order to overcome this gap. Authors in [2] propose Trust and Risk Aware Access Control, the proposed policy is based on different zone levels and decision are made according to the trusted zones. A Policy Based Access Control [3] includes context information and applies constraints as a rule to perform authorization. Policy definition is separated from system implementation, allowing access rules to be changed without affecting the system. BiLayer Access Control [4] separates the subject layer and policy layer, and access policies are matched with pseudo roles. Security Privacy Access Control Model [5] proposed for policy integration and reconciliation among collaborating healthcare organizations to provide access based on ST constraints. Hierarchy Similarity Based Access Control [6] proposes that improved interoperability in EHR access can be achieved by matching the policies with the different hospitals, and calculates a

© Springer International Publishing AG, part of Springer Nature 2018
A. Abraham et al. (Eds.): IBICA 2017, AISC 735, pp. 350–360, 2018.
https://doi.org/10.1007/978-3-319-76354-5_32

hierarchy similarity score index for the requested attributes and resources. Based on the score, the system decides access to the participating healthcare entity.

In distributed environments, each organization has its own local Access control policy. In order to regulate access in collaborative systems, a global policy have to be defined. A global policy represents the combination of the collaborator's access policies. However, the enforcement of such a global policy may lead to conflicting decisions. The challenge in such systems is how to preserve collaboration goals without loosing security control. The paper proposes a new approach based on organization's weight to prioritize authorization policies taking into account some important criteria such as collaboration level between organization, policies similarity level and trust level. This weighted combination leads organizations to set a number o access policy rules base on their weight in the collaboration. In this paper, we propose a new formula to calculate organization's weight. Besides, we present a new algorithm to resolve potential detected conflicts occurring during the composition of the global Access Control policy. This algorithm is based on a set of important parameters which are organization weight, object owner and object sensitivity.

The paper is organized as follows: Sect. 2 presents the main access control requirements in distributed systems followed by related works in Sect. 3. Section 4 details our proposed approach to create the global access control policy. Section 5 presents illustrative example. The last section concludes the paper with future works.

2 The Main Access Control Requirements in Distributed Systems

Many access control requirements have to be considered. Among the main needs are privacy, confidentiality, interoperability, trust and policy cohabitation [8]:

1. Trust: Distributed environments have uncertainties and are unreliable. Trust is composed of many attributes including reliability, dependability, honesty and competence [9]. Including trust into access control will significantly increases security.
2. Policies cohabitation: Policies cohabitation aim to conciliate different or even conflicting access control policies, and resolve detected conflicts. In fact, each organization must be able to protect its own data while respecting the global security policy.
3. Privacy preserving: Privacy arises as a major concern in todays collaboration. In fact, how to use suitable control to preserve privacy and perform cooperation needs.
4. Interoperability: Access control in distributed environments needs to be able to support different policies, resources and users. One of the primary goals of standardization is to maximize interoperability. Ontologies can be used to specify access control vocabularies adopted by others [10].

5. e-contracts: To govern collaborations, security contracts are established. They aim to detail the objectives and tasks to be achieved, to attribute responsibilities to each party and to specify penalties in case of abuses.

In order to guarantee a secured environment for collaborating organizations, the above-mentioned requirements should be considered.

3   Related Work

In collaborative environments, each organization has it's own access control policy, thus the combination of multiple policies is a necessary process to create a new global access control policy. However, the combination of different polices may create authorization conflict. Negotiation and conflict resolution are the most important mechanism to this end. Many works have been proposed in policy conflict detection and policy composition algorithm [11–14]. Authors in [11] present an automated policy combination for data sharing across multiple organizations. They made it possible by adopting bottom-top approach in the decomposition of the policy rules into different classes based on the subject constraints; each rule in a class that has the same subjects is combined by the condition-based attribute combination based algebraic operations. Their proposed approach makes the combined policy more restrictive, that is, the combined policy permits a request when all the policies permit it, denies a request when any one of policies denies it. Authors in [12] propose a novel access control policy composition method that can detect and resolve policy conflicts in cloud service composition. However, there proposed conflict detection algorithm did not include all cases of policy conflicts also their proposed conflict resolution algorithm is not based on a convenient solution. In [13], a purpose-based access control model was presented in order to preserve privacy, also a new algorithm was described to resolve conflicting policies based on purpose in order to preserve policies privacy. However, authors did not present how to resolve inter-organizational policies conflict. In the context of Online Social Networks, authors in [14], present a multiparty access control model to capture the essence of multiparty authorization requirements, also they propose Multiparty Policy Evaluation Process that include two steps and propose systematic conflict resolution mechanism to resolve conflicts during multiparty policy evaluation. Several solutions were proposed to resolve potential conflict such Threshold-based conflict resolution, Strategy-based conflict resolution with privacy and Decision Voting.

4   Weighted Access Control Policies Cohabitation

In this section, we introduce Weighted Access Control policies cohabitation approach. Our approach is based on different steps in order to conciliate conflicting access control policies in whole one.

4.1   Main Steps for the Proposed Approach

Our approach is basically based on the definition of the shared resources. Different steps to create new global access control policy are presented hereafter (Fig. 1).

1. Collaboration: In this step we identify the participating organizations and the set of resources and services assigned to each partner.
2. Public access control policy: Each partner defines how to access to his resource and under which condition. In this steps organization can define the sensibility levels of their shared data (defined hereafter).
3. Security policy standardization: In distributed systems, organization's policies are described by different policy languages, semantics or formats. Thus, it's mandatory step to unify policies in order to detect potential conflicts and proceed to compose global policy free of conflicts.
4. Organization's weight: based on trust level, collaboration level and policy similarity level, we calculate each organization weight in order to conciliate different policies (definition hereafter).

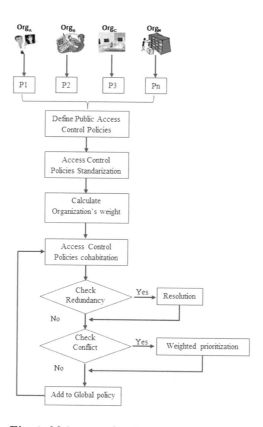

Fig. 1. Main steps for the proposed approach

5. Access control policies cohabitation: this step aims to check different access control policy rules issued by participating organizations, detect redundancy and conflict in order to resolve them. At the end of this step new global access control policy free of conflict is created.
6. E-contract: in order to govern the collaboration between different organization, contact is established. It contains the objective of the collaboration, the organizations responsibilities and the global access control policy.

At the end of this process, a new global access control policy is created. In order to resolve potential detected conflict we use the organization's weight. This parameter is calculated based on a set of specific criteria. Detailed method is explained hereafter.

4.2   Proposed Organization's Weight Calculation Method

Our proposed approach to resolve conflict between different access control policies is based on the organization's weight. Organization's weight α_i is determinated based on trust level, collaboration level, policy similarity level.

Organization's Weight. To calculate the organization's weight, it is mandatory to determine the most important factors in distributed relations. In fact, organization's weight is calculated based on Trust level, collaboration level and policy similarity level. Trust remains a paramount requirement in inter-organizational collaboration. It measures the trustworthiness of a particular organization in the collaboration. The collaboration level define the degree of collaboration between organization. We add an important factor in order to respect the security and autonomy of each organization, it's policy similarity level.

We use the weighted average to calculate α_i value based on the three parameters that are the TL, CL and PSL. We suggest the following formula:

$$\alpha_i = x_i * TL_i + y_i * CL_i + z_i * PSL_i \tag{1}$$

Where:

- $TL_i \in [0,1]$ is trust level, we use the scale of 0 to 1 where 0 reflects complete distrust
- $CL_i \in [0,1]$ is collaboration level, it represents the degree of collaboration
- $PSL_i \in [0,1]$ is policy similarity level, it gives the similarity score for any two given security policies
- x_i is the weight of TL_i, x_j is the weight of CL_i, z_i is the weight of PSL_i
  $x_i + z_i + z_i = 1$

1. Determination of the Trust Level (TL_i): In cross-organizational collaboration trust represents the relationship between parties in which one or more parties have the confidence [9]. Trust models can be divided into two types: identity-based and behavior-based. Identity-based trust model uses trust certificates to

verify the reliabilities of components. Behavior-based models observe and take the cumulative historical transaction behavior and also feedback of entities to evaluate the reliability. Trust is evaluated by three major methods: Direct trust, Recommendation Trust, and Integrated Trust. Direct trust is derived from the historical transaction. Recommendation trust is where the organization consults a third party to quantify the trust level of an organization [16]. Integration trust is a combination of both direct and recommendation trust. This is usually done by a weighted approach [17]. In [16] authors propose an integrated trust metric combining direct trust and recommendation trust using a weighted approach. In our approach we recommend to use the integrated trust adapted to our context and defined as below:

$$TL_i = a_i * DT_i + (1 - a_i) * RT_i \tag{2}$$

where, DT_i is the direct trust, RT_i is the recommendation trust by other organization and a_i is the weight of DT_i and RT_i to calculate trust level.

2. Collaboration Level (CL_i): this parameter aims to evaluate the contribution of the organization O_i in the collaboration. This metric can be calculate based on the number of services executed by the organization in the coalition and the importance of these services compared with others. For example we can calculate CL as:

$$CL_i = \frac{\sum_{j=1}^{k} a_j * S_j}{\sum_{j=1}^{n} a_j * S_j} \tag{3}$$

where, k = number of services executed by O_i, n = total number of services in the collaboration, a_j service weight in the collaboration.

3. Policy Similarity Level (PSL): this parameter is very important because it gives the similarity score for any two given security policies, which approximates the percentage of the rule pairs having the same decision. Authors in [18] propose an algorithm that iterates through all rules in policy set of each organization and calculates similarities between each rule attribute of both policies. It determines the closeness of two policies concluding the probability that the two policies can securely integrate with each other. The paper [19] proposes an access control framework that applies a Hierarchy Similarity Analyzer (HSA) on the policies that need to be merged. It calculates a Security Level (SL) and assigns it to each user and resource attribute in the given rule set. Authorizations ensure secured sharing of resources using simple security rule where a user at lower level can access high-level resource only under the authorization of a higher level user. The formal definition to calculate SPL_i is given in Eq. (4), where $Num(sameDecision(AR_1i; AR_2j))$ denotes the quantity of the rule pairs having the same decision and $Num(allDecision(AR_1i; AR_2j))$ denotes the amount of the total decision pairs.

$$SPL(P_1; P_2) = \frac{Num(sameDecision(AR_{1i}; AR_{2j}))}{Num(allDecision(AR_{1i}; AR_{2j}))}; AR_{1i} \in P_1; AR_{2j} \in P_2 \tag{4}$$

Object Sensitivity Level: each organization should specify the object sensitivity level of his data that refers to her privacy concern. The OSL has range between [0, 1] based on the object confidentiality level.

For example if the confidentiality is: High → OSL = 1, Medium → OSL = 0.5, Low → OSL = 0.

The Global Object Sensitivity Level GOSL is defined as follows:

$$GOSL\,(R_i, O_i) = \alpha_i \times OSL_i \tag{5}$$

Algorithm 1. Redundancy and Conflict rule detection

Input:
n is the number of organizations in the collaboration
AR_{ih} **is Access Control Policy Rule h in organization** O_i **where** $AR_{ih} = \{S_{ih}, Obj_{ih}, A_{ih}, E_{ih}\}$
AR_{km} **is Access Control Policy Rule m in organization** O_k **where** $AR_{km} = \{S_{km}, Obj_{km}, A_{km}, E_{km}\}$
$nb_{AR}\,(O_i)$ **is the number of Access Control Policy Rule in organization** O_i
$nb_{AR}\,(O_k)$ **is the number of Access Control Policy Rule in organization** O_k
        Output: RedundantRule is a set of redundant rules. ConflicRule is a list
of conflicting rules

1: **for** $i \leftarrow 1$ to $n - 1$
2:     **for** $k \leftarrow i + 1$ to n
3:         **for** $h \leftarrow h$ to $nb_{AR}\,(O_i)$
4:             **for** $k \leftarrow m$ to $nb_{AR}\,(O_k)$
5:                 **if** $(S_{ih} = S_{km}) \wedge (Obj_{ih} = Obj_{km}) \wedge (A_{ih} = A_{km}) \wedge (E_{ih} = E_{km})$ **then**
6:                     $RedundantRule \leftarrow (AR_{ih}, AR_{km})$
7:                         **if** $(S_{ih} \cap S_{km} \neq \varnothing) \wedge (Obj_{ih} \cap Obj_{km} \neq \varnothing) \wedge (A_{ih} \cap A_{km} \neq \varnothing) \wedge$ $(E_{ih} \neq E_{km})$ **then**
8:                             $ConflictRule \leftarrow (AR_{ih}, AR_{km})$
9:                         **else**
10:                            $m \leftarrow m + 1$
11:                        **end if**
12:                    **end if**
13:                **end for**
14:                $h \leftarrow h + 1$
15:            **end for**
16:            $k \leftarrow k + 1$
17:        **end for**
18:        $i \leftarrow i + 1$
19: **end for**
20: **return** $RedundantRule \leftarrow (AR_{ih}, AR_{km})$
21: $ConflictRule \leftarrow (AR_{ih}, AR_{km})$

4.3 Our Proposed Conflict Resolution Algorithm

Global Access Control Policy. We define access control rules as $AR = \{S, Obj, A, E\}$. Where S is subject, Obj is object, A is action and E is effect. The $AR_i = \{S_i, Obj_i, A_i, E_i\}$ indicates that any subject in organization $O - i$ is authorized to exercise action A on the Object Obj_i.

Let as $AR_1 = \{S_1, Obj_1, A_1, E_1\}$ Access rule in Organization O_1,
$AR_2 = \{S_2, Obj_2, A_2, E_2\}$ Access rule in Organization O_2

Conflict Detection. As we defined previously, Access Control Policy Rule in organization O_i is $AR_{ih} = \{S_{ih}, obj_{ih}, A_{ih}, E_{ih}\}$. Also in organization O_k, we define $AR_{km} = \{S_{km}, Obj_{km}, A_{km}, E_{km}\}$, AR_{km}. The Algorithm 1 hereafter describes the conflict rule detection between two different access control rules.

In our Algorithm 1, we compare the Access Control Policy Rule two by two for all participating organizations in the collaboration.

Conflict Resolution. Our proposed Algorithm 2, if the organization with higher weight is the owner of the resource, we prioritize its access policy rule.

Algorithm 2. Global Access Control Policy Composition

    Input:
α **is the organization weight, we have** $\alpha_i > \alpha_j$
$GOSL(Obj, O_i)$, $GOSL(Obj, O_i)$ **is Global Object Sensitivity Level of** O_i, O_j
    Output:
AR_g **is Global Access Control Policy Rule,** $AR_g = \{S_g, Obj_g, A_g, E_g\}$
$list \leftarrow ConflictRule(AR_i, AR_j)$
 1: **if** $list \neq \varnothing$ **then**
 2:    **for** each $(AR_i, AR_j) \in list$
 3:        **if** $R \in O_i$ **then**
 4:            $AR_g \leftarrow AR_i$ || if the organization with higher weight is the object owner then we prioritize its rule and put it in the global policy
 5:        **else**
 6:            **if** $Obj \in O_j$ **then** || if the Obj belong to the organization with lower weight, we have two case
 7:
 8:                **if** $GOSL(Obj_i, O_i) \geqslant GOSL(Obj_j, O_j)$ **then**
 9:                    $AR_g \leftarrow AR_i$ || The organization with higher weight have to preserve the privacy and confidentiality, we prioritize restrictive rule.
10:                **else** $GOSL(Obj_j, O_j) \geqslant GOSL(Obj_i, O_i)$
11:                    $AR_g \leftarrow AR_j$ || if the object is more sensitive and confidential in O_j we prioritize it's rule to preserve privacy.
12:            **end if**
13:        **end if**
14:        **end if**
15:    **end for**
16: **end if**

However, if the resource belongs to the organization with lower weight, we compare GOSL in order to decide which policy is privileged.

5   Illustrative Example

Healthcare organizations are currently facing pressure to improve productivity and to reduce costs while at the same time the demand for more hospital services is increasing. In order to provide optimal care for patients, Healthcare organizations decide to establish inter-organizational collaboration. However, these systems are facing security threads and are requiring a security mechanism, particularly access control.

5.1   E-Health Example

Hospital A would like to create collaboration other hospitals and health centers. We have the flowing AR defined by each organization. Potential conflicts can be detected. For example, the doctor in O_2 can read and write in the object F_1 based on the AR_{11}, but he cannot write in F_1 based on A_{21}. How we can resolve this conflict? (Tables 1 and 2).

Table 1. Global specifications

Organization O_i	Subject S	Actions A	Object Obj	Effect E
$O_1 = Hospital A$	$S_1 =$ doctor	$A_1 =$ read	$Obj_1 = F_1$	Permit
$O_2 = Hospital B$	$S_2 =$ nurse	$A_2 =$write	$Obj_2 = F_2$	Deny

Table 2. Access control policy rules

O_1	O_2
$AR_{11} = O_1, S_1, (A_1 \cup A_2), F_1$	$AR_{21} = O_2, S_1, Op_1, F_1$
$AR_{12} = O_1, S_2, Op_1, F_1$	$AR_{22} = O_2, S_1, (A_1 \cup A_2), (F_1 \cup F_2)$

5.2   Global Access Control Policy Composition

Based on the information given by O_1 and O_2, we assume that $OSL(F_1, O_1) = 1$, $OSL(F_1, O_2) = 0.5$, $\alpha_1 = 0.5$, $\alpha_2 = 0.3$.

We calculate the GOSL based on Eqs. (1) and (2) we have:
$GOSL(F_1, O_1) = 0.5, GOSL(F_1, O_2) = 0.15$

Based on Global Access Control Policy Composition algorithm, we have:

1. If $F_1 \in O_1\,(O_1)$ then $AR_g \leftarrow AR_{11}$
2. If $F_1 \in O_2\,(O_2)$ then we compare GOSL $O_1,\,O_2$
   In our case $GOSL\,(F_1, O_1) > GOSL\,(F_1, O_2)$. Then $AR_g \leftarrow AR_{11}$

The proposed algorithm helps as to combine multiple access control policy rules in order to compose global access control policy free of conflicts. However, potential conflict may persist if the participating organization did not accept the proposed solution to resolve it. In this case, we can go to the negotiation step where other specific criteria should be considered.

6   Conclusion and Future Work

In this paper, we propose a new weighted approach to resolve detected conflict when composing global policy in distributed systems. The new approach uses a set of parameters that are paramount in collaborative systems such as organization weight and object privacy. Besides, we present a new formulas to calculate organization weight. Finally we propose an algorithm to resolve potential detected conflicts occurring during the composition of the global Access Control policy.

The next stage of our work is the implementation of our approach using the eXtensible Access Control Markup Language (XACML) standard. Moreover, we look to propose a new approach based on automated negotiation.

References

1. Sandhu, R.S., Coyne, E., Feinstein, H.L., Youman, C.E.: Role-based access control models. IEEE Comput. **29**(2), 38–47 (1996). Arch. Rat. Mech. Anal. **78**, 315–333 (1982)
2. Burnett, C., Chen, L., Edwards, P., Norman, T.J.: TRAAC: trust and risk aware access control. In: Twelfth Annual International Conference on Privacy, Security and Trust, pp. 371–378 (2014)
3. Rashid, A., Kim, I.K., Khan, O.A.: Providing authorization interoperability using rule based HL7 RBAC for CDR (Clinical Data Repository) framework. In: Proceedings of the 2015 12th International Bhurban Conference on Applied Sciences and Technology, IBCAST 2015, pp. 343–348 (2015)
4. Alshehri, S., Raj, R.K.: Secure access control for health information sharing systems. In: 2013 IEEE International Conference on Healthcare Informatics, pp. 277–286 (2013)
5. Chi, H., Jones, E.L., Zhao, L.: Implementation of a security access control model for inter-organizational healthcare information systems. In: Proceedings of the 3rd IEEE Asia-Pacific Services Computing Conference, APSCC 2008, pp. 692–696 (2008)
6. Bhartiya, S., Mehrotra, D., Girdhar, A.: Proposing hierarchy-similarity based access control framework: a multilevel electronic health record data sharing approach for interoperable environment. J. King Saud Univ. Comput. Inf. Sci. (2015)

7. Yanhuang, L., Nora, C.B., Jean-Michel, C., Frdric, C., Vincent, F.: Reaching agreement in security policy negotiation
8. Elkandoussi, A., Elbakkali, H.: On access control requirements for inter-organizational workflow. In: Proceedings of the 4th Edition of National Security Days (JNS4), pp. 1–6 (2014)
9. Deepak, P., Mohsen, A.S., Kotagiri, R., Rajkumar, B.: A taxonomy and survey of fault-tolerant workflow management systems in cloud and distributed computing environments
10. Sabrina, K., Alessandra, M., Stefan, D.: Access control and the resource description framework: a survey
11. Duan, L., Chen, S., Zhang, Y., Liu, R.P.: Automated policy combination for data sharing across multiple organizations. In: IEEE International Conference on Services Computing, pp. 226–233 (2015)
12. Lin, L., Hu, J., Zhang, J.: Packet: a privacy-aware access control policy composition method for services composition in cloud environments. J. Front. Comput. Sci. **10**(6), 1142–1157 (2016)
13. Wang, H., Sun, L., Varadharajan, V.: Purpose-based access control policies and conflicting analysis. In: Rannenberg, K., Varadharajan, V., Weber, C. (eds.) Security and Privacy – Silver Linings in the Cloud. IFIP Advances in Information and Communication Technology, vol. 330, pp. 217–228. Springer, Heidelberg (2010)
14. Hu, H., Ahn, G.J., Jorgensen, J.: Multiparty access control for online social networks: model and mechanisms. IEEE Trans. Knowl. Data Eng. **25**, 1614–1627 (2012)
15. Elkandoussi, A., Elbakkali, H., Elhilali, N.: Toward resolving access control policy conflict in inter-organizational workflows. In: 2015 IEEE/ACS 12th International Conference of Computer Systems and Applications (AICCSA), pp. 2161–5330 (2015)
16. Li, W., Wu, J., Zhang, Q., Hu, K., Li, J.: Trust-driven and QoS demand clustering analysis based cloud workflow scheduling strategies. Cluster Comput. **17**(3), 1013–1030 (2014)
17. Tan, W., Sun, Y., Li, L.X., Lu, G., Wang, T.: A trust service-oriented scheduling model for workflow applications in cloud computing. IEEE Syst. J. **8**(3), 868–878 (2014)
18. Lin, D., Rao, P., Bertino, E., Lobo, J.: An approach to evaluate policy similarity. In: Proceedings of 12th ACM Symposium on Access Control Models and Technologies, pp. 1–10 (2007)
19. Shalini, B., Deepti, M., Anup, G.: Proposing hierarchy-similarity based access control framework: a multilevel electronic health record data sharing approach for interoperable environment. J. King Saud Univ. Comput. Inf. Sci. **29**, 505–519 (2017)

A Comparative Study on Access Control Models and Security Requirements in Workflow Systems

Monsef Boughrous[✉] and Hanan El Bakkali

Information Security Research Team (ISeRT), ENSIAS,
Mohammed V University, Rabat, Morocco
monsef.boughrous@gmail.com, h.elbakkali@um5.net.ma

Abstract. Workflow systems handle data and ressources that often require integrity preserving and may also need a high-level of confidentiality. Thus, they should be protected against unauthorized access. Organizations, use workflow management systems to manage, control and automate their business processes. Likewise, they adopt access control models to express their security needs and establish thier access control policies. Therefore, organizations have to choose a flexible access control model that corresponds to their security requirements, without sacrificing the resiliency of their workflow system. The contribution of this paper is to provide a study on access control models and comparing them according to a set of criteria and requirements that we believe are necessary to ensure security and resiliency in workflow systems.

Keywords: Workflow · Access control · Security requirements

1 Introduction

Workflow has emerged as a new technology in the 1990s. It allows enterprises to automate, analyze, and control their business processes, in order to improve the quality of their services. Actually, it's widely used in almost all domains, including finance, healthcare, distributed systems, eScience, cloud computing, manufacturing, and production. The Workflow Management Coalition (WfMC) has proposed a model that takes a broad view of the workflow system. They define workflow as the partial or complete automation of a business process, it allows to define and coordinate tasks and activities, in order to achieve or contribute to overall business goals in accordance with a defined set of rules [1]. Thus, the WfMC specifies that the Workflow Management Systems (WFMS) is a software system that execute workflows and provides procedural automation of business processes, by managing sequences of work activities and invoking appropriate human and/or IT resources [1,2].

To achieve a successful execution of the workflow, organizations manage and coordinate their activities using WFMS. However, executing workflows requires

© Springer International Publishing AG, part of Springer Nature 2018
A. Abraham et al. (Eds.): IBICA 2017, AISC 735, pp. 361–373, 2018.
https://doi.org/10.1007/978-3-319-76354-5_33

performing series of tasks and activitys that needs a set of access control permissions, to allow subjects (workflow users) get access to objects (resources) needed to get the work done at the right time by the correct person.

From a security perspective, access control is one of the most important security measures, which allow users to get physical or logical access to various resources in workflow systems. The access control policy applied in those systems is very important, and it must be both strict and flexible. Strict to ensure a high level of security, especially for confidential and sensitive data, where threats against integrity, authorization, and availability are very high. As well as flexible to facilitate the allocation of access permissions for authorized users. Moreover, to grant or refuse access permission several security requirements are inspected following the security strategy of the organization [3].

All this puts security and resiliency as essential and integral part of the workflow systems, so that the WFMS can manage and execute workflows in a flexible and secure way. Therefore, the major problem is to find a robust access control model (ACM) that ensure efficiency of management and administration of access rights, as well as not affecting the resiliency of the system.

In this paper we present a comparative study of different access control models and the necessary security requirements that can be applied in workflow systems. In Sect. 2 we describe access control issues in WFMS. In Sect. 3 we define the concepts and access control requirements to enforce security in workflow systems. Section 4 outlines a state of the art of various access control models, and Sect. 5 present a comparative table of models according to the main security criteria, finally we conclude this paper in Sect. 6.

2   Access Control Issues in Workflow Systems

Security is an essential part of any workflow system principally if the WFMS handle critical and sensitive information. In such case, the workflow security, reliability, and resiliency issues become more important. In order to protect the workflow systems it's necessary to protect the affected entities which are tasks, roles, subjects and objects. In [4], the WfMC published a white paper attempting to standardize security consideration and mechanisms that are important within a workflow system. Also, the WfMC in [2], discuss how to integrate the appropriate security services into the architecture and standards specified in the reference model. The security consideration for workflow systems identified are authentication, authorization, access control, audit, data privacy, data integrity, non-repudiation, security management and administration [4].

Security requirements in a workflow system are not limited only to authentication, access control, integrity, privacy and non-repudiation [5]. Other requirements exist made the security policies difficult and complicated to establish, especially if the workflow contains conflicting entities that can break down the execution of the workflow. The authors in [5] specify a set of security measures that are needed to build secure workflow systems, which are confidentiality, integrity, availability, authentication, authorization, audit, anonymity and separation of duties. In [6], authors surveyed distinct security requirements of multi-user collaboration systems (Computer-supported cooperative work (CSCW) systems and workflow systems), and their security policies. Also, they discussed the issues underlying the use of role-based security models and their limitations to express security policies in workflow systems. Moreover, they specify that traditional security requirements of integrity, availability, and confidentiality are naturally present, and the security requirements that represent unique aspects of workflow systems are static and dynamic separation of duties, Chinese wall security policy, confidentiality and privacy, context-sensitive access control policies, metalevel administrative security policies, fine-grain access control [6].

In relation with our purpose to identify the needed security requirements to secure workflow systems, and based on the surveyed papers, we believe that the essential requirements to ensure security and flexibility in workflow systems are static and dynamic separation of duties (SoD), binding of duties (BoD), Least privilege principle (LP), delegation of capabilities, granularity of specification and more other criteria that we will describe in Sect. 3.

3 Security Requirements and Evaluation Criteria

The adopted ACM must supports several security requirements and mechanisms. In [7] they specify three types of requirements. First type are requirements with administrative properties including Auditing, Flexibility, Delegation of capabilities and Team collaboration. Second are requirements with enforcement properties which are Static and Dynamic Separation of Duties (SoD), Binding of Duties (BoD), Least Privilege principle (LP), Granularity of specification, Context awareness, Privacy preservation and Resiliency. The third requirements are related to performance properties including Cardinality and Response time.

3.1 Separation of Duties

Separation of duties (SoD) prevent a user to hold enough permissions to commit a fraudulent act. SoD is supported by the least privilege principle, organizations use SoD to ensure the integrity of their business processes and properly address conflict of interest situation [6]. Two categories of SoD, are used in workflow systems [8]: Static separation of duties (SSoD), the simplest version of SoD, also called strong exclusion, it's applied during the administration time [8]. In role based access control (RBAC) policies, SSoD prevents users from assignment to incompatible roles or from being members of any exclusive roles at the same time [9]. Dynamic separation of duties (DSoD), also known as weak exclusion. Therefore the user can be a member of any exclusive roles at the same time, but he cannot activate both of them at the same time [9]. DSoD uses more policies than SSoD to limit the availability of privileges for a user even during work sessions [9]. DSoD provides operational flexibility, it's applied at the run-time (workflow execution) [8].

3.2   Least Privilege Principal

Least Privilege principal (LP) aims to ensure that each user has only the needed permissions to execute the task he is assigned to and no more [10,11]. In RBAC, role hierarchy and role delegation can cause a violation of the security policy. Therefore, LP restrictions must be enforced in such cases to prevent a user to hold more privileges than what he needs to perform his tasks [3].

3.3   Binding of Duties

BoD for Binding of Duties is a state where tasks are linked by a BoD relationship so they must be performed by the same subject. Two types of BoD constraints are distinguished: the Role-Binding constraint, based on role and Subject-Binding constraints, based on the subject (user, machine, etc.). The Subject-Binding requires that the same individual who performed a task should also perform the tasks linked to it. While Role-Binding constraint implies that related tasks must be performed by members of the same role, and it is not necessary to be the same individual [12].

3.4   Cardinality

Cardinality (Card) constraints define the maximum number of assignments for a single role, and the maximum number of roles assignment that a single user can have, also it can be static or dynamic [10]. Another definition by Jason Crampton in [12], specify that cardinality requires that the task is executed a number of times by a number of different users. Moreover, to enforce this type of constraint, we have to know the maximum number of tasks that can be handled at the same time, and which user and role have executed the task instance.

3.5   Delegation of Capabilities

The delegation process is where a user A (the delegator) delegates (grants or transfers) his rights to user B (the delegate) and then user B becomes authorized to perform a task on behalf of user A [13]. Delegation of Capabilities (DoC) provides organizational flexibility and enhances workflow systems resiliency. It ensures a flexible management of permissions in ACMs, allowing the delegate to get all needed permissions to perform a conflicting task instance [8]. However, the arbitrary application of this mechanism can lead to a breach of security policies and/or a lack of quality in work [8]. More information about delegation mechanism presented in [13].

3.6   Flexibility

Flexibility (Flex) aims to provide efficient and flexible configuration for ACM. Therefore, the administration and management of access rights must be flexible and transparent, to deal with unexpected situations in dynamic environments.

such as workflow systems, where authorizing or preventing users must be flexible and not creating a conflictual situation, especially while the execution of the workflow [8, 10].

3.7   Granularity of Specification

Granularity of Specification (GoS) is the measure where a task or permission is decomposed into small pieces, allowing a fine-grained control of access rights [10]. WFMS often manage workflows with multiple kind of tasks that need access to various resources and permissions. Therefore, the ACM must be able to protect information and resources of all types and at different levels of granularity.

3.8   Resiliency

Resiliency (Res) is the ability of a WFMS to use suitable strategies to bypass Workflow Satisfiability Problem Situation (WSPS) which refers to the situation where the WFMS is unable to assign a user to a task in a workflow instance for availability or security reasons, without violating the security policy [8].

According to E. Sahafizadeh et al. [11], ACMs are not suitable in enterprise settings, if they cannot accommodate unexpected changes. They must be flexible and supports changes at run-time in order to ensure workflow resiliency, but it is very risky to override WSPS, without applying the LP principle.

3.9   Team Collaboration

Team collaboration (TmC) is an important concept that permit to assist team members while they work together, on a complex task, which need the collaboration of different users to complete its execution. Generally, ACMs support team collaboration, but the most known model that support these criteria is the Team Based Access Control model (TMAC), proposed to take into account this concept in collaborative environments [10].

3.10   Context Awareness

Context Awareness (CnA) specify the dependance of the authorization decisions on context and conditions, in order to enhance the level of protection. They can be simple, such as using time, date or location [11], or made on the base of the subject and object information and attributes. Likewise, they are used in the operation of authentication to ensure the identity of the subject or to take access decision. Moreover, context based AC may be related to the physical environment, events and user actions (e.g. signing a contract) [6].

3.11 Privacy Preservation

Privacy preservation (PrP) can be achieved by ensuring confidentiality and privacy policies, which are related to the information flow constraints. Privacy policies protect the interest of individual users, and they are expressed in terms of consent, obligation, data category and context [6]. Where confidentiality policies express the interest of organizations [6]. The authors in [14], specify that conditions and obligations are necessary components in privacy laws. Conditions let policy writers specify fine-grained privacy policies to meet privacy law requirements, while obligations regulate subject's actions.

4 State of the Art on Access Control Models

Workflow security has become increasingly important and challenging, to meet these challenges several ACMs have been proposed in the literature.

4.1 Classical Access Control Models

Discretionary Access Control (DAC) is an access control (AC) model where the owner defines the access permissions on his own resources, deciding who has access to what, and the type of access to allow. Authorizations are expressed into an AC matrix, which represents subjects, objects and the type of access permission, it consists of two dimensions, the first dimension identify subjects (users, network equipment, etc.), and the second dimension list objects (resources) [15, 16]. The decomposition by columns creates AC lists (ACLs) where each object has a list of subjects and their access rights, while the decomposition by rows creates, the AC capability lists (ACCLs) that specify operations that each subject can perform on his authorized objects [15, 16]. The AC matrix provides a strong flexibility to DAC model [16].

Mandatory Access Control Model (MAC) takes its mandatory name from the way of labeling information, where ordinary subjects cannot change their labels unless a security officer (administrator) change their access authorizations [11, 15]. Its access strategy is very strict, which allows controlling the information flow of the system [16]. The definition of authorisation take into consideration the access policy and the sensitivity of information. The authorization decision compare security classes of the subject and the requested object, with consideration of the access policy and the sensitivity of information [11, 15, 16]. MAC model improves the system's security and prevent the damages from security problems (false programs, users mistakes, etc.). However, the security control scheme is strict and lacks flexibility, the security classes and rules description is difficult and complex in deployment. Therefore, MAC model is adopted in highly hierarchical organizations and environments where the security strategy have a definite security policy rules, (e.g. banks and military domain) [16].

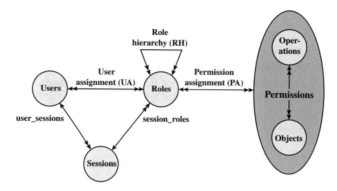

Fig. 1. RBAC model.

4.2   Role Based Access Control Model

Since 2004 RBAC is adopted as an ANSI/INCITS standard [17], it's widely considered the most appropriate AC model for organizations. RBAC introduce role concept to manage access rights which are defined by an authority responsible [3]. It grants permissions to roles, rather than users. Therefore, the access decision is taken by examining permissions assigned to the role that the user is assigned to. The benefit of using role concept appears when a person joins or leaves the organization or a role within the organization, it is not necessary to modify access rights every time [8] (Fig. 1).

According to [10,15,17], the RBAC model consists of four related conceptual models, RBAC0 contains the core concept and the minimum functionalities. RBAC1 adds roles hierarchie to RBAC0 features. RBAC2 adds static and dynamic constraints (not related/related to session). RBAC3 include all features of RBAC0, RBAC1, and RBAC2. Roles hierarchie reflect the hierarchy of responsibilities within an organization, in which roles of the first rank inherit all privileges of the roles in lower levels [15]. Generally, a major role will not necessarily need all the privileges of inferior roles, and this can cause a breach of the least privilege principle.

RBAC offers great granularity of security, but it requires more work in the implementation. It ensures system integrity and availability, allows much easier management, and it supports static and dynamic separation of duty. However, RBAC lacks decentralized management of policies [10]. Also, traditional RBAC lacks the ability to specify a fine-grained control on individual users in certain roles and on individual object instances. The definition of roles during the lifetime of a session lacks flexibility in changing environments [10,11,15,18].

4.3   Attribute Based Access Control Model

ABAC model is a logical model that controls access to objects by evaluating policy rules against the attributes of entities (subject and object) relevant to a request [19]. Attributes represent any security characteristics that could distinguish subjects (e.g. age, size, etc.) [15,19].

ABAC is dynamic and more flexible than RBAC for fine-grained changes or adjustments to subjects access profile. Its implementation is simple, and the policy rules definition can take in consideration semantic context [10]. It provides a security control point preventing unauthorized users [3,20]. Thus, it's limited by the richness of the available attributes for objects and subjects, and the computational power of the system (performance issue) [15,19].

4.4   Task Based Access Control Model

TBAC model controls access by comparing permissions of a task and those of the role assigned to it. Moreover, it grants permissions only to users in active execution of the task [18]. Therefore, TBAC requires an authorization management with constant monitoring of permission to decide what to activate and deactivate [21]. Thus, it controls subjects activities inside the organization, and gives certain granularity. It's flexible and dynamic, but it still an extended form of RBAC [21].

4.5   Team Based Access Control Model

Developed by R. S. Sandhu as an extension of RBAC, it addresses the problem of deploying RBAC policy in a collaborative environment, and it allows controlling access of activitys that will be better accomplished by the collaborators [18]. TMAC introduces the concept of team as an abstraction for group of users with specific roles in order to accomplish tasks or a specific objective in collaboration. Moreover, permissions are associated with roles and teams, thus permissions of a team member are shared with other members of the same team, this combination of permissions may breach the LP principle.

4.6   Task Role Based Access Control Model

TRBAC model is an improved model based on tasks and roles. It contains the basic features of RBAC and TBAC. In TRBAC model tasks are not only considered as sub-roles but they have a distinct sense of role. Their characteristics are the basis of the AC strategy, they link between roles and privileges [20]. TRBAC assign privileges to tasks and users to roles, thus a user cannot get permissions to access resources of a task unless he is assigned to the correct role and performing the task [20]. In TRBAC the allocation of users privileges through roles is described using task property, task state and constraints rules [22]. The task concept allows describing clearly the LP principle [22]. However, it's hard to identify small differences between roles, as consequence users needed privileges are very hard to equal privileges granted by roles [20]. Its flexibility is poor and the system administrator privileges represent security risks [20].

4.7   Task Attribute Based Access Control Model

The basic idea of TABAC is the description of tasks using attributes which are related to tasks and tasks are related to privileges, thus tasks allow to dynamically manage privileges [20]. TABAC architecture operates as follow. Work is decomposed into atomic tasks according to its procedure, and when a user start the execution process of a task, privileges are associated to the task according to its status, then privileges are withdrawn immediately after the end of the user mission [20]. TABAC supports fine-grained specification, separation of account management, and the privilege verification process is strict, and it enforces attributes and permissions constrain [20].

4.8   Role Attribute Based Access Control Model

RABAC model combine features offered by both models RBAC and ABAC. In [23] Kuhn et al., identify three alternatives to integrate attributes into RBAC. First alternative is Dynamic Roles, which use attributes context to dynamically assign roles to users. Second alternative is Attribute Centric where roles are just another attributes of users, this method largely discards the advantages of role concept in RBAC. Third alternative is Role Centric in which available permissions are limited in each session by the activated roles [23].

RABAC models proposed in the literature aims to preserve the best features of ABAC (flexibility, and scalability) and RABAC (the administrative convenience). In [24] Rajpoot et al., enhance attributes and RBAC model (AERBAC), in a novel way of individual objects to deal with role-permission explosion problem, and the object expressions enable content-based AC, also they provide fine-grained AC, policy modification and visualization, etc. In [25] Hui Qi et al., present an interesting work which analyze the existing RABAC models and they propose an enhanced model more fine-grained, flexible and efficient.

4.9   Organizational Based Access Control

OrBAC model, introduces the concept of organization as the most important entity in the model [26]. Authors in [26], specify that OrBAC entities are subjects, roles, objects, views, actions and activities. Similarly to role entity in managing and structuring subjects, they introduce the entity view to structure objects and add new objects to the system. The entity Action contain computer actions, while the entity activity is used to abstract actions, otherwise it joins actions that partake the same principles [26]. Besides, they extend the model with an entity contexts to specify the concrete circumstances where organizations grant roles permissions to perform activities on views. This new entity [26].

OrBAC aims to allow modeling a variety of security policies, and to reduce the complexity of managing access rights. Otherwise, it tries to take into account delegation, but the design of AC policie is much more complex, and building organization hierarchy or setting up teams is a difficult work as it multiplies the access decision rules. Moreover, its flexibility is not obvious [18].

4.10   Privacy-Aware Role Based Access Control

P-RBAC extends RBAC model to integrates and express complex privacy-aware AC policies, as well conditions under which a requester may perform his actions, and obligations that subjects must perform before and after the granted action [14]. P-RBAC model is designed on the base of several important privacy policy guidelines derived from privacy laws and public privacy policies of well-known organizations [14].

P-RBAC consists of conceptual models with different modeling capabilities [14]. The base model is Core P-RBAC, it provides basic components with a compromised design that should represent public privacy policies, privacy statements and privacy notice on the AC policies derived from privacy related acts [14]. Advanced models extends Core P-RBAC, such as Hierarchical P-RBAC with the hierarchical organizations, introducing role, object and purpose hierarchy [14]. Thus, Conditional P-RBAC extending Core P-RBAC to express more complex conditions introducing permission assignment sets and context [14].

4.11   Role-Based XACML Administration and Delegation Profile

XACML-ADRBAC [27], combine and enhance the XACML v3.0 administration and delegation profile (XACML v3.0), and the XACML-ARBAC profile [28]. In order to enforce and improve its scalability and delegation mechanism they incorporate a delegation model (PBDM), which extends RBAC to include user-to-user/role-to-role delegations, and provide single/multiple step delegation into the Administrative-RBAC Model (ARBAC) [28]. Thus, the role based approach facilitates the delegation of permissions to a large number of users, and it allows delegators to delegate or modify any subset of their assigned permissions [27]. Moreover, they extend the XACML-ARBAC profile to cover the XACML v3.0 profile and benefit from its features (policy administration and dynamic delegation).

5   Comparative Study

In this section, we present a comparative study of the ACMs that we presented in Sect. 3. The comparative study is based on the requirements presented in Sect. 3, which they must be supported by the adopted ACM, in order to ensure security and resiliency in workflow systems. Table 1 shows this comparison, the first column of the table shows criteria, and the first row shows ACMs. The intersection cell indicates if the model supports the criterion, "YES" if its supported, "NO" if not, and "Partial" means that the criterion is partially supported, but it can be enforced under some conditions or configuration of the ACM, "N.S" means that the criterion is not specified.

In order to understand what we mean by "Partial" which can be ambiguous. This is an example for LP principle in the RBAC model, which is partially supported, but it's possible to enforce it. Roles hierarchy in RBAC ensure that

Table 1. Comparative table

	SoD	BoD	LP	Card	DoC	GoS	Flex	Res	TmC	CnA	PrP
DAC	NO	NO	NO	NO	YES	NO	YES	YES	Partial	NO	NO
MAC	Partial	NO	Partial	YES	NO	NO	NO	NO	NO	NO	NO
RBAC	YES	YES	Partial	YES	N.S	Partial	YES	Partial	N.S	Partial	NO
TBAC	YES	YES	Partial	YES	N.S	Partial	YES	Partial	N.S	Partial	NO
TRBAC	YES	YES	YES	YES	N.S	NO	NO	NO	N.S	NO	NO
TMAC	Partial	Partial	NO	YES	YES	Partial	Partial	YES	YES	Partial	NO
ABAC	Partial	Partial	YES	Partial	N.S	YES	YES	Partial	N.S	YES	NO
TABAC	YES	YES	YES	YES	NO	YES	YES	NO	NO	YES	NO
RABAC	YES	YES	YES	YES	N.S	YES	YES	Partial	NO	YES	NO
OrBAC	YES	YES	N.S	YES	YES	N.S	NO	Partial	NO	YES	Partial
Privacy-RBAC	YES	YES	Partial	YES	N.S	Partial	YES	Partial	N.S	YES	YES
XACML-ADRBAC	YES	YES	Partial	YES	YES	Partial	YES	YES	N.S	Partial	N.S

roles hold only the minimum needed privileges, but a role of the first rank in roles hierarchy inherit all the privileges of the roles in lower levels, which will not necessarily need, and that can cause a violation of the LP principle.

An other example is the Resiliency requirement which ensure that the workflow system can override WSPS. Supporting resiliency require that the model is flexible in managing access rights and the most important thing is supporting delegation criterion, in order to delegate privileges when a conflicting situation appear especially in the run-time of the workflow. For RABAC model Resiliency requirements is partially supported because RABAC supports SoD, BoD and LP, which is good, but increase the possibility of creating conflicting situations, moreover the delegation criterion is not specified for RABAC.

6   Conclusion and Future Work

Workflow systems enhance enterprises efficiency, and make them more competitive, thus they improve business processes quality. The workflow management system provides powerful tools to manage, control and execute workflows. However, there are several security issues in workflow systems, as the problems that result from the strict application of the security policy, like resiliency problem. That's why enterprises should adopt an ACM that meets their strategy and security requirements, in order to control access on their resources, and ensure the security of their system. Therefore, the main issue is to find an access control model from a wide range of access control models that offer different features.

The objective of this paper is to give a comparative study of access control models and security requirements that must fulfill the security needs in workflow systems. We can use this comparative study to select an adequate model that responds to our needs and security strategy. As future work, we will propose an approach that fulfills those requirements, especially delegation to enhance workflow resiliency and override the workflow satisfiability problem.

References

1. Workflow Management Coalition Terminology & Glossary, June 1996. http://www.aiai.ed.ac.uk/project/wfmc/ARCHIVE/DOCS/glossary/glossary.html
2. Hollingsworth, D.: Workflow management coalition the workflow reference model. The Workflow Management Coalition Specification, no. TC00-1003, January 1995
3. Andress, J.: The Basics of Information Security: Understanding the Fundamentals of InfoSec in Theory and Practice, 2nd edn. Syngress, Waltham (2014)
4. Workflow Security Considerations - White paper. Workflow Management Coalition, vol. 1.0, no. WFMC-TC-1019, February 1998
5. Atluri, V., Warner, J.: Security for workflow systems. In: Gertz, M., Jajodia, S. (eds.) Handbook of Database Security. Springer, Boston (2008)
6. Ahmed, T., Tripathi, A.R.: Security policies in distributed CSCW and workflow systems. IEEE Trans. Syst. Man Cybern. Part A Syst. Hum. **40**(6), 1220–1231 (2010)
7. Hu, V.C., Scarfone, K.: Guidelines for Access Control System Evaluation Metrics. NISTIR 7874, September 2012
8. El Bakkali, H.: Enhancing workflow systems resiliency by using delegation and priority concept. J. Digital Inf. Manage. **11**(4), 267–276 (2013)
9. Younis A., Kifayat K., Merabti M.: A novel evaluation criteria to cloud based access control models. In: 11th International Conference on Innovations in Information Technology. IEEE (2015)
10. El Kandoussi, A., El Bakkali, H.: On access control requirements for inter-organizational workflow. In: The 4th Edition of National Security Days. IEEE, May 2014
11. Sahafizadeh, E., Parsa, S.: Survey on access control models. IEEE (2010)
12. Crampton, J., Gagarin, A., Gutin, G., Jones, M., Wahlstrom, M.: On the satisfiability of constraints in workflow systems. In: ACM Transactions on Private Security, vol. 19 (2016)
13. Ali, A., Habiba, U., Shibli, M.A.: Taxonomy of delegation model. In: 12th International Conference on Information Technology - New Generations (2015)
14. Ni, Q., Bertino, E., Lobo, J., Calo, S.B.: Privacy-aware role-based access control. IEEE Secur. Priv. **7**(4), 35–43 (2009)
15. Stallings, W., Brown, L.: Computer Security: Principles and Practice, 3rd edn. Pearson (2015)
16. Bai, Q., Zheng, Y.: Study on the access control model in information security. IEEE, July 2011
17. American national standard for information technology: Role based access control. ANSI INCITS 359 (2004)
18. Smari, W.W., Clemente, P., Lalande, J.-F.: An extended attribute based access control model with trust and privacy: application to a collaborative crisis management system. Future Gener. Comput. Syst. **31**, 147–168 (2014)
19. Hu, V.C., Kuhn, D.R., Ferraiolo, D.F.: Attribute-based access control. IEEE Comput. **48**(2), 85–88 (2015)
20. Yi, L., Ke, X., Junde, S.: A task-attribute-based workflow access control model IEEE (2013)
21. Mallare, I.J.G., Pancho-Festin, S.: Combining task and role based access contro with multi-constraints for a medical workflow system. IEEE (2013)
22. Sainan L.: Task-role-based access control model and its implementation. In: 2nc International Conference on Education Technology and Computer. IEEE (2010)

23. Kuhn, D.R., Coyne, E.J., Weil, T.R.: Adding attributes to role-based access control. IEEE Comput. **43**(6), 79–81 (2010)
24. Rajpoot, Q.M., Jensen, C.D.: Attributes enhanced role-based access control model. In: Proceedings of the 12th International Conference, TrustBus 2015, Valencia, Spain, p. 317, 1–2 September 2015
25. Qi, H., Luo, X., Di, X., Li, J., Yang, H., Jiang, Z.: Access control model based on role and attribute and its implementation. In: International Conference on Cyber-Enabled Distributed Computing and Knowledge Discovery. IEEE (2016)
26. Kalam, A.A.E., et al.: Organization based access control. In: Proceedings POLICY 2003, IEEE 4th International Workshop on Policies for Distributed Systems and Networks, pp. 120–131 (2003)
27. Xu, M., Wijesekera, D.: A role-based XACML administration and delegation profile and its enforcement architecture. In: SWS 2009. ACM (2009)
28. Xu, M., Wijesekera, D., Zhang, X., Corray, D.: Towards session-ware RBAC administration and enforcement with XACML. In: The IEEE International Symposium on Policies for Distributed Systems and Networks (POLICY) (2009)

Author Index

© Springer International Publishing AG, part of Springer Nature 2018
A. Abraham et al. (Eds.): IBICA 2017, AISC 735, pp. 375–376, 2018.
https://doi.org/10.1007/978-3-319-76354-5

Printed in the United States
By Bookmasters